Environmental Challenges and Solutions

The Environmental Challenges and Solutions series aims to improve our understanding of the Earth's most important environmental challenges, and how we might more effectively solve or at least mitigate these challenges. Books in this series focus on environmental challenges and solutions in particular geographic regions ranging from small to large spatial scales. These books provide multidisciplinary (technical, socioeconomic, political, etc.) analyses of their environmental challenges and the effectiveness of past and present efforts to address them. They conclude by offering holistic recommendations for more effectively solving these challenges now and into the future.

We are welcoming proposals specifically on the following topics: -Environmental sciences and social aspects -Environmental sciences and human conflicts -Boosting de-carbonization of energy system to face climate and economic emergency -Climate emergency and health: sources, science-based analysis and global solutions -Indoor and outdoor chemicals, environmental quality and public health in a changing world -Demographic, climate and environmental change: evidence and recommendation for global sustainability -Technological, industrial and corporate change for supporting environmental and social sustainability -Nanotechnology applied to water decontamination/purification processes -Sorbents and other materials in environmental protection -Raw materials supply and circular economy -Strategies to boost clean mobility

All books are written in a concise and readable style, making them suitable for both specialists and non-specialists starting at first year graduate level. Proposals for the book series can be sent to the Publishing Editor, Éva Lörinczi, eva.loerinczi@springer.com.

Smita S. Kumar • Vivek Kumar
Editors

Nature-Based Solutions for Decentralised Liquid Waste Management

Springer

Editors
Smita S. Kumar
Department of Environmental Studies
Central University of Haryana
Mahendergarh, Haryana, India

Vivek Kumar
Centre for Rural Development and Technology
Indian Institute of Technology Delhi
New Delhi, Delhi, India

ISSN 2214-2827 ISSN 2214-2835 (electronic)
Environmental Challenges and Solutions
ISBN 978-3-032-08361-6 ISBN 978-3-032-08362-3 (eBook)
https://doi.org/10.1007/978-3-032-08362-3

This Springer imprint is published by the registered company Springer Nature Switzerland AG
The registered company address is: Gewerbestrasse 11, 6330 Cham, Switzerland

Preface

The demand for effective and sustainable wastewater treatment solutions is growing rapidly, particularly in decentralized systems where conventional, large-scale systems are often impractical or too costly. As urbanization continues to accelerate and the global population increases, decentralized liquid waste management is becoming essential for safeguarding water resources, improving public health, and ensuring environmental sustainability. Nature-based solutions (NbS) are at the forefront of this paradigm shift, offering cost-effective, eco-friendly alternatives that work in harmony with natural processes to treat wastewater.

This edited book aims to explore a wide range of nature-based solutions for decentralized liquid waste management, focusing on innovative, sustainable approaches to address the challenges of wastewater treatment at the local level. Each chapter has been meticulously curated to provide insights into the various technologies and methodologies that are reshaping decentralized wastewater treatment systems. From constructed wetlands and waste stabilization ponds to anaerobic baffled reactors and phytoremediation, the contributions highlight cutting-edge research, case studies, and real-world applications of NbS in managing wastewater.

The book begins with an **Introduction** that provides a broad overview of decentralized wastewater management, setting the stage for the subsequent chapters. The following sections explore the diverse array of nature-based technologies and their potential for treating both domestic and industrial wastewater. Key chapters on **Constructed Wetlands** and **Waste Stabilization Ponds** explore how these systems can offer sustainable solutions for wastewater treatment, while chapters on **Biochar-Based Water Treatment** and **Anaerobic Baffled Reactors** provide insights into novel approaches that are gaining attention for their efficiency and low environmental impact.

Emerging contaminants, such as microplastics, are addressed in the chapters on **Microbial Technologies** and **Constructed Wetlands**, showcasing how nature-based solutions are being adapted to meet the challenges posed by new pollutants. The inclusion of **Case Studies** brings practical examples of decentralized wastewater treatment solutions in diverse geographic and industrial contexts, offering valuable lessons for practitioners and policymakers alike.

The book also emphasizes the importance of circularity and sustainability, with a dedicated chapter on **Circularity and Nature-Based Solutions**, illustrating how NbS contribute to a more sustainable approach to wastewater management. A key theme that runs through the book is the integration of science, technology, and nature to create solutions that are not only effective but also socially, economically, and environmentally beneficial.

It is our hope that this book will serve as a comprehensive resource for researchers, engineers, policymakers, and anyone interested in exploring the future of decentralized wastewater treatment. By presenting the latest advancements, real-world applications, and case studies, we aim to inspire further research, innovation, and the widespread adoption of nature-based solutions for wastewater management worldwide.

We extend our gratitude to all the authors and contributors for their invaluable expertise and insights. Their dedication and commitment to advancing the field of decentralized wastewater treatment have made this book possible. We also thank the readers for their interest, and we trust that this book will provide meaningful guidance in the pursuit of sustainable and innovative solutions for managing liquid waste in decentralized settings.

Mahendergarh, Haryana, India Smita S. Kumar
New Delhi, Delhi, India Vivek Kumar

Contents

Part I
Introduction

Chapter 1
Nature-Based Solutions for Decentralized Liquid Waste Management

Smita S. Kumar, Renu Meena, Jyoti Yadav, and Vivek Kumar

Abstract Nature-based solutions (NbS) for decentralized wastewater treatment are gaining recognition due to their sustainability, cost-effectiveness, and environmental benefits. These solutions, such as constructed wetlands, waste stabilization ponds, soak pits, and hybrid aerobic/anaerobic systems, offer an alternative to traditional, centralized wastewater treatment, particularly in rural, peri-urban, and resource-limited areas. This chapter discusses the types of NbS for wastewater treatment, the challenges encountered in their implementation, and offers suggestions for improvement, such as incorporating smart technologies, promoting hybrid systems, and enhancing public engagement. It also explores future prospects, including the integration of resource recovery, resilience to climate change, and the growing role of NbS in addressing global water and sanitation issues. By addressing these challenges and capitalizing on emerging trends, NbS can play a central role in sustainable wastewater management.

Keywords Constructed wetlands · Nature-based solutions · Decentralized liquid waste management

S. S. Kumar (✉) · R. Meena · J. Yadav
Department of Environmental Studies, Central University of Haryana, Mahendergarh, Haryana, India
e-mail: drsmita@cuh.ac.in

V. Kumar
Centre for Rural Development and Technology, Indian Institute of Technology Delhi, Delhi, India

S. S. Kumar, V. Kumar (eds.), *Nature-Based Solutions for Decentralised Liquid Waste Management*, Environmental Challenges and Solutions,
https://doi.org/10.1007/978-3-032-08362-3_1

1.1 Nature-Based Strategies for Decentralized Treatment of Liquid Waste

Liquid waste, in the form of wastewater, is generated from a variety of sources, including residential, commercial, and industrial activities. Wastewater that is no longer fit for human consumption or other uses is classified as liquid waste, which, if improperly managed, can lead to significant environmental and public health issues. Wastewater is typically categorized into gray water (wastewater from bathing, kitchen use, and laundry) and blackwater (wastewater from toilets), each requiring different treatment approaches. In both rural and urban areas, the improper disposal of wastewater results in water stagnation, foul odors, and the spread of diseases such as dengue, malaria, and cholera, as it becomes a breeding ground for disease-carrying vectors like mosquitoes (Anh et al. 2002). Effective liquid waste management is crucial for maintaining environmental sustainability and public health. Proper disposal, treatment, and reuse of wastewater not only reduce the risk of contamination but also contribute to water conservation. A key challenge in wastewater management, particularly in rural and peri-urban areas, is the need for decentralized and simple solutions that can be easily implemented and maintained by local communities (Li et al. 2024). Nature-based strategies offer promising alternatives for the treatment of liquid waste at a decentralized level. These strategies utilize natural processes and ecosystems to treat wastewater effectively and sustainably. Techniques such as constructed wetlands, waste stabilization ponds, reed bed systems, and phytoremediation harness the power of plants, microbes, and natural filtration to purify water, making it safe for reuse (Zhang et al. 2014). Nature-based solutions for wastewater treatment not only provide effective sanitation but also contribute to the overall health of the ecosystem by promoting biodiversity and enhancing the resilience of water systems Dalwani and Gopal (2020). These decentralized systems are often cost-effective, require minimal energy, and can be tailored to suit local needs, making them ideal for rural and peri-urban areas with limited infrastructure.

This chapter summarizes various nature-based strategies for decentralized wastewater treatment, highlighting their benefits, applications, and the role they play in sustainable liquid waste management. By exploring these innovative and environmentally friendly solutions, communities can develop appropriate wastewater management systems that improve health outcomes, promote water reuse, and protect local ecosystems.

1.2 Liquid Wastes in India

Liquid waste, commonly referred to as wastewater, is an inevitable by-product of various industrial, agricultural, and domestic activities. The increasing population and rapid industrialization in India have led to an overwhelming increase in the generation of liquid waste. Managing and treating this waste is crucial for preserving water resources, maintaining public health, and promoting sustainable development. Liquid waste can be categorized into several types, each originating from different sectors, and the management strategies for each type vary accordingly. This write-up explores the different types of liquid waste in India, their sources, characteristics, and potential environmental impacts.

1.2.1 Domestic Liquid Waste

Domestic liquid waste primarily refers to wastewater generated from households, which includes sewage, gray water, and wastewater from various household activities. The types of domestic liquid waste are as follows:

Sewage: Sewage is the waste generated from toilets, which contains human excreta, urine, and other biological matter. It often contains harmful pathogens, including bacteria, viruses, and parasites, posing a significant threat to public health if not treated adequately.

Gray Water: Gray water is the wastewater generated from activities such as bathing, washing dishes, and laundry. It is typically less contaminated than sewage but may contain organic matter, detergents, and other chemicals. Although it is not directly hazardous, untreated gray water can lead to water pollution if discharged untreated into natural water bodies (Sanwal 2021).

Stormwater Runoff: Rainwater runoff from urban areas carries pollutants such as oil, chemicals, debris, and plastics. Though rainwater is generally considered clean, urban runoff is a significant source of pollution when combined with various contaminants from streets and surfaces.

1.2.2 Industrial Liquid Waste

India's industrial sector is a major contributor to liquid waste, which varies depending on the type of industry and the processes involved. Common types of industrial liquid waste include:

Chemical Wastewater: Chemical industries generate wastewater containing harmful chemicals, heavy metals, and toxic compounds. This type of liquid waste is a major concern due to its potential to contaminate water sources and harm aquatic life. Industries such as textiles, dyes, pharmaceuticals, and food processing produce significant amounts of chemical wastewater.

Mining Wastewater: Mining operations, including coal and mineral extraction, generate liquid waste containing suspended solids, heavy metals (like lead, mercury, and arsenic), and other pollutants. If untreated, mining wastewater can lead to the contamination of nearby rivers, lakes, and groundwater sources.

Oil and Petroleum Wastewater: Refineries and industries dealing with petroleum products release wastewater containing oil, grease, and hydrocarbons. These substances are toxic to aquatic ecosystems and can lead to bioaccumulation in the food chain.

Textile and Dyeing Effluents: The textile industry, especially in regions like Tamil Nadu and Gujarat, produces large volumes of wastewater from dyeing and washing processes. These effluents often contain dyes, salts, and other toxic chemicals that can significantly degrade the quality of receiving water bodies.

1.2.3 Agricultural Liquid Waste

The agricultural sector contributes significantly to liquid waste generation in India. Agricultural runoff and wastewater typically contain:

- Pesticides and Fertilizer Runoff: The overuse of pesticides and chemical fertilizers in agriculture results in runoff that contains harmful chemicals like nitrates, phosphates, and toxic substances. This type of waste contributes to water pollution and can cause eutrophication in water bodies (Hariprasad and Dayananda 2013).
- Irrigation Runoff: Water used for irrigation often carries soil particles, nutrients, and chemicals into nearby rivers and lakes. Excessive irrigation can lead to salinity and contamination of freshwater resources.
- Livestock Wastewater: Animal farming generates wastewater containing animal manure, urine, and other organic waste. This type of liquid waste can lead to nutrient overload and contamination of water resources if not managed properly.

1.2.4 Hospital and Healthcare Liquid Waste

Healthcare facilities, including hospitals, clinics, and laboratories, generate significant amounts of liquid waste, often classified as biomedical waste. The types include:

Infectious Wastewater: Liquid waste generated from hospitals, including blood, body fluids, and wastewater from surgical procedures, may contain pathogens. This type of waste requires special treatment and disposal methods to prevent the spread of diseases.

Pharmaceutical Wastewater: Hospitals and pharmaceutical industries may generate wastewater containing residues of medicines and chemicals. Improper disposal of pharmaceutical waste can lead to contamination of water supplies and may contribute to antibiotic resistance.

1.2.5 Wastewater from Slaughterhouses and Food Processing Units

The food processing industry, including slaughterhouses, dairies, and beverage production, generates wastewater containing organic matter, fats, oils, and other contaminants. The main issues related to this type of waste include:

Organic and Nutrient Load: The wastewater from food processing industries contains a high concentration of organic matter, which, when discharged untreated, can deplete oxygen levels in water bodies and harm aquatic life.

Slaughterhouse Wastewater: This waste contains blood, fats, and other animal by-products, often carrying high levels of BOD (biochemical oxygen demand). The discharge of untreated slaughterhouse wastewater into water bodies can lead to severe environmental damage (Kothari et al. 2024).

1.2.6 Municipal Liquid Waste

Municipal liquid waste includes water discharged by public facilities such as schools, government buildings, and public washrooms. These wastewater streams often combine sewage and gray water, adding to the strain on municipal treatment systems.

Liquid waste in India, originating from domestic, industrial, agricultural, healthcare, and municipal sources, poses a significant environmental and public health challenge (Mawari et al. 2022). To mitigate the adverse impacts, effective wastewater treatment, proper disposal practices, and sustainable management strategies are essential (Inyinbor Adejumoke et al. 2018). Public awareness, regulatory measures, and technological innovations in wastewater treatment can help address the growing concern of liquid waste management and contribute to environmental sustainability in India. Broadly, different sources of waste water and its type have been enlisted in Table 1.1.

Table 1.1 Sources of waste water and its type

No.	Source of waste water	Types of waste water	Quantity/day/person
1.	Toilets	Blackwater	3 liters
2.	Bathing	Gray water	20–30 liters
3.	Kitchen	Gray water	5–10 liters
4.	Washing cloth	Gray water	15–20 liters
5.	Animals	Gray water	10–15 liters

1.3 Criteria for Selection of Technology

Sustainability assessment of the technologies for wastewater treatment solutions is of vital importance for helping the decision-makers to select the most sustainable technology among multiple alternatives. Some of the criteria have been proposed for sustainability assessment of the technologies for wastewater treatment solutions (Table 1.2, Fig. 1.1).

Table 1.2 Criteria for sustainability assessment of wastewater treatment technologies

Consideration	Requirement
Quality of treated effluent	Production of treated effluent of stipulated quality without interruption
Power requirement	Reduce energy consumption
Land required	Minimize land requirement
Capital cost of plant	Optimum utilization of capital
Operation & maintenance costs	Lower recurring expenditure
Maintenance requirement	Simple and reliable
Operator attention	Easy to understand procedures
Reliability	Consistent delivery of treated sewage
Resource recovery	Production of quality water and manure
Load fluctuations	Withstand variations in organic and hydraulic loads

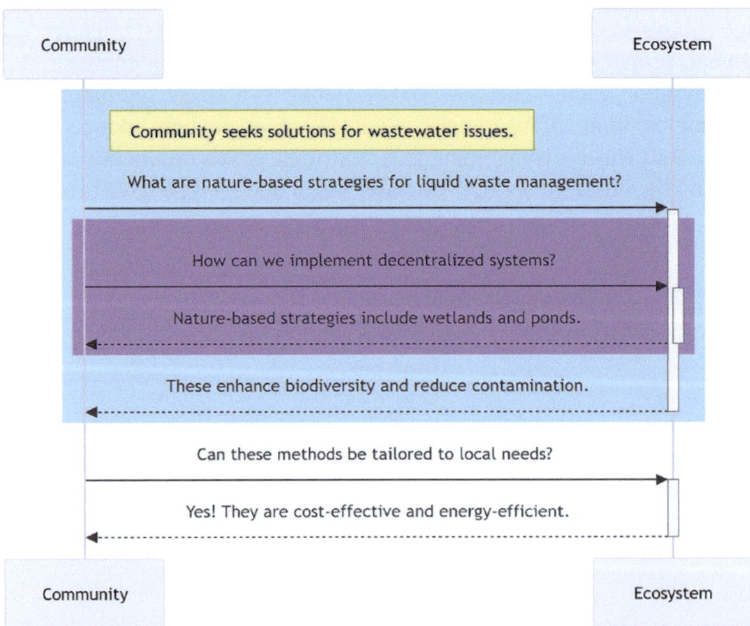

Fig. 1.1 Selection of nature-based strategies for liquid waste management

1.4 Detailed Account of Natural Liquid Waste Management Techniques

Natural liquid waste management techniques, also known as sustainable or eco-friendly methods, focus on leveraging natural processes to treat wastewater. These methods typically rely on biological and physical processes to degrade contaminants, allowing for wastewater treatment that integrates well with the environment. The following natural liquid waste management techniques are widely used in India and around the world:

1.4.1 Soak Pit/Leach Pit/Magic Pit

A soak pit is a simple and commonly used method for the disposal of liquid waste, especially domestic sewage. It involves digging a pit that allows the wastewater to percolate through the soil and gradually filter out contaminants. The pit is usually filled with gravel and sand, creating a filtration system that helps in removing solid particles from the wastewater.

1.4.1.1 Working Principle

The wastewater is directed into the soak pit, where it is gradually filtered by the surrounding soil and the materials used within the pit. The pit's porous structure helps the water percolate slowly, allowing natural filtration and absorption into the ground. The soil, along with microbial organisms, plays a crucial role in breaking down organic matter and filtering out pollutants.

1.4.1.2 Advantages

Low cost and simple to construct.

Effective for small-scale, household sewage treatment.

Requires minimal maintenance if properly designed and located in an area with suitable soil.

1.4.1.3 Disadvantages

May not be effective for large volumes of wastewater or high-strength waste.

Can lead to groundwater contamination if not designed or maintained properly.

1.4.2 Waste Stabilization Ponds (WSPs)

Waste stabilization ponds (WSPs) are a form of natural wastewater treatment that uses shallow ponds or lagoons to treat sewage through biological processes, primarily involving algae, bacteria, and other microorganisms.

1.4.2.1 Working Principle

Wastewater is allowed to flow into a series of ponds, with the first pond being primarily anaerobic (without oxygen), where organic matter is degraded by anaerobic bacteria. The subsequent ponds are typically aerobic (with oxygen), allowing aerobic microorganisms to further treat the wastewater. Algae in the aerobic ponds play a crucial role in oxygenating the water, as well as in absorbing nutrients like nitrogen and phosphorus, which helps to reduce pollution.

1.4.2.2 Types of Ponds in WSP Systems

Anaerobic Pond: Degrades organic matter in the absence of oxygen.

Facultative Pond: Contains both aerobic and anaerobic zones, where facultative bacteria break down pollutants.

Maturation Pond: Aerobic treatment takes place here, allowing the removal of pathogens.

1.4.2.3 Advantages

Cost-effective and simple to design and operate.

Suitable for small to medium communities.

Does not require chemical additives or complex machinery.

Effective in treating domestic and industrial wastewater with moderate organic content.

1.4.2.4 Disadvantages

Requires a large land area, which can be a limitation in urban areas.

Treatment can be slow and dependent on climatic conditions (temperature and sunlight).

Can have unpleasant odors if not maintained properly.

1.4.3 Constructed Wetlands

Constructed wetlands (CWs) are man-made systems designed to mimic the natural processes of wetlands in treating wastewater. They use plants, soil, and microbial communities to remove contaminants from wastewater (Massoud et al. 2009).

1.4.3.1 Working Principle

Constructed wetlands are typically divided into two types: free-water surface (FWS) wetlands and subsurface flow (SSF) wetlands. Free-water surface (FWS) wetlands: Water flows over the surface of the wetland and is treated by plants, microorganisms, and substrate. Subsurface flow (SSF) wetlands: Water flows below the surface, where it is filtered by the substrate and roots of plants. Plants such as reeds, bulrushes, or cattails are used to absorb excess nutrients and provide oxygen through their root systems. Microorganisms in the soil and on plant roots help break down organic materials and pathogens.

1.4.3.2 Advantages

Effective in removing nutrients, suspended solids, and pathogens.

Suitable for domestic, industrial, and agricultural wastewater.

Provides an aesthetically pleasing and natural treatment option.

Low operational and maintenance costs.

1.4.3.3 Disadvantages

Requires adequate space for construction.

Limited treatment capacity in areas with heavy pollution or high wastewater flows.

Seasonal variations, such as cold temperatures, may reduce the efficiency of treatment.

1.4.4 Soil Bio-technology

Soil bio-technology utilizes the natural biological processes within the soil to treat wastewater. The focus is on harnessing the ability of soil organisms, such as bacteria, fungi, and protozoa, to break down pollutants in wastewater.

1.4.4.1 Working Principle

In this method, wastewater is applied to soil in a controlled manner. The soil acts as a natural filter, while microorganisms in the soil degrade organic pollutants, convert toxic compounds into nontoxic forms, and remove excess nutrients. The soil provides a medium for filtration, adsorption, and microbial activity, which purifies the water. Additionally, the application of organic waste, such as compost or bio-solids, enhances the microbial activity in the soil, improving the overall wastewater treatment process.

1.4.4.2 Advantages

Low cost, sustainable, and energy-efficient.

Suitable for small-scale wastewater treatment, such as for farms or rural households.

Enhances soil fertility by introducing organic matter.

1.4.4.3 Disadvantages

Limited capacity for treating high-strength or industrial wastewater.

The treatment process can be slow and requires careful management of the soil system.

Can lead to soil salinization if not managed properly.

1.4.5 Aerobic and Anaerobic High Rate Systems

These systems use aerobic (oxygen-rich) and anaerobic (oxygen-deprived) processes to treat liquid waste rapidly through microbial activity. The high-rate systems aim to accelerate the biological treatment processes by increasing microbial growth and activity.

1.4.5.1 Aerobic Systems

Aerobic treatment units (ATUs) are designed to introduce oxygen into the wastewater, which enhances the growth of oxygen-loving bacteria. These bacteria break down organic pollutants more efficiently.

Common aerobic processes include activated sludge and oxidation ditches.

1.4.5.2 Advantages

Rapid treatment of wastewater.

Effective for removing organic matter, nitrogen, and phosphorus.

Produces high-quality effluent that can be safely discharged or reused.

1.4.5.3 Disadvantages

High energy consumption due to aeration requirements.

Maintenance-intensive and may require skilled personnel to manage the system.

1.4.5.4 Anaerobic Systems

Anaerobic digesters are used to treat wastewater in the absence of oxygen. This process is particularly useful for treating high-strength wastewater, such as from industrial processes (e.g., food processing and distilleries). Anaerobic systems convert organic matter into biogas (methane) that can be used as an energy source.

1.4.5.5 Advantages

Low energy requirements as aeration is not needed.

Production of biogas, which can be used for energy generation.

Suitable for high-strength waste, including organic waste from industries.

1.4.5.6 Disadvantages

Longer hydraulic retention time.

Potential for odor production.

Requires careful management to maintain the balance of anaerobic microorganisms.

Natural liquid waste management techniques, such as soak pits, waste stabilization ponds, constructed wetlands, soil biotechnology, and high-rate aerobic and anaerobic systems, offer sustainable solutions for treating wastewater in India. These methods utilize biological processes and natural resources, providing environmentally friendly alternatives to traditional, energy-intensive treatment systems (Harikumar and Bindhya 2012). However, the selection of a treatment method depends on factors such as wastewater quality, volume, land availability, and cost. While these techniques are highly effective in many cases, regular maintenance and proper design are crucial for ensuring that they function optimally and do not result in environmental degradation. The selection of a process requires analysis of all factors. In order to provide additional factors for the final considerations, key parameters need to be evaluated and weighed as shown in Table 1.3 to reach a final recommendation.

1.5 Problems and Challenges Faced in Nature-Based Solutions (NbS) for Decentralized Treatment of Wastewater

Nature-based solutions (NbS) for decentralized wastewater treatment offer sustainable, cost-effective, and environmentally friendly alternatives to conventional methods (Cohen-Shacham et al. 2016). However, despite their many advantages, these solutions face a range of challenges and issues that can hinder their implementation and effectiveness. Below is a detailed overview of the problems and challenges associated with nature-based solutions for wastewater treatment:

1.5.1 Limited Treatment Capacity and Efficiency

High Pollutant Loads: Nature-based systems like constructed wetlands, soak pits, or waste stabilization ponds are often less effective at handling high-strength wastewater, especially from industrial or commercial sources. Wastewater with a high concentration of chemicals, heavy metals, or pathogens may overwhelm the natural systems' ability to filter and purify the water.

Inconsistent Treatment Efficiency: The efficiency of natural treatment systems, such as constructed wetlands, varies depending on seasonal changes (e.g., temperature and rainfall) and pollutant types (Ventura et al. 2024). For example, cold weather can slow down microbial activity in systems like wetlands or waste stabilization ponds, reducing their effectiveness during the winter months.

1.5.2 Space Requirements and Land Availability

Large Land Area Requirements: Many NbS, such as constructed wetlands, waste stabilization ponds, and soil-based technologies, require significant amounts of land. In urban areas where land is scarce and expensive, finding space to implement such solutions becomes challenging.

Land Use Conflicts: In densely populated areas, there may be competition for land use between wastewater treatment systems and other urban infrastructure, such as housing, commercial areas, and agricultural land. This can make it difficult to implement decentralized wastewater treatment solutions in areas where space is limited.

Table 1.3 Various treatment technologies for liquid waste management

Household level management						
	Technical description	O &M	Cost	Tech. Know.	Applicability	Advantage
1. Kitchen garden with piped root zone system	Grease trap: (450 mm × 350 mm × 300 mm) Perforated nonpressure PVC pipe (50 mm diameter and length as per requirement) Trench (150 mm–200 mm depth and 200 mm width) Gravel of size. (20 mm–25 mm size)	Periodical cleaning of the grease trap (every week) Cleaning of perforated pipes (once in a year)	Rs. 80/– per meter length including labor cost (2006)	Trained mason	Houses with adequate courtyard.	Simple and cost-effective technology Cent percent utilization of water to produce vegetables and fruits Prevents water stagnation Prevents vector breeding
2. Kitchen garden without piped root zone system	Grease trap (450 mm × 350 mm × 300 mm) Simple bed of appropriate size to absorb the available water.	Periodical cleaning of grease trap (every week).	Rs. 50/– per square meter including labor cost (2006)			
3. Leach pit	Distance between the wall and the pit as 1 m. Pit (a diameter of 1.75 m and a depth of 1 m). Pit volume: 0.75 cubic meters.	Periodical cleaning of the P-trap and removal of the sludge	Rs. 1000/– per leach pit		Houses without adequate space for kitchen garden where waste water discharge is relatively more.	
4. Soak pit	Excavation of 1 m × 1 m × 1 m pit. Filling of pit by boulders from bottom 250 mm by 125 mm to 150 mm boulders; second 250 mm layer by 100 mm to 125 mm size boulders; third 250 mm layer by 50 mm to 75 mm size boulders. 225 mm earthen pot (or plastic container) over the last layer of the boulders	Filter to be cleaned every fortnight or month A hook of thick wire is required to take filter media out and clean/ wash it and dry and replace it in the earthen pot	For a rural family of five persons = Rs. 270/– (i.e., Rs. 54/– per person)	House owner can construct pit himself by getting the information of the design		Cheapest technology for management of water at household level Prevents gray water stagnation Prevents vector breeding

(continued)

Table 1.3 (continued)

Household level management						
	Technical description	O &M	Cost	Tech. Know.	Applicability	Advantage
5. Household level gray water treatment and reuse system	A three-stage gray water filtration system. It involves screening (grease and silt removal), soap froth removal, equalization and filtration.	Very frequent cleaning and user attention	Rs. 850/–	Individual households may construct and operate	In water scarce areas, water can be cleaned and reused not only for gardening but for other use also.	
Community level management						
	Technical description	O &M	Cost	Tech. Know.	Applicability	Advantage
1. Plantation with intercepting chamber	Gray water can be reused conveniently for plantation. Channel should be built from that point to the plantation area in such a way that the water does not stagnate on the platform. Water can be given to plantation either by piped root zone system or without piped root zone.	The platform, channel, and slit chamber will have to be cleaned periodically		Trained mason		Beneficial return from plantation. The application of the technology will help in preventing water stagnation around such public places.

Household level management						
	Technical description	O &M	Cost	Tech. Know.	Applicability	Advantage
2. Community leach pit	Depth: 1.2 m; diameter: More than 1.5 m.	Silt chamber should be periodically cleaned and sludge to be removed		Required		Stagnation of spilled over water Prevents vector breeding Main water source will not be contaminated
3. Soakaway channel	A. Sludge tank: 93 × 53 × 45 cm (deep) pit B. Soakaway channel: 4.5 m long and 0.6 m broad, 0.6 m deep in the beginning and 0.75 m in the end	The sludge tank should be cleaned once or twice a month.			More useful in rocky terrain where leach pit may not function properly.	Large quantities of community gray water can be absorbed without any open stagnation of gray water
3. Sludge stabilization ponds	The gray water is passed to large shallow ponds and placed serially in a stabilization system in which it is stabilized. A. Anaerobic ponds: Pond depth: 8–10 ft., solids settle at the bottom are digested anaerobically, hydraulic retention time: 12 days B. Facultative ponds: Pond depth: 4–5 ft. Upper layer—Aerobic (oxidation takes place) Lower layer—Anaerobic Hydraulic retention time: 3–5 days C. Maturation pond: Wholly aerobic, destruction of pathogens				Stabilized water pollution due to untreated gray water is avoided. Surface water pollution, due to untreated gray water is avoided.	The process is a natural process. The pathogenicity is reduced and the stabilized water becomes useable. Capital cost is very low. O&M cost is also very low & affordable.

1.5.3 Maintenance and Operation Challenges

Ongoing Maintenance Needs: While nature-based systems are generally low-maintenance, they still require periodic maintenance, such as vegetation management, sludge removal, and system monitoring. Without proper care, the systems can become less effective, leading to reduced treatment efficiency and potential failure.

Lack of Technical Expertise: In many areas, especially rural and underserved regions, there may be a lack of local technical expertise or skilled personnel to operate and maintain these systems. This gap can hinder their effectiveness and lead to system failure if not adequately managed.

Clogging and Sedimentation Issues: In soak pits and leach pits, the filtration medium (such as sand, gravel, and soil) can become clogged over time, reducing the percolation capacity. Similarly, constructed wetlands and waste stabilization ponds may accumulate sludge or organic matter, requiring periodic dredging or cleaning, which can be a labor-intensive process.

1.5.4 Environmental and Ecological Risks

Eutrophication and Overloading of Water Bodies: Improperly designed or maintained natural treatment systems may discharge treated effluent with excess nutrients, such as nitrogen and phosphorus, into nearby water bodies. This can lead to nutrient pollution, causing eutrophication, algal blooms, and oxygen depletion, which adversely affects aquatic ecosystems (Novotny 2011).

Disruption of Local Ecosystems: In constructed wetlands, there is a risk of introducing non-native plant species that may outcompete indigenous flora and disrupt local ecosystems. Additionally, the introduction of wastewater can alter local hydrology and water quality, impacting surrounding ecosystems if not managed properly.

Pathogen Removal Limitations: Nature-based solutions, especially those relying on biological treatment, may not be effective in completely removing pathogens such as bacteria, viruses, and protozoa, especially in systems that are not properly maintained. Without appropriate disinfection steps, treated wastewater may still pose a health risk.

1.5.5 Climate and Seasonal Variability

Temperature Sensitivity: Many nature-based solutions, like constructed wetlands and waste stabilization ponds, are highly sensitive to temperature variations. Cold climates can significantly reduce microbial activity and slow down the treatment process, leading to lower treatment efficiency during colder months. Similarly, extreme heat can cause high evaporation rates, affecting water levels and the treatment capacity of systems like ponds.

Rainfall and Flooding Issues: In areas with seasonal monsoons or heavy rainfall, natural systems can be overwhelmed by high volumes of wastewater or stormwater runoff, leading to inefficiencies in treatment. The influx of large volumes of water can lead to flooding of wetlands or waste stabilization ponds, and excess nutrients and solids may be washed away before the system can effectively treat them.

1.5.6 Water Reuse and Effluent Quality

Inconsistent Effluent Quality: While natural systems can remove many pollutants, they may not consistently produce effluent that meets stringent standards for water reuse, particularly in high-density or industrial areas. Effluent from NbS may still contain residual nutrients, organic matter, and pathogens that make it unsuitable for high-quality reuse applications, such as drinking water or irrigation for crops.

Health and Safety Concerns: In decentralized wastewater treatment systems, particularly when effluent is reused for agricultural or domestic purposes, the risk of contamination remains a concern. Without adequate disinfection, wastewater from natural systems may still pose a health hazard, particularly when it is used for crop irrigation or as a source of water for livestock or humans.

1.5.7 Funding and Financial Constraints

High Initial Setup Costs: Though the operational costs of nature-based solutions tend to be lower than conventional treatment methods, the initial capital investment in designing and constructing systems such as constructed wetlands or waste stabilization ponds can be significant. In developing countries like India, securing the necessary funding for large-scale implementation can be a challenge, especially in rural areas or small communities.

1.5.8 Social and Cultural Barriers

Public Perception and Awareness: People may have misconceptions or a lack of understanding about the benefits of nature-based solutions, leading to reluctance in adopting them. In some cultures, there may also be resistance to using wastewater treatment systems that involve direct contact with the environment, such as constructed wetlands or soak pits.

Behavioral Change and Community Engagement: Successful implementation of decentralized wastewater treatment systems often requires a shift in community behavior and attitudes. Without active participation and engagement, these systems may not be operated and maintained properly, resulting in failure or underperformance.

1.5.9 Scaling Challenges

Scalability in Urban Areas: While nature-based solutions are effective for small-scale or rural applications, scaling them up to handle the vast volumes of wastewater generated in urban settings can be challenging. The amount of land, space, and resources needed to treat large volumes of wastewater in dense urban environments may be impractical or too costly.

Integration with Existing Infrastructure: Integrating nature-based solutions into existing wastewater treatment infrastructure may be difficult. In urban areas, decentralized systems may need to be retrofitted into existing sanitation frameworks, which can be complex and costly.

1.6 Suggestions for Improving Nature-Based Solutions (NbS) for Decentralized Treatment of Wastewater

Nature-based solutions (NbS) for decentralized wastewater treatment have significant potential to address wastewater challenges sustainably. However, several improvements can be made to enhance their efficiency, scalability, and overall effectiveness. Below are key suggestions to improve NbS for decentralized wastewater treatment:

1.6.1 Improve System Design and Adaptation

1.6.1.1 Tailor Solutions to Local Contexts

NbS should be customized to suit local conditions, including climatic factors (temperature, rainfall), wastewater characteristics (domestic vs. industrial), and available space. For example, constructed wetlands or waste stabilization ponds should be designed with locally available plant species and soil types to maximize treatment efficiency and resilience.

1.6.1.2 Hybrid Systems (Integrated Approach)

Combine natural systems with low-energy mechanical systems to create hybrid solutions. For example, integrating constructed wetlands with microfiltration or ultraviolet (UV) disinfection can improve pathogen removal (Capodaglio et al. 2021). Hybrid systems can enhance overall treatment capacity, especially for higher-strength wastewater or areas with higher pollution loads.

1.6.1.3 Design for Flexibility and Scalability

NbS should be designed to be scalable to accommodate future population growth or increased wastewater volumes. For instance, modular wetlands can be expanded over time, and decentralized systems like waste stabilization ponds can be designed to operate in stages.

1.6.2 Enhance Operational and Maintenance Support

1.6.2.1 Training and Capacity Building

Local communities should be trained in the operation and maintenance of nature-based systems. Capacity building for local engineers, technicians, and community members can help ensure the system's longevity. Additionally, governments and NGOs can play a key role in providing technical support and knowledge-sharing platforms.

1.6.2.2 Regular Monitoring and Data Collection

Implement systems for regular monitoring of the wastewater quality, system performance, and potential ecological impacts. Monitoring should include key parameters like biochemical oxygen demand (BOD), chemical oxygen demand (COD), nutrient levels (nitrogen, phosphorus), pathogens, and sludge accumulation. Data can be used to improve system performance and inform maintenance schedules.

1.6.2.3 Automation of Monitoring Systems

While NbS are low-tech, integrating simple sensors or automated systems for key parameters (e.g., water level, flow rate, effluent quality) can help identify system failures early and improve maintenance procedures. Automated alerts can assist operators in taking timely corrective actions, reducing the risk of system malfunction.

1.6.3 Increase the Efficiency of Treatment Processes

1.6.3.1 Optimized Plant Selection for Constructed Wetlands

Careful selection of plant species is crucial to enhance the pollutant-removal efficiency of constructed wetlands. Certain plants, like cattails or reed species, are highly effective in removing nutrients, suspended solids, and organic matter. Research and experimentation to identify locally adaptable and effective plants can increase the treatment capacity of these systems.

1.6.3.2 Enhance Microbial Activity

Introduce methods to enhance the microbial activity in nature-based systems, such as bioaugmentation (introducing beneficial microorganisms) or organic amendments (e.g., compost). These methods can help break down organic pollutants faster and more efficiently, improving treatment performance.

1.6.3.3 Use of Aerobic Zones

In systems like waste stabilization ponds and constructed wetlands, creating or expanding aerobic zones can enhance the breakdown of organic material. In waste stabilization ponds, for instance, optimizing aeration can significantly improve the oxidation of nutrients and organic matter, leading to faster treatment times.

1.6.4 Address Environmental and Ecological Risks

1.6.4.1 Mitigate Eutrophication Risks

To prevent nutrient overload and eutrophication in receiving water bodies, nature-based solutions should be carefully managed. One strategy is to incorporate nutrient removal technologies (e.g., denitrification processes in wetlands) or integrate stormwater management practices to control nutrient levels in effluent (Triest et al. 2016).

1.6.4.2 Integrate Aquatic and Terrestrial Ecosystems

When designing constructed wetlands or other nature-based systems, integrate them with natural ecosystems such as buffer zones and riparian corridors. These areas can further filter and stabilize the effluent before it reaches open water bodies, reducing the risk of ecological damage.

1.6.4.3 Implement Regular Sludge Management

Regular removal of accumulated sludge in constructed wetlands or waste stabilization ponds is necessary to prevent overloading the system, as well as to avoid potential contamination of groundwater or surface water. The management of sludge and its reuse (e.g., as organic fertilizer) can further contribute to the sustainability of the system.

1.6.5 Optimize Resource Recovery from Wastewater

1.6.5.1 Biogas Production from Anaerobic Systems

In decentralized systems such as anaerobic digesters, implementing systems for capturing and utilizing biogas (methane) produced during the treatment process can create valuable energy resources. This can offset operational costs and contribute to energy sustainability in off-grid areas.

1.6.5.2 Nutrient Recovery

Nature-based systems can be optimized to recover valuable nutrients like nitrogen, phosphorus, and potassium, which can be used as fertilizers. This approach can close the nutrient loop, reduce the need for chemical fertilizers, and promote circular economy principles (Puchongkawarin et al. 2015). For example, constructed wetlands can be designed to accumulate nutrients that can later be extracted and used in agriculture.

1.6.5.3 Water Reuse for Irrigation

Ensure that treated wastewater from nature-based systems is reused in agriculture, reducing the pressure on freshwater resources. Properly treated effluent can be used for irrigation in nonfood crop areas or for landscape watering, contributing to water conservation.

1.6.6 Enhance Public Engagement and Community Participation

1.6.6.1 Community-Based Design and Management

Involving local communities in the design, operation, and maintenance of decentralized treatment systems is critical for long-term success. Participatory approaches help ensure the system aligns with local needs and ensures greater acceptance and accountability. Engaging communities also fosters a sense of ownership and responsibility, leading to better care and reduced chances of system failure.

1.6.6.2 Public Awareness Campaigns

Raise awareness about the benefits of nature-based solutions through educational programs, workshops, and outreach activities. Promoting the importance of wastewater treatment and the environmental, health, and economic benefits of NbS can help garner public support and encourage wider adoption.

To improve the effectiveness of nature-based solutions for decentralized wastewater treatment, a holistic approach is needed that combines optimal system design, better operational practices, environmental considerations, community engagement, and supportive policy frameworks. By addressing the challenges and making targeted improvements in these areas, NbS can become a more reliable and sustainable solution for wastewater treatment, offering a promising alternative to traditional, energy-intensive methods.

Hybrid systems, which combine nature-based solutions with traditional treatment technologies, will become more prevalent. For instance, constructed wetlands or waste stabilization ponds can be integrated with membrane filtration, UV disinfection, or solar-powered technologies (Justino et al. 2023). This combination will allow decentralized systems to handle higher pollution loads, improve pathogen removal, and meet stricter regulatory standards, making NbS viable even in urban or industrial settings. Furthermore, integrating NbS with urban green spaces, parks, and recreational areas could provide both ecological and social benefits while treating wastewater (Oral et al. 2020). Nature-based solutions are inherently more adaptable and resilient to climate variability (Stefanakis et al. 2021). Advances in biotechnology and microbiology will continue to improve the effectiveness of NbS. For example, engineered or bio-enhanced systems that use specific microbial consortia or algae to accelerate nutrient removal or pathogen degradation could become commonplace (Kaszycki et al. 2021).

1.7 Conclusions

Nature-based solutions (NbS) represent a promising and sustainable approach to decentralized wastewater treatment, offering a range of benefits, including low energy consumption, resource recovery, and enhanced environmental resilience. As the demand for effective and affordable wastewater treatment grows, especially in rural and urban fringe areas, NbS can offer a flexible, scalable solution. However, several challenges remain, including ensuring system efficiency, scalability, and long-term sustainability, as well as overcoming issues related to maintenance, public awareness, and policy support. By integrating smart monitoring systems, improving system design through hybrid approaches, and focusing on community engagement, these challenges can be addressed effectively. Furthermore, NbS can play an important role in resource recovery, water reuse, and climate change adaptation, helping to close the loop in wastewater management. As research and technology continue to evolve, the future of NbS looks promising, and they will likely become integral components of sustainable water management strategies worldwide. By supporting NbS with appropriate policies, regulatory frameworks, and innovative designs, their role in wastewater treatment will continue to expand, contributing to a more circular, resilient, and sustainable future for water resources globally.

References

Anh NV, Ha TD, Nhue TH, Heinss U, Morel A, Moura M, Schertenleib R (2002) Decentralized wastewater treatment-new concept and technologies for Vietnamese conditions. In: 5th specialised conference on small water and wastewater treatment systems, Istanbul, Turkey, pp 24–26

Capodaglio AG, Bolognesi S, Cecconet D (2021) Sustainable, decentralized sanitation and reuse with hybrid nature-based systems. Water 13(11):1583

Dalwani R, Gopal B (2020) Nature-based solutions for restoration of freshwater ecosystems: Indian experiences. In: Nature-based solutions for resilient ecosystems and societies. Springer, Singapore, pp 231–245

Harikumar PS, Bindhya MK (2012) A synoptic study on the preparation of a liquid waste management plan for Kerala State, India. Environ Nat Resour Res 2(2):74–83

Hariprasad NV, Dayananda HS (2013) Environmental impact due to agricultural runoff containing heavy metals–a review. Int J Sci Res Publ 672

Inyinbor Adejumoke A, Adebesin Babatunde O, Oluyori Abimbola P, Adelani Akande Tabitha A, Dada Adewumi O, Oreofe Toyin A (2018) Water pollution: effects, prevention, and climatic impact. In: Water challenges of an urbanizing world, vol 33, pp 33–47

Justino S, Calheiros CS, Castro PM, Gonçalves D (2023) Constructed wetlands as nature-based solutions for wastewater treatment in the hospitality industry: a review. Hydrology 10(7):153

Kaszycki P, Głodniok M, Petryszak P (2021) Towards a bio-based circular economy in organic waste management and wastewater treatment–the Polish perspective. New Biotechnol 61:80–89

Kothari R, Azam R, Bharti A, Goria K, Allen T, Ashokkumar V, Tyagi VV (2024) Biobased treatment and resource recovery from slaughterhouse wastewater via reutilization and recycling for sustainable waste approach. J Water Process Eng 58:104712

Li X, Zhang X, Zhao M, Zheng X, Wang Z, Fan C (2024) Application of decentralized wastewater treatment technology in rural domestic wastewater treatment. Sustainability 16(19):8635

Massoud MA, Tarhini A, Nasr JA (2009) Decentralized approaches to wastewater treatment and management: applicability in developing countries. J Environ Manag 90(1):652–659

Mawari G, Kumar N, Sarkar S, Frank AL, Daga MK, Singh MM et al (2022) Human health risk assessment due to heavy metals in ground and surface water and association of diseases with drinking water sources: A study from Maharashtra, India. Environ Health Insights 16:11786302221146020

Novotny V (2011) The danger of hypertrophic status of water supply impoundments resulting from excessive nutrient loads from agricultural and other sources. J Water Sustain 1(1):1–22

Oral HV, Carvalho P, Gajewska M, Ursino N, Masi F, Hullebusch EDV et al (2020) A review of nature-based solutions for urban water management in European circular cities: a critical assessment based on case studies and literature. Blue-Green Syst 2(1):112–136

Puchongkawarin C, Gomez-Mont C, Stuckey DC, Chachuat B (2015) Optimization-based methodology for the development of wastewater facilities for energy and nutrient recovery. Chemosphere 140:150–158

Sanwal CS (2021) Ill effects of untreated household Grey water used in agricultural irrigation. In: Multidimensional approaches to impacts of changing environment on human health. CRC Press, pp 247–261

Stefanakis AI, Calheiros CS, Nikolaou I (2021) Nature-based solutions as a tool in the new circular economic model for climate change adaptation. Circ Econ Sustain 1:303–318

Triest L, Stiers I, Van Onsem S (2016) Biomanipulation as a nature-based solution to reduce cyanobacterial blooms. Aquat Ecol 50:461–483

Ventura JRS, Tulipan JU, Banawa A, Umali KD, Villanueva JAL (2024) Advancements and challenges in decentralized wastewater treatment: a comprehensive review. Desalin Water Treat 100830

Zhang DQ, Jinadasa KBSN, Gersberg RM, Liu Y, Ng WJ, Tan SK (2014) Application of constructed wetlands for wastewater treatment in developing countries–a review of recent developments (2000–2013). J Environ Manag 141:116–131

Part II
Decentralized Technologies for Wastewater Treatment

Chapter 2
Constructed Wetland as a Sustainable Option to Decentralized System for Industrial Wastewater Treatment

Abhilasha Yadav, Deepak Kumar, Priya Saharan, Sashi Yadav, Sunil Kumar Chhikara, Smita S. Kumar, and Jyoti Rani

Abstract Wetlands constitute a vital source of habitat life for both people and wildlife by enhancing our standard of living and providing shelter to wildlife. Wetlands served as the foundation for both basic and complex discoveries. For instance, papyrus was used to make paper and ships, while dried reeds were used to build houses. Wetlands provided a range of economic and life-sustaining resources; therefore, early civilisations frequently developed around marshes and rivers. In ancient times, wetlands were used for wastewater treatment as a sustainable technology.. Constructed wetlands have been used to treat wastewater for more than five decades. The idea of using constructed wetlands for wastewater treatment emerged from the observation of natural wetlands' ability to filter and purify water, inspiring the design of artificial systems that mimic these ecological processes. Constructed wetlands have traditionally been used to treat residential or municipal wastewater, they are currently successfully used to treat a range of wastewater treatment processes, including the wastewater from wineries, dairies, and textile mills, and many other industries. Our water resources are on the verge of danger due to the disposal of untreated wastewater from the industries located around the water bodies. On the other hand, constructed wetlands (CWs) are of the unimpeded water surface flow type or the type of subsurface flow, are natural treatment systems that without the aid of mechanical or energy-intensive machinery, stabilize waste through the actions of bacteria, media, or plants. Conventional wastewater treatment methods have some drawbacks like high energy consumption, formation of disinfection by-

A. Yadav · D. Kumar · P. Saharan · J. Rani (✉)
Deenbandhu Chhotu Ram University of Science and Technology, Murthal, Haryana, India
e-mail: jyoti.energy@dcrustm.org

S. Yadav · S. K. Chhikara
Maharshi Dayanand University, Rohtak, Haryana, India

S. S. Kumar
Department of Environmental Studies, Central University of Haryana, Mahendergarh, Haryana, India

S. S. Kumar, V. Kumar (eds.), *Nature-Based Solutions for Decentralised Liquid Waste Management*, Environmental Challenges and Solutions,
https://doi.org/10.1007/978-3-032-08362-3_2

products, create environmental impacts, and concentrate disposal and limited treatment capacity. Therefore, the creation and maintenance of the artificial wetlands should be tailored to the specific pollutants of concern for optimal treatment performance. Industrial wastewater has been constructed wetlands that were treated employing both surface and subsurface flow. For subsurface flow created wetlands, both horizontal and vertical flow systems have been devised. It has also been claimed that other hybrid built wetlands have been used to cleanse industrial wastewater. However, vertical flow built wetlands are quite effective at treating a variety of wastewater pollutants.

Keywords Constructed wetlands · Wastewater · Nutrients removal · BOD · TSS · Decentralized system

Abbreviations

CW	Constructed Wetland
BOD	Biological Oxygen Demand
COD	Chemical Oxygen Demand
PE	Population Equivalent
TP	Total Phosphate
AOX	Absorbable Organic Halogen
BHT	Butylated Hydroxytoluene
BHA	Butylated Hydroxyanisole
TSS	Total Suspended Solids
PCBs	Polychlorinated Biphenyls
BTEX	Benzene, Toluene, Ethylbenzene, Xylene
PAHs	Polycyclic Aromatic Hydrocarbons
HRT	High Retention Time
LCW	Laboratory Constructed Wetland
SFCW	Surface Flow Constructed Wetland
VFCW	Vertical Flow Constructed Wetland
SBC	Sludge Biochar
CBC	Cattail Biochar
VSSF	Vertical Subsurface Flow
MFC	Microbial Fuel Cell
CWM	Constructed Wetland Microorganism
TN	Total Nitrogen
HSSF	Horizontal Subsurface Flow
HCW	Hybrid Constructed Wetland
SS	Suspended Solid
FWS	Free Water Surface
SFCW	Surface Flow Constructed Wetland

SSF	Subsurface Flow
HF	Horizontal Flow
VF	Vertical Flow
FTW	Floating Treatment Wetland
SMFC	Sediment Microbial Fuel Cell
EAB	Electroactive Bacteria ThC: Total Hydrocarbon

2.1 Introduction

A wetland is a natural or artificial area where water covers the soil or is present at or near the surface for varying periods, creating an environment that supports water-tolerant vegetation. These ecosystems are crucial for biodiversity, water purification, and flood control. Wetlands are vital for maintaining ecological balance, as they filter pollutants, store floodwaters, and provide habitats for diverse species. They also play a crucial role in carbon sequestration, helping to mitigate climate change. Wetlands are places that get flooded or regularly soaked in surface or ground water and for long enough to keep the soil moist. The Convention imposes general responsibilities on Contracting Parties to conserve wetlands over their whole territory as well as additional obligations for wetlands that have been listed on a "List of Wetlands of International Importance" to achieve these goals. A framework for international cooperation in the conservation of wetland habitats is provided by the Ramsar Convention, which was adopted in Iran in 1971 and is also known as the Convention on Wetlands of International Importance, particularly as a waterfowl habitat. The main goals of the Convention are to stop the loss of wetlands and to make sure that they are conserved because wetlands are crucial for ecological processes as they are having a diverse array of flora and fauna. Also they serve as filters, trapping contaminants from the water as it passes through and travels to oceans, lakes, and rivers. Natural wetlands provide a variety of purposes that are valuable to society. For instance, wetland plants have ability to fix carbon dioxide is good for the environment, and supporting food chains makes human activity (like fishing) more valuable. Ecological, sociocultural, and economic qualities can be three key categories that collectively make up the total value of wetland.

2.1.1 Services Provided by Natural Wetlands

- Improving water quality and quantity in groundwater aquifers: Including safe drinking water, supporting traditional agricultural practices, and preserving cultural landscapes, agricultural productivity, reduces the costs associated with water treatment, and sustains industries reliant on clean water sources.

- Flood incident control (protective buffers): It prevents damage to infrastructure, reduces recovery costs, and ensures the continuity of local economies by protecting agriculture, property, and livelihoods from flood impacts.
- Sediments and other contaminants are retained in wetland soil and plants: Wetlands retain sediments and contaminants in their soil and plants, providing economic services by reducing the need for costly water treatment processes. This natural filtration improves water quality, benefiting industries reliant on clean water and supporting sustainable agriculture by preventing soil degradation and maintaining ecosystem health.
- Carbon dioxide fixation: By mitigating climate change, reducing the financial impact of carbon emissions through natural carbon sequestration. This helps industries meet carbon regulations and offsets, while promoting sustainable practices that can lead to long-term cost savings and investment opportunities in carbon credit markets.
- Storing and releasing heat: Wetlands' ability to store and release heat provides economic services by moderating local climates, reducing energy costs associated with heating and cooling. This natural temperature regulation supports agriculture by stabilizing growing conditions and can lower infrastructure maintenance costs by mitigating extreme weather impacts.
- The assimilation of solar radiation and corresponding assistance to food chains: Sustainable farming maintains ecosystems that contribute to tourism and other nature-based industries, ensuring food security and economic resilience.

2.1.2 Constructed Wetlands

A "constructed wetland" is a wetland that has been created particularly for the purpose of pollution reduction through wastewater treatment process. Hence, planned CWs (constructed wetlands) are created to replicate original wetlands by adding plant life, soils, and microbial communities that support them in order to power the wastewater treatment process. In the past few decades, CWs, a green wastewater treatment method that mimics natural wetlands have been widely used to treat several forms of wastewater, including domestic sewage, agricultural wastewater, industrial effluent, mine drainage, landfill leachate, storm water, contaminated river water, and urban runoff (Saeed and Sun 2012; Sen et al. 2013; Badhe et al. 2014). The CWs are reportedly the best technology for developing nations because they are typically less expensive to build, run, and maintain. Also, they offer excellent pollutant removal efficiency with low energy consumption, and have capacity to treat a variety of effluent types from different sources. Wastewater treatment facilities have been closely observed for many years, with a focus on the heavy metals, removal of organic materials, nutrients, and faecal microorganisms. Heavy metals are practically non-biodegradable, in contrast to organic substances, and they prefer to assemble in the receiving water's sediments and aquatic life. CWs are slow-moving water bodies that are relatively shallow (usually less than 0.6 m), where dense stands of water-tolerant vegetation, such as reeds, cattails, or bulrushes, are grown. CWs are designed by humans in systems that often lengthy, narrow canals or pits and many other forms. When a wetland is created where none previously existed, it can be designed for the best possible wastewater treatment while avoiding several environmental problems and user disputes that arise with organic wetlands. Organic wetland is a type of wetland characterized by the accumulation of organic matter, such as peat or muck, in its soil. These wetlands are rich in decomposed plant material, which forms thick, carbon-rich layers, making them important for carbon storage and supporting unique ecosystems. Constructed wetlands (CWs) can be erected pollutant practically anywhere including on sites with limited uses in contrast to organic wetlands which are constrained by the location and accessibility of the sewerage source. Typically, a CW performs more efficiently than a natural wetland of similar size. It frequently has a sloped bottom and a regulated hydraulic regime. Because the vegetation and other system elements may be handled as needed, process reliability is also increased (Reed 1988). As a result during the past decades the use of CWs has greatly increased from the typical secondary and tertiary residential sewage treatment to the treatment of a variety of industrial effluents (Breitholtz et al. 2012; Mburu et al. 2012; Zhang et al. 2021; Verlicchi and Zambello 2014).

2.2 Status of Industrial Pollution

There are often very high quantities of organics, suspended particles, ammonia, or other contaminants in industrial effluent and as a result per treatment is almost always necessary before using built wetlands. Numerous industrial wastewater have compositions that are significantly different from municipal sewage and also from one another. BOD/COD is a ratio that a metric that infers the biological biodegradability speculatively. The wastewater is easily biodegradable whenever this ratio exceeds 0.5, such as effluent from slaughterhouses, breweries, the food industry, dairies, or yeast- and starch-manufacturing facilities. These wastewater typically have values lying between 0.6 and 0.7 for the BOD/COD ratio but it can be as high as 0.8 (Vymazal 2014). On the other hand, wastewater from the pulp and paper industry for instance are examples of wastewater with low levels of biodegradability due to low BOD/COD ratios. Population equivalent (PE) commonly referred to as unit per capita loading or equivalent person is one of the parameters. It compares the polluting capacity of an industry (as measured by biodegradable organic matter) to the population, or the number of people, who would produce the same amount of pollution. PE of 60 g BOD daily per individual could be used to make a rough comparison between the strength of industrial wastewater and municipal sewage (Pitter et al. 2009). The effluent of different industries have been explored for their treatment in CW. The higher the value of (40–70) PE the more treatment is required for that wastewater. Strength of industrial WW in terms of PE is shown in Table 2.1 (Vymazal 2014).

Table 2.1 Strength of industrial wastewater expressed in term of population equivalent according to BODs

Type of wastewater	Name of product	Population equivalent/ Production amount (ton)
Sugar mill industry	Sugar beet	45–70
Dairy industry	Milk	40–230
Paper mill industry	Paper	200–900
Brewery industry	Beer	150–350
Laundry industry	Laundry	350–900
Tannery	Leather	1000–5000
Petrochemical industry	Oil and petroleum	950–1500
Aquaculture industries	Fish farm effluent	250–350
Textile industry	Dye	850–1000

2.2.1 *Petrochemical Industry*

In petrochemical industries, oil and other hydrocarbon-rich petroleum sources, such as oil and natural gas sands, are refined into a variety of finished items and intermediate components (Speight 2015). In operations processes for making lubricating oil, topping, cracking, blowing down cooling towers, and dripping water and sludge from tanks, and storm water runoff and drainage all produce wastewater. Organics, oil and grease, suspended particles, ammonia, phenolics, H_2S, and heavy metals are all present in petroleum refineries as typical wastewater pollutants (Kadlec 1997).

2.2.2 *Tannery Industry*

Tannery industries generate/produce typically 25 m^3 and 80 m^3 of wastewater per tonne of processed raw material. The classification of tanning techniques depends on the substance (tannin or chromium) used to bond the collagen fibres. As much as 50 mg/L of chromium may be present in wastewater generated from tanneries (Keeley et al. 2016). Although chromium is not present in the wastewater from vegetable tanning procedures, there is a large concentration of resistant COD because tannin maximum 3000 mg/L are present (Lin and Liu 1994). Additionally, excessive salt concentrations in tannery wastewater can cause NaCl levels as high as 80 g/L (Hussain et al. 2018). Most anaerobic treatment technologies, such as anaerobic filters and up flow anaerobic sludge blanket (UASB) reactors are employed for vegetable tannery effluent (Mannucci et al. 2010).

2.2.3 Aquaculture Recirculating Fish and Prawn Industry

In recent years have seen a rapid expansion of aquaculture that is intense and uses recirculating the systems, and this growth has brought with it the necessity for reliable treatment. Biological treatment is thought to be the most economically viable method for enabling water reuse in these systems. The difficult to treat farm effluents are characterized by low waste concentrations at high volumes (Moore et al. 2001). Typically, the BOD5, COD, ammonia, and total phosphate contents in trout farm effluents are about 15, 12, 1.3, and 0.3 mg/L, respectively. This is diluted 20–25 times more than medium strength city wastewater. Phosphorus removal from the effluents of freshwater fish farms is the most difficult task in order to meet water quality targets (Comeau et al. 2001).

2.2.4 Paper and Pulp Industry

In terms of freshwater the paper and pulp industry comes in third place worldwide behind the chemical and primary metal industries (Alayu and Yirgu 2018). The composition of wastewater from the pulp and paper large volumes of effluent are produced by the pulp and paper industries nearly 75–225 m^3 of wastewater per unit of product. The process type, wood material kind, and technology employed, management techniques, internal recovery of the effluent and the volume of water consumed in the particular process (Pokhrel and Viraraghavan 2004; Jagaba et al. 2022). While some xenobiotic pollutant produce during the manufacturing paper and pulp process (dioxins, furans, phenol, resin acids, and chlorinated lignin), along with are naturally occurring wood extractive like tannin, resin acids, stillborn, and lignin (Ali and Sreekrishnan 2001).

2.2.5 Textile Industry

Two crucial elements in the textile manufacturing process are the dyeing and finishing operations (Lin and Liu 1994). The most common textile processing configuration includes size, removing impurities, cleaning, bleaching, mercerizing, dying, rinsing, and finishing (Armaroli and Balzani 2007). Moreover, the fabric wastewater pH ranges from 2 to 12 and can be very wide (Can et al. 2006). Pointed out that there are more than 10,000 dyes that can be purchased commercially that are synthetic in origin and chemically complicated structure, fad-resistant in the presence of water, light, and various other things. Additionally, the reduction of azo dyes produces aromatic amines, a class of chemicals widely regarded as hazardous to the environment and many of which are poisonous or carcinogenic (Pinheiro et al. 2004). The most recent techniques for removing colour from textile wastewater.

These technologies cover a wide range of chemical, physical, and biological (microbial and fungal cultures, as well as anaerobic cultures) processes.

The different type of pollutants are released by the industries are various classes with elevated organic parameters like COD, BOD, TOC, and heavy metals are typical wastewater pollutants create impact on environment. Table 2.2 shows that ammonia concentrations are highest in the paper and pulp sectors; BOD levels are higher in the fish, dairy, paper and pulp, and distillery industries; and TSS concentrations are highest in the distillery industries. The distillery, textile, and dairy sectors emit the majority of pollutants. A significant amount of heavy metals was removed from the landfill leachate.

Table 2.2 Principal pollutants of different industrial wastewater

Sr. No.	Industries	Primary pollutant	Other pollutants	References
1.	Paper and pulp mill	Highly intense colour, chlorophenolic compounds	Volatile organic compounds, fatty acids, lignin, AOX, or resins, BOD, COD, and suspended particles	Sumathi et al. (2014)
2.	Textile industries	Harmful residual dyes and highly coloured, often alkaline	Dyes and pigments BOD, COD, TOC, AOX, organic halogens, metals, chloride, sulphate, sulphide, and nitrogen	Birame (2012)
3.	Coffee processing industries	Acidic, rich in nutrients and organic matter, resultant phenolic compounds	Peptides, phenols, sugars, pectins, nutrients, Sulphur, nitrogen, and phosphorous	Subba Rao et al. (2020)
4.	Tannery industries	High salt content, up to 80 gdm^3 of NaCl, high organic loading	Chromium salts, phenolic, tannin, organic matter, BOD, TOC, COD	Mannucci et al. (2010)
5.	Oil refinery industries	Low biodegradability, include coloured, often alkaline	Polychlorinated biphenyls (PCBs), pesticides, dioxins, including mercury, pathogens, oil and grease, BTEX (benzene, toluene, ethylbenzene, xylene), polycyclic aromatic hydrocarbons (PAHs)	Radelyuk et al. (2021)
6.	Winery industries	Variable flow and loading, high amount of organic matter and suspended solids, acids	Butylated hydroxyanisole (BHA) and butylated hydroxytoluene (BHT) p-hydroxybenzoic acid, tannin acid, vanillic acid, gallic acid derivatives and syringic acid, flavanols, anthocyanins, flavonols, flavones, proanthocyanidins	Ahmad et al. (2020)
7.	Distillery industries	High Organic load	Phenolic compounds, mainly gallic acid, lactic acid, p-coumaric acid and gentisic acid, tartaric acid, succinic acid, acetic acid and malic acid	Rani et al. (2013)

2.3 Challenges in Industrial Wastewater Treatment

Industrial wastewater treatment poses several challenges due to the complexity and variability of pollutant present in it. Here are some common challenges in industrial wastewater treatment.

2.3.1 High Levels of Contaminants

Different type of industrial wastewater throughout paper and pulp, dairy, winery, textile, and distillery can contain high concentrations of contaminants like heavy metals, organic compounds, and toxic chemicals as discussed in Sect. 2.2. These contaminants can be difficult to remove and can pose a risk to human health and the environment if not treated properly.

2.3.2 High Volumes of Wastewater

Industries can generate large volumes of wastewater the treatment of which can be expensive and energy intensive. The cost of treatment can be further increased if the wastewater is highly concentrated. Different industries like winery, oil refinery, coffee processing industries, distillery, and textile create high volume of wastewater.

2.3.3 Variability of Wastewater Composition

The composition of industrial wastewater can vary widely according to the industry and the production processes involved. This variability can make it challenging to design and operate an effective treatment system in Table 2.3.

2.3.4 Regulatory Compliance

Industries are often subject to strict regulations governing the discharge of wastewater into the environment. Compliance with these regulations can be challenging and can require significant investment in treatment infrastructure development of the wastewater treatment.

Table 2.3 Comparison of wastewater pollutants released by the different industries

Parameters	pH	TSS	BOD	COD	Ammonium	Nitrate	Phosphate	Heavy metals	References
Paper and pulp industries	8.05	644	1300	–	2900	–	4.12	–	Alayu and Yirgu (2018)
Distillery industries	–	6337	2539	1386	–	11.9	17	–	Billore et al. (2001)
Dairy industries	8	644	1300	2920	–	3.96	–	–	Dipu et al. (2011)
Olive oil industries	7.02	577	–	264	124	96.33	9.45	–	Bruno (2021)
Winery industries	12.9	129	942	1558	28	–	17	–	Skrzypiecbcef and Gajewskaad (2017), Vymazal (2014)
Fish processing industries	10	–	1250	–	95	–	–	–	Vymazal (2014)
Meat industries	8	–	–	239	122	129	17	–	Zheng et al. (2022)
Textile industries	–	324	190	493	–	–	–	–	Hussain et al. (2018)
Landfill leachate	6.5	143	10–185	584.19	275–753	1.83	2.3	Mn removal of 35.27%, 69.48%, and 94%	Bulc (2006)

2.3.5 Energy Consumption

Running a wastewater treatment facility involves a lot of energy usage. For main and secondary stages of the wastewater treatment process seeing the most frequent use. The cost of energy may make up as much as 30% of the entire cost of operations. Therefore, controlling energy costs is a key concern for water treatment plants (Musabandesu and Loge 2021).

2.3.6 Operational Cost

Another significant consequence is high operational costs. To maintain regulation many chemicals are used for the wastewater treatment. During the process sludge is produced, and it needs to be adequately treated before disposal. Chemicals and sludge treatment have both seen considerable price increases recently (Keeley et al. 2016). Learning how to utilize correct chemical dosages and reduce sludge volumes is essential for minimizing cost.

2.3.7 Skilled Labour

Besides operation of wastewater treatment workforce is required for design, engineering, construction, maintenance, and troubleshooting, which required specialized persons in respective fields. The personnel employed in the wastewater treatment sector need to have extensive training and experience in order to be familiar with the technical aspects of the operation of the wastewater treatment plants. Because the entire treatment process depends on machines, the technicians must be available round the clock to manage any mishaps that may arise at the treatment plant.

2.3.8 Sludge Production

The leftover solid particles from the wastewater treatment process are referred to as sludge. Many chemicals are being used for wastewater treatment resulting in generation of hazardous sludge. Multiple treatments, including chemical, biological, and mechanical ones, are used in the therapeutic procedure. The disposal of extra sludge created as a by-product of the wastewater treatment process is the main issue. Untreated disposal can seriously damage the ecosystem.

Overall, the difficulties in treating industrial wastewater call for a customized and creative strategy to ensure that the wastewater is treated efficiently, in accordance with legal standards, at the lowest possible operational cost.

Technology complexity including the presence of toxic contaminants and irregular rate of wastewater generation is the major challenges in industries.

2.4 The Need for Sustainable and Possible Solutions

Wastewater treatment has been carried out by traditional centralized facilities for almost a century in the developed countries. These large installations are energy-intensive, necessitating additional investment in order to replace the ageing facilities, and construct new ones after 25–30 years (Armaroli and Balzani 2007). However, places with low income typically can't afford to build such substantial centralized facilities and appoint the skilled labour for operation and maintenance of facilities. As a result alternative treatment methods are needed in both developed and emerging locations as the sustainable solution. Decentralized wastewater system is economical as they reduce the energy demand for pumping the wastewater from its sources of generation to the treatment plants. Different type of CW may be developed to treat the different type of pollutants. Small CW may be developed to treat the industrial wastewater generated from small-scale industries. Decentralized CW has been used for wastewater treatment by different researchers. Decentralized CW is a good option of domestic wastewater treatment in developing countries that are struggling with food and energy supply. The majority of resources used in CW are renewable, so no synthetic raw materials are required. Instead, they generally need natural ones, including plants and aggregate materials, offering an ecological solutionBataille et al. 2018). Additionally, CWs have several difficulties due to the increasingly strict water safety standards for wastewater management and water recycling across the globe. More research and innovation activity are required.

The selection of appropriate plants and substrates highlights the crucial role of substrates and wetland phytoplankton in enhancing the performance of CWs. Strong focus should be placed on the careful selection of macrophytes in order to maximize the supply of oxygen and carbon compounds, increase toxin absorption, particularly for emerging toxin types, and decrease contaminant overload, heavy metals (Poindexter 2014). A detail of different phytoplankton used by different researchers in CW is shown in Table 2.4. Further, artificial ventilation, phase feeding, extrinsic carbon incorporation, an increase in microorganisms, the dispersal of different plants, a combination of various substrates, and hybrid CWs (Eslamian et al. 2019) may be employed to increase the efficiency of pollutant removal.

The utilization of plants to filter wastewater is always seen as a novel and appealing idea by society, which can raise their social acceptability. Natural wetland serves as decentralized system for wastewater treatment. The need for decentralised CWs is driven by several factors, including reduced environmental impact, efficient resource conservation, enhanced climate change resilience, greater social

Table 2.4 Summary of constructed wetlands and its performance for wastewater treatment

Sr. No.	Location of CW	Wastewater type	Types/Dimensions of wetland	HRT	Plant species	Pollutants removal efficiency	Remarks	References
1.	Barcelona, Spain	Winery wastewater	Vertical CW, horizontal CW	–	*Typha latifolia, Phragmites australis, Schoenoplectus validus*	More than 90% of the organic pollutants	No chemicals consumption and avoidance of off-site management and transportation practices in WW	Flores et al. (2023)
2.	South Florida, United States	Landfill leachate	Hybrid constructed wetlands	–	*Typha latifolia, Spartina alterniflora*	Removal of nitrogen removal up to 79%	Wood chip-CW reduced nitrate accumulation	Yang et al. (2023)
3.	Ljubljana, Slovenia	Domestic wastewater	Laboratory constructed wetland (LCW) with horizontal subsurface flow (117 cm × 30 cm × 50 cm)	4 days	*Water iris, Iris pseudacorus*	N-NO_2 removal was increased, but P-PO_4^{3-} removal was lowered	Microbes were more abundant in the effluent than the fibres	Rozman et al. (2023)
4.	Queensland, Australia	Wastewater treatment	(SFCW) surface flow-constructed wetland	2($m^3 \cdot m^{-2} \cdot day^{-1}$)	*Eleocharis dulcis, Schoenoplectus validus, Baumea articulata,* and *Bolboschoenus caldwellii*	Removal of NOx and ammonium up to (92–100%) respectively	–	Goukeh and Alamdari (2024)

(continued)

Table 2.4 (continued)

Sr. No.	Location of CW	Wastewater type	Types/Dimensions of wetland	HRT	Plant species	Pollutants removal efficiency	Remarks	References
5.	Italy	Landfill leachate	Hybrid constructed wetland	–	*Phragmites australis, Arundo donax, and A. plinii*	N_2, Cl^- (80–90%) respectively	The SPF process was carried out in a raceway pond reactor (RPR) as polishing step	Spiniello et al. (2023)
6.	Qingdao, China	Wastewater	Hybrid CW	–	*Typha latifolia*	Removing organic matter up to 76%	Electron donor supplementation and clogging repair	Wu et al. (2023)
7.	Ibadan, Nigeria	Sewage effluent	Vertical-flow constructed wetlands (VFCW)	(HRT) of 0.5, 1, and 2 days	*Coix lacryma-jobi, L (CL)*	Removal of ammonia and phosphate up to 96%	Due to their shorter more fibrous roots, which minimizes preferential flow paths in the system	Ewemoje et al. (2023)
8.	Ben Guerir, Morocco	Olive mill wastewater	Vertical-flow constructed wetlands (VF-CWs)	–	*Typha latifolia, Phragmites australis,* and *Arundo donax aquatic*	COD up to 91.44%, phenolic compounds up to 79.47%	High performance of hybrid system (sand filter + VF CWs) in reducing the organic pollution	Achak et al. (2023)
9.	Chongqing, China	Wastewater	SBC/CBC CW	–	*Typha latifolia*	Removal of nitrogen (91%), ammonia (93.5)	The sludge biochar had a stronger electron transfer capacity (ETC) than the cattail biochar	Zheng et al. (2022)

10.	Palwal, Haryana	Textile industry	VSSF CW	3 and 6 cm/day	*I. pseudocarus, S. secundatum, T. latifolia,* and *P. australis*	Removal of BOD (66%) and COD (84%)	The process is 42% more economical than the activated sludge process	Shenoy et al. (2022)
11.	Nanjing, China	Municipal wastewater treatment plants	MFC-CW	–	*Water hyacinth*	Nitrate removal up to 99.3%	Utilize biomass carbon sources in MFC-CW to treat secondary effluent for simultaneous nitrogen removal and energy recovery	Tao et al. (2022)
12.	Kharagpur, India	Wastewater treatment plants	Vertical flow CWs	–	*I. pseudocarus*	Removal and recovery of nitrogen up to 89%	Nutrient pollution increases exponentially with time	Chakravarthi et al. (2021)
13.	Rouyn-Noranda, Canada	Mine effluent	–	HRT of 47.4	*Typha Latifolia*	Removal of ammonia 97.2%, Se up to 85%	–	Etteieb et al. (2021)
14.	Omu-Aran, Nigeria	Piggery effluent treatment	(HSSF CW)	3 days, 0.014 m^3/day	*Canna indica* and *Oryza sativa L*	Removing heavy metals and nutrients up to 92%	Indicating a higher metal translocation potential	Olawale et al. (2021)
15.	Chongqing, China	Municipal wastewater	VFCWs	–	*C. pyrenoidosa and D. magna*	80% of the total nitrogen removed	Effluent could damage the anti-oxidative capabilities in enhanced ROS level	Liu et al. (2020)

(continued)

Table 2.4 (continued)

Sr. No.	Location of CW	Wastewater type	Types/Dimensions of wetland	HRT	Plant species	Pollutants removal efficiency	Remarks	References
16.	Penang, Malaysia	Palm oil mill effluent (POME)	Subsurface flow CTWS	12 days	*Cyperus alternifolius*	95% removal of suspended solids	Removal of pollutant with higher HRT	Davis et al. (2019)
17.	Lavras, Brazil	Swine wastewater	(VFCWS), (HFCWS)	4.7, 3.1, and 2.3 days for (HFCWS1)	*Cynodon* spp.	Remove up to 96.4–99.0% for TC, 94.2% and 97.6% for ThC	–	Amorim et al. (2019)
18.	UP, India	Domestic wastewater treatment	Constructed wetland microcosms (CWMs)	–	*Phragmites australis, Typha latifolia*	Removal of nitrogen up to 93%	–	Kumar and Dutta (2019)
19.	Chongqing, China	Wastewater treatment plant effluent	Vertical flow constructed wetland (VFCW)	–	*Oryza sativa L*	PO_4^{3-}-P (89.7–99.4%), NO_3^- -N (82.7–99.6%), and TN (51.9–93.7%)	Removed in the sacrificial iron anode chamber by co-precipitation and adsorption	Wang et al. (2019)
20.	Karnataka, India	Wastewater treatment	(HSSF-CW) and (VFCW)	12 h and 24 h	*Pennisetum pedicellatum* and *Cyperus rotundus*	PO_4-P, 85% and 90%, BOD 77% and 83%	Sustainable alternative to the tertiary conventional treatment of domestic wastewater	Thalla et al. (2019)
21.	Catania, Italy	Storm water runoff	Hybrid constructed wetland (H-CW)	–	*Clostridium perfringens, Typha latifolia*	Removal of TSS, BOD up to 89%	–	Ventura et al. (2019)

22.	Lahore, Pakistan	Textile effluent	VFCWs	31 h to 76 h	*Brachiaria mutica*	Decreased COD and BOD up to 79% and 71% and Cr (97%)	Smaller size with high efficiency is an advantage for field-scale applications	Hussain et al. (2018)
23.	Ljubljana, Slovenia	Landfill leachate	Vertical flow and horizontal flow	0.5 cm d^{-1}	*Phragmites australis* and *Typha latifolia*	Removal of COD was 50%, BOD (59%), and Fe (84%)	The ratio between N and P showed a limited level of P for biological processes	Bulc (2006)
24.	Irapuato Gto., México	Tannery industries	Artificial wetlands	–	*Scirpus americanus,*	The 16S rRNA phylogenetic study of SOB isolates demonstrates that they belong to the genera specific	The majority of the isolates can oxidize reduced Sulphur compounds and are organotrophic	Avila-Vega et al. (2014)

acceptance, economic feasibility, regulatory compliance, and the promotion of research and innovation. Overall, the need of sustainable solutions in terms of constructed wetlands is essential to address the environmental, social, and economic challenges associated with industrial wastewater treatment using chemical and physical methods. By adopting sustainable practices, constructed wetlands can provide long-term benefits while minimizing their ecological footprint and contributing to a more sustainable future.

Several researchers have explored the efficiency of various types of construction wetlands for the treatment of different types of industrial wastewater. A detailed summary of the research work done on constructed wetland is presented in Table 2.4.

2.5 CW as Sustainable Option for a Decentralized Wastewater Treatment System

Decentralized wastewater treatment systems, commonly referred as onsite or cluster systems, have a several advantages over conventional centralized systems. The following understated justifies the requirement for decentralized wastewater treatment systems.

2.5.1 Cost-Effectiveness

Decentralized systems may be more economical, especially in rural areas or places with dispersed populations. They lessen the expenses involved in infrastructure development do not need to transfer wastewater to a centralized treatment facility.

2.5.2 Environmental Sustainability

Decentralized CW have a reduced environmental footprint. They reduce the energy needed for pumping and transportation by treating wastewater closer to the source. Thus lowering greenhouse gas emissions. Decentralized systems can also be created to include sustainable practice like reusing water for irrigation or employing natural treatment methods.

2.5.3 Reliability and Resilience

Centralized wastewater treatment systems are susceptible to breakdowns, such as natural catastrophes or infrastructure failures, which can result in extensive sewage backups or contamination. On the other side, decentralized systems offer redundancy and resilience by distributing treatment capacity over numerous locations. The whole effluent treatment process isn't significantly impacted if one system malfunctions because the others can keep working. Wastewater from these systems can be treated and reused for non-potable purposes such as toilet flushing, irrigation, or industrial processes. It also aids in water conservation in areas with acute shortage of freshwater resources.

2.5.4 Scalability and Flexibility

Decentralized systems are scalable and flexible. They are competent appropriate for both small towns and single-family homes since they are made to handle a variety of effluent volumes and types. Decentralized system's modular design enables simple expansion or change in response to shifting requirements.

2.5.5 Protection of the Public's Health

Controlling the spread of waterborne infections and safeguarding the public's health depends on effective wastewater treatment. Decentralized systems move wastewater treatment closer to the point of origin, lowering the risk of contamination and thus enhancing community cleanliness in general. It's vital to remember that not all scenarios may call for decentralized wastewater treatment systems. When choosing the best wastewater treatment strategy for a particular area, it is important to consider elements including regional rules, population density, land availability, and water quality factors etc.

2.6 Classification of Constructed Wetland

CWs are classified depending upon the purpose and wetland hydrology.

2.6.1 Classification of Wetland Depending on Purpose

Constructed wetlands can be divided into three primary categories according to their purpose served by wetlands (ElZein et al. 2016).

2.6.1.1 Habitat Creation

Use of Constructed wetlands for Water and sewage treatment systems improve current ecological advantages by attracting birds, generating green areas, and providing habitats for animals. There are four distinct CW types: swamps, ephemeral wetlands, marshes, and ponds.

2.6.1.2 Flood Control

Systems that increase capacity to store storm water, particularly in urban locations that act as runoff catchers when there are floods.

2.6.1.3 Wastewater Treatment

The CW is developed for treatment of wastewater coming from a number of sources. CWs are based on the water flows direction and the type of plant that is present (Fig. 2.1). There are two general types based on the CW system's flow path (Xu et al. 2022).

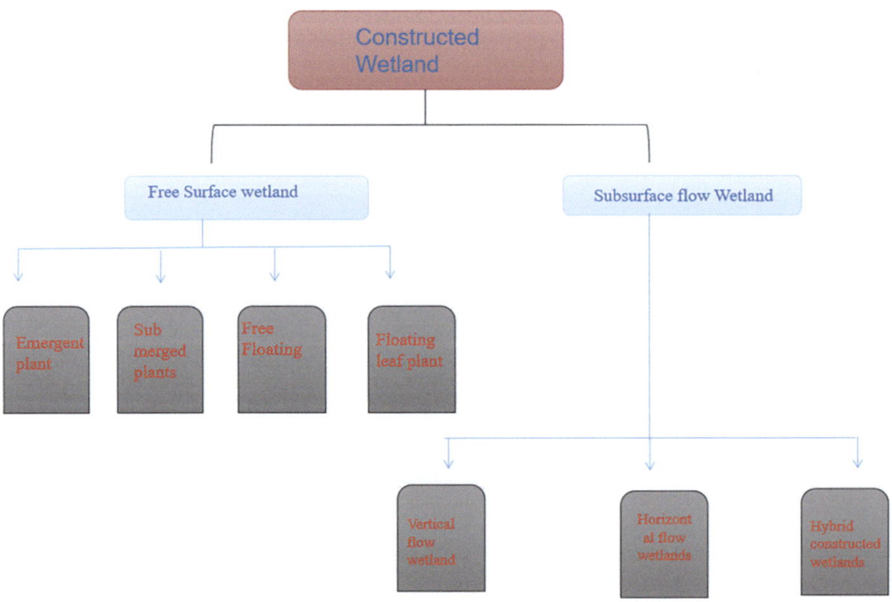

Fig. 2.1 A system for classifying of CWs utilized in wastewater management

2.6.2 *Classification of CW Based on Wetland Hydrology*

2.6.3 *On the Bases of Wetland Hydrology*

2.6.3.1 Free Water Surface (FWS)/Surface Flow Constructed Wetlands (SFCWs)

FWS systems resemble natural wetlands because they have shallow wastewater flows over saturated substrates. This kind of CW consists of a shallow basin or channel with a soil layer that is between 30 cm and 40 cm thick and is used to cultivate macrophytes. As geotextile, geomembrane, or clay material covers the bottom of the basin to prevent sewage/wastewater from penetrating groundwater (Patel et al. 2023). Above it is a water column beneath the earth layer is 10–50 cm deep, meaning that water is exposed to the atmosphere and solar radiation. The system's exit enables the water level to be adjusted (Fig. 2.2). In order to remove pollutants by the water flows horizontally through the plant stems and rhizomes, undergoing many physical, biological, and chemical processes before coming into contact with the various plant parts, and the accompanying biofilm. Common plant species employed include bulrush (*Scirpus* spp.), common reed (*Phragmites australis*), cattails (*Typha* spp.), and herbs (*Juncus* spp.) (Colares et al. 2020)

If the water inside the system is left nearly stagnant owing to poor design or construction, FWS CWs may breed mosquitoes. This type of wetland performs well in eliminating pathogens, biochemical oxygen demand, and suspended solids (SS), but phosphorus (P) removal is often limited (Valipour and Ahn 2016). FWS CWs have been used to treat primary and secondary municipal effluents, but they are mostly used to polish treated effluents, storm water runoff from roads and highways, and agricultural effluents.

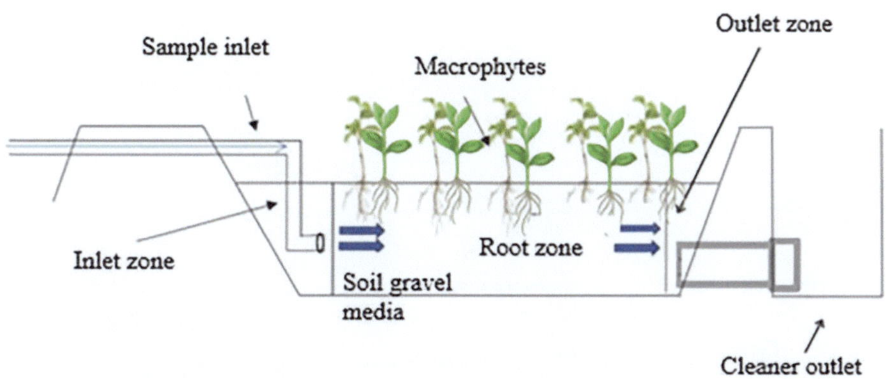

Fig. 2.2 A schematic illustration of a typical cross-section of a wetland with a free water surface

2.6.3.2 Subsurface Flow Constructed Wetlands (SSF CWs)

Depending on the direction of the flow, wastewater in SSF systems travels either horizontally or vertically through the substrate that promotes plant growth. SSF CWs could be further separated horizontal flow (HF) and vertical flow (VF).

Vertical Flow Constructed Wetlands (VFCWs)

The bed's bottom is covered in a geotextile/geomembrane material and slopes slightly (1–2%) (Stefanakis 2018). The most typical setting is a basin with many layers of progressively coarser sand and gravel from top to bottom. The substrate ranges in total thickness between 30 cm and 180 cm. Usually, the bed's top layer is made of sand. These aeration tubes provide for improved aeration of the deeper areas of the bed. Intermittent loading is preferred in VFCW. In the VFCW, the gravitational force is used for the flow of wastewater in vertical direction less surface area is needed for VFCWs than for HSF and FWS CWs (Bosnina 2021). VFCWs are particularly more successful because of their better aeration powers, eliminating organic matter (BOD) and ammonia nitrogen. Phosphorus removal is still limited, and many changes have been suggested to boost performance (Moore et al. 2001). The two most commonly used plant species are cattails (*Typha latifolia*) and common reeds (*Phragmites australis*) (Pedescoll et al. 2013). Influent enters the wetland system from the top layer of VFCW, as shown in Fig. 2.3, and then moves downwards with the help of vertical perforated aeration and is collected at the bottom using the drainage-collection pipeline system. Vertical flow beds offer higher initial purification than horizontal flow beds because the water travels from the surface to the bottom (Table 2.5).

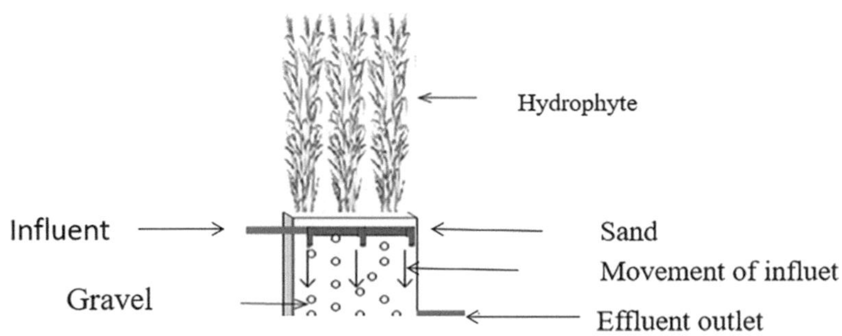

Fig. 2.3 A schematic illustration of a typical cross-section of vertical flow constructed wetland

Table 2.5 Advantages and disadvantages of VFCWs and HFCWs

Types of CW	Advantages	Disadvantages
VFCWs	It decreases area demand Excellent nitrification Better organics and suspended particles removal Straightforward hydraulics Enhancing oxygen mixing	Minimal flow distances Low nitrate removal Poor denitrification Greater technical requirement Performance decline, particularly in the removal of phosphorus
HFCWs	Extended flowing distance Established a nutrient gradients Effective at removing suspended particles and organics Possible denitrification Humic acid production for the elimination of nitrogen and phosphorus	High area demand and clogging issues Sulphur transformation that may impair nitrification sensitivity Decreased phosphorus removal efficiency Hydraulic calculations must be done carefully for optimum oxygen supply and minimal ammonium oxidation It is challenging for wastewater to move uniformly through the packed material (due to the possible presence of dead zones)

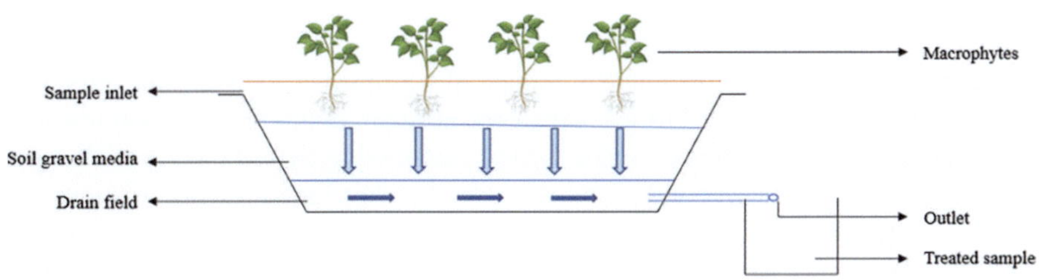

Fig. 2.4 A typical horizontal subsurface flow built constructed wetland

Horizontal Subsurface Flow Constructed Wetlands (HSF CWs)

In this type of constructed wetland, the water surface remains unexposed to the atmosphere because the water level is maintained approximately 5-10 cm below the gravel surface (Fig. 2.4). The substrate used is made up of several types of rock and gravel. Water enters the substrate media's pores horizontally, contacting the substrate grains, plant roots, and associated biofilm. The substrate layer can be anywhere between 30 cm and 100 cm thick. The bed's bottom typically has a 1–3% slope and is covered in an impermeable geomembrane (Eslamian et al. 2019).

Basins composed of gravel are used to cultivate common reeds (*Phragmites australis*) and other wetland plants species like *Typha* (e.g., *latifolia, angustifolia*) and *Scirpus* (e.g., *lacustris, californicus*) (Vymazal 2013). Although this CW type has been demonstrated to be very effective in treating municipal wastewater, removing SS and organic matter (BOD) at high rates. The system design enhancement is suggested to maximize performance in effluent recirculation, including wastewater step-feeding, water level increase, and effluent treatment with gravity filters

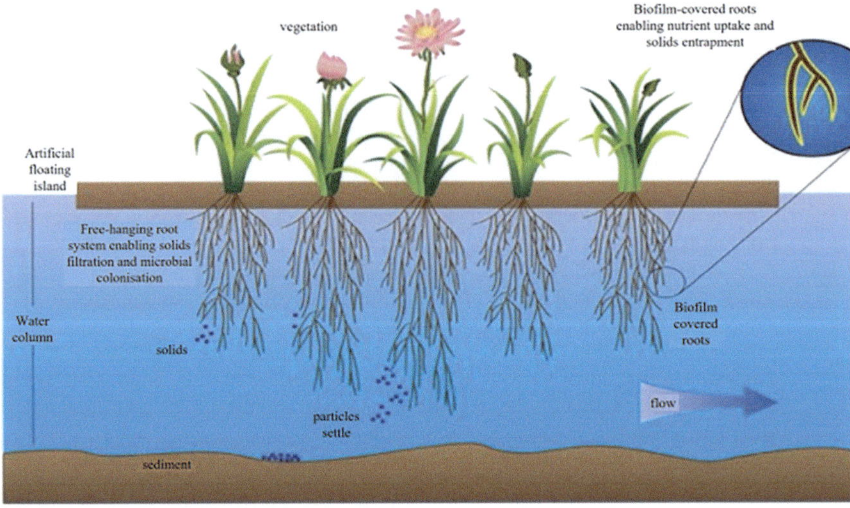

Fig. 2.5 An illustration of a typical floating treatment wetlands (FTWs)

containing particular substrates for efficient removal of nutrients (McCarty et al. 2011). HSF CWs treat a wide range of industrial wastewater, such as mine drainage, dairy, swine, olive mill, landfill leachate, cork effluent, contaminated groundwater, hydrocarbons, etc. SSF CWs could be further classified as hybrid CWs, which combine multiple different wetland systems, were also developed for the treatment of wastewater (Ávila et al. 2014). This design typically included two stages of multiple parallel CWs, such as VF-HF, HF-VF, HF-FWS, and FWS-HF CWs, in series (Vymazal 2013). It has several advantages and disadvantages as discussed below.

Floating Treatment We4tlands (FTWs)

In FTWs the plants are installed atop a floating plastic component. Because of this, some systems appear to be floating islands. The plants develop deep, dense roots in the underlying water column, comparable to those of other CW species (Headley et al. 2008). Water level changes have no impact on the combined system because plants and the porous plastic mat float on the water's surface (Fig. 2.5). FTWs are used to treat storm water, residential, and municipal wastewater and purify water in rivers, channels, lakes, etc.

2.7 Media and Vegetation in CW

Media and vegetation play crucial roles in treating industrial wastewater in constructed wetlands. Here is how they are used.

2.7.1 Media (Substrate)

The media, also known as the substrate, is a vital component of the constructed wetland system. It provides a support structure for the vegetation and also facilitates the treatment processes. Different types of media can be used, including gravel, sand, soil, or a combination of these materials (Hussain et al. 2018). The media selection depends on elements like hydraulic conductivity, pollutant removal efficiency, availability, and cost (Fassman-Beck et al. 2015). The media provides a surface area for microbial attachment and biofilm development. Microorganisms present in the biofilm help break down organic matter and metabolize pollutants through various biological processes, such as aerobic and anaerobic degradation, nitrification, and denitrification. The media also acts as a physical filter, trapping suspended solids and facilitating sedimentation (Tripathi and Goshisht 2022).

2.7.2 Vegetation

Vegetation, such as emergent plants and wetland grasses, plays a vital role in constructed wetlands. They contribute to the treatment process through the following mechanism: nutrient uptake, oxygenation, filtration and absorption, and hydraulic redistribution. Approximately 70% removal of suspended particles occurs by filtration methods (Pelissari et al. 2014). The selection of appropriate plant species depends on factors such as tolerance to wastewater pollutants, hydraulic conditions, and climate. Commonly used plant species include cattails (*Typha* spp.), reeds (*Phragmites* spp.), bulrushes (*Scirpus* spp.), *Pennisetum purpureum*, *P. australis* and *T. latifolia*, *Canna indica*, *Schoenoplectus validus*, *Carex rostrata*, *Phragmites*, and *Typha*. *Carex rostrata* and *T. lotifolia* performed better in removing ammonium from water in the summer, whereas *Typha* removed all of the nitrogen in the winter. When completely removing nitrogen in the summer, *Canna indica* and *Pennisetum purpureum* performed, but different grasses performed better in the winter (Das et al. 2021).

The combination of media and vegetation in constructed wetlands creates a favourable environment for microbial activity, physical filtration, and chemical processes, resulting in removing and transforming pollutants in industrial wastewater (Brix 2020).

2.8 Factors Affecting Constructed Wetland

Constructed wetlands are artificial ecosystems created to mimic the functions of wetlands in order to purify and treat wastewater. Various factors and components have an impact on performance and output efficiency before, during, and after the operation of the technology. Numerous component elements, including design parameters, wetland macrophytes, electrode materials and substrate, microbes, and hydraulic retention time (HRT), have been shown to have a significant impact on the performance of constructed wetlands (Vymazal 2010). Several elements, as listed below, affect how well-engineered wetlands work.

2.8.1 Design Parameters

A constructed wetland's performance is greatly influenced by its design parameters. These factors include the wetland's size and form, the water's depth, the kind of substrate used, and the flora. The precise requirements of the treated wastewater, the amount of available land, and the regional climate typically define the design parameters (Sumathi and Hung 2004).

2.8.2 Hydraulic Loading Rate (HLR)

It is the volume of wastewater introduced per unit area per day in CW. The hydraulic loading rate measures how quickly wastewater moves through a wetland. Since the HLR controls how long water stays in the wetland, which in turn impacts how much pollutants are being eliminated. It is a significant component in how well the wetland performs (Kadlec 2009).

2.8.3 Substrate Type

Substrate offers a surface on which microorganisms can grow and degrade the contaminants in the wastewater. Gravel, sand, and dirt are a few examples of possible materials for the substrate (Ohore et al. 2022). The substrate selection will depend on the type of pollutants to be removed and the treated wastewater's regulatory requirements. In order to remove pollutants substrates perform a variety of integrated physical, chemical, and more significant biological functions (Tao et al. 2022). They also provide storage for various toxins and act as a support system for living things. The effectiveness of the system as a whole is significantly influenced by the substrate or filtration medium selection in CW. More importantly, in order to enhance the system's performance, it is necessary to take into account the substrate's physical, chemical, and biological characteristics like (e.g., surface charge, chemical stability, toxicity, particle size, hydraulic and electrical conductivity, porosity, area, and mechanical strength and specific surface area) properties. For instance, both the hydraulic properties and the porosity are influenced by the sizes of the substrate particle. Its permeability influences how wastewater moves through the system, and microorganisms use it to undergo chemical and biological changes (Doherty and Zhao 2015).

2.8.4 Vegetation

A constructed wetland's effectiveness can be impacted by the plants employed in it in a number of different ways. They provide microorganisms with a surface to adhere to and take nutrients out of the water and release oxygen into it, supporting aerobic bacterial growth (Tiner 2016). Aquatic macrophytes flourish in wetlands or damp environments very efficiently (Brix and Schierup 1990). Some are planted in the ground, while others float on the water's surface without any support. Macrophytes are categorised into four types based on their growth environment: submerged, emergent, floating, and free-floating leafed macrophytes. (Brix and Schierup 1990). The type of macrophytes utilized in CWs decides the system's effectiveness may be considered during selection of vegetation in CWs.

(i) Effective natural climate adaption: Researchers found that the region's popular aquatic plants performed better in the CWs (Oodally et al. 2019).
(ii) Excessive biomass production and quick growth.
(iii) Their ability to absorb nutrients.
(iv) Adaptability and dissemination simplicity.
(v) Strong root development: New roots could release more oxygen than old or ageing roots in their experiment using a sediment microbial fuel cell (SMFC) and wetland plants. This suggests that while choosing a macrophyte, it is vital to consider the root's maturity and development (Can et al. 2006).
(vi) The ability to transmit oxygen to roots by fostering an aerobic environment. However, because of the wide variety of plant species, more research is required to assess and choose plant species that could be used in CW to simultaneously produce electricity and wastewater treatment (Stefanakis 2018).

2.8.5 Temperature

The water's temperature can influence the activity and proliferation of microorganisms. Greater temperatures generally speed up biological activities, but exceptionally high temperatures can also harm a wetland's ability to function by affecting the population of microbial flora growing on the substrate (Sumathi and Hung 2004).

2.8.6 pH

In a built wetland, the pH of the water can also impact the development and activity of microorganisms. Some wetland plants and microbes can withstand only within a range of pH levels, although most of the plants and microbes prefer a neutral pH (Quagraine 2018). Sphagnum moss (*Sphagnum* spp.), cranberry (*Vaccinium acrocarpon*), and tamarack (*Larix laricina*) are plants that can survive in an acidic environment; however, cattail (*Typha* spp.), bulrush (*Scirpus* spp.), common reed (*Phragmites australis*), and saltgrass (*Distichilis spicaqtao*) are plants usually grow in alkaline habitat (Makopondo et al. 2020).

2.8.7 Hydraulic Retention Time

The total time (in days) water remains in a wetland is known as the retention time. Longer retention times enable more thorough wastewater treatment but may also cause blockage and decreased performance of constructed wetlands.

2.8.8 Inlet Wastewater Characteristics

The wetland's performance may also be impacted by the characteristics of the wastewater entering it. The development and activity of microorganisms in the wetland can be influenced by several variables, including the kind and concentration of pollutants, temperature, and pH (Yan et al. 2023).

2.9 Recent Challenges in the Treatment of Industrial Wastewater by Using Constructed Wetlands

Recent challenges in the treatment of industrial wastewater using constructed wetlands includes the following.

2.9.1 Emerging Pollutants

Industrial operations introduce different pollutants in the wastewater, like antibiotics, personal care items, industries-specific pollutants, and microplastics. These contaminants might not be efficiently removed by conventional designed wetland systems, necessitating additional treatment methods or changes (Kataki et al. 2021).

Antibiotic-resistant microorganisms and antibiotics are frequently found in industrial wastewater. If the wetland system is not built correctly or managed, constructed wetlands may encourage the growth and spread of antibiotic-resistant genes in the environment (Niu et al. 2022).

Various industries produce particular types of pollutants that call for specialized treatment strategies. For instance, developed wetland systems have particular difficulties due to oil and grease in effluent from the petrochemical industry. To efficiently treat industrial-specific pollutants, developing specialized wetland designs and incorporating per-treatment procedures may be required (Butler-Minor 2016).

2.9.2 Climate Change and Extreme Weather Events

Droughts and extreme precipitation events can become more frequent and intense due to climate change. Built-in wetlands must be planned to withstand climate change and keep their best treatment capabilities (Mitsch and Gosselink 2015). Furthermore, variations in temperature and precipitation patterns can impact the development and well-being of wetland plants, which may impact pollutant removal.

2.9.3 Wastewater Resource Recovery and the Circular Economy

As the idea of the circular economy gets popular, it is becoming more important to recover valuable resources from it. Built-in wetlands can be used for resource recovery, such as extracting metals or removing nutrients for use as fertilizers (Morsy et al. 2020). However, it is difficult and call for creative solutions to maximize resource recovery while ensuring efficient pollutant removal.

2.9.4 Integrating Constructed Wetlands with Other Treatment Methods

Constructed wetlands are frequently employed as a component of a more extensive wastewater treatment system that uses various treatment methods. It is possible to increase treatment effectiveness by combining built wetlands with other treatment methods like membrane filtration or activated sludge systems (Brix 2020). Developing and optimizing the integration of several treatment systems presents difficulties in terms of hydraulic compatibility, process control, and operation management.

2.10 Conclusion

The industries are releasing several million litres of wastewater per day. If not treated, this wastewater may impact aquatic and terrestrial life. Decentralized constructed wetlands may offer a quick, easy, and cost-effective solution for treating some particular types of industrial wastewater. Several factors, including design parameters, hydraulic loading rate, substrate type, vegetation, temperature, pH, retention time, and inlet wastewater characteristics, influence the performance of a constructed wetland. Understanding these factors and designing the wetland accordingly is essential for achieving optimal performance and efficient wastewater treatment. Compared to conventional treatment techniques, constructed wetlands require less energy. They rely on organic procedures like photosynthesis and microbial activity to remediate wastewater. They are more energy-efficient and have a smaller carbon imprint. One of the main benefits of constructed wetlands is their potential to manage industrial wastewater while encouraging environmental sustainability. A sustainable and affordable option for decentralized industrial wastewater treatment is to build wetlands. They offer advantages for the environment, energy efficiency, and the flexibility to alter treatment procedures. In order to get beyond these obstacles and realize the full potential of artificial wetlands for industrial wastewater treatment, there is a constant need for study, innovation, and cooperation between academics, engineers, and business professionals.

References

Achak M, Boumya W, Elamraoui S, Asdiou N, Taoufik N, Barka N et al (2023) Performance of olive mill wastewater treatment using hybrid system combining sand filtration and vertical flow constructed wetlands. J Water Process Eng 53:103737. https://doi.org/10.1016/j.jwpe.2023.103737

Ahmad B, Yadav V, Yadav A, Rahman MU, Yuan WZ, Li Z, Wang X (2020) Integrated biorefinery approach to valorize winery waste: a review from waste to energy perspectives. Sci Total Environ 719:137315

Alayu E, Yirgu Z (2018) Advanced technologies for the treatment of wastewaters from agro-processing industries and cogeneration of by-products: a case of slaughterhouse, dairy and beverage industries. Int J Environ Sci Technol 15:1581–1596

Ali M, Sreekrishnan TR (2001) Aquatic toxicity from pulp and paper mill effluents: a review. Adv Environ Res 5(2):175–196

Amorim FD, Silva JRM, Fia R, Oliveira LFCD, Campos CMM (2019) Coliform removal in a constructed wetland system used in post-swine effluent treatment. Rev Ambient Água 14

Armaroli N, Balzani V (2007) The future of energy supply: challenges and opportunities. Angew Chem Int Ed 46(1–2):52–66

Ávila C, Matamoros V, Reyes-Contreras C, Piña B, Casado M, Mita L et al (2014) Attenuation of emerging organic contaminants in a hybrid constructed wetland system under different hydraulic loading rates and their associated toxicological effects in wastewater. Sci Total Environ 470–471:1272–1280

Avila-Vega DE, Alvarez-Mayorga B, Arvizu-Medrano SM, Pacheco-Aguilar R, Martinez-Peniche R, Hernandez-Iturriaga M (2014) Microbiological profile and incidence of salmonella and

listeria monocytogenes on hydroponic bell peppers and greenhouse cultivation environment. J Food Prot 77(11):1904–1910
Badhe N, Saha S, Biswas R, Nandy T (2014) Role of algal biofilm in improving the performance of free surface, up-flow constructed wetland. Bioresour Technol 169:596–604
Bataille C, Åhman M, Neuhoff K, Nilsson LJ, Fischedick M, Lechtenböhmer S et al (2018) A review of technology and policy deep decarbonization pathway options for making energy-intensive industry production consistent with the Paris agreement. J Clean Prod 187:960–973
Billore SK, Singh N, Ram HK, Sharma JK, Singh VP, Nelson RM, Dass P (2001) Treatment of a molasses based distillery effluent in a constructed wetland in Central India. Water Sci Technol 44(11–12):441–448
Birame CS (2012) Development of a low-cost alternative for metal removal from textile wastewater. CRC Press. Balkema
Bosnina M (2021) Characterising and modelling pollutant dynamics in urban stormwater constructed wetlands. Doctoral dissertation, Cardiff University
Breitholtz M, Näslund M, Stråe D, Borg H, Grabic R, Fick J (2012) An evaluation of free water surface wetlands as tertiary sewage water treatment of micro-pollutants. Ecotoxicol Environ Saf 78:63–71
Brix H (2020) Wastewater treatment in constructed wetlands: system design, removal processes, and treatment performance. In: Constructed wetlands for water quality improvement. CRC Press, pp 9–22
Brix H, Schierup HH (1990) Soil oxygenation in constructed reed beds: the role of macrophyte and soil-atmosphere interface oxygen transport. In: Constructed wetlands in water pollution control. Pergamon, pp 53–66
Bruno, M. R. (2021). Sustainable use and practical application in the industrial sector of permanent crops biomass: Orange tree (C. sinensis L.), Apricot tree (P. armeniaca L.) and Olive tree (O. europea L.) 11563/149565
Bulc TG (2006) Long term performance of a constructed wetland for landfill leachate treatment. Ecol Eng 26(4):365–374
Butler-Minor C (2016) Clear Creek estuary restoration: establishing an ecological monitoring program and baseline conditions. Environmental Science and Management Professional Master's Project Reports, 52
Can OT, Kobya M, Demirbas E, Bayramoglu M (2006) Treatment of the textile wastewater by combined electrocoagulation. Chemosphere 62(2):181–187
Chakravarthi VP, Ratri A, Masumi S, Borosha S, Ghosh S, Christenson LK et al (2021) Granulosa cell genes that regulate ovarian follicle development beyond the antral stage: the role of estrogen receptor β. Mol Cell Endocrinol 528:111212
Colares GS, Dell'Osbel N, Wiesel PG, Oliveira GA, Lemos PHZ, da Silva FP et al (2020) Floating treatment wetlands: a review and bibliometric analysis. Sci Total Environ 714:136776
Comeau Y, Brisson J, Réville JP, Forget C, Drizo A (2001) Phosphorus removal from trout farm effluents by constructed wetlands. Water Sci Technol 44(11–12):55–60
Das MT, Kumar S, Ghosh P, Shah G, Malyan SK, Bajar S et al (2021) Remediation strategies for mitigation of phthalate pollution: challenges and future perspectives. J Hazard Mater 409:124496
Davis LG, Madsen DB, Becerra-Valdivia L, Higham T, Sisson DA, Skinner SM et al (2019) Late upper paleolithic occupation at Cooper's Ferry, Idaho, USA,~ 16,000 years ago. Science 365(6456):891–897
Dipu S, Kumar AA, Thanga VSG (2011) Phytoremediation of dairy effluent by constructed wetland technology. Environmentalist 31:263–278
Doherty L, Zhao Y (2015) Operating a two-stage microbial fuel cell–constructed wetland for fuller wastewater treatment and more efficient electricity generation. Water Sci Technol 72(3):421–428
ElZein ZAKAA, Abdou A, Abd ElGawad I (2016) Constructed wetlands as a sustainable wastewater treatment method in communities. Procedia Environ Sci 34:605–617

Eslamian S, Okhravi S, Eslamian F (2019) Constructed wetlands: hydraulic design. CRC Press
Etteieb S, Zolfaghari M, Magdouli S, Brar KK, Brar SK (2021) Performance of constructed wetland for selenium, nutrient and heavy metals removal from mine effluents. Chemosphere 281:130921
Ewemoje OE, Semprini L, Wood BD, Sangodoyin AY, Radniecki TS (2023) Effectiveness of the tropical plants Rhynchospora corymbosa and Coix lacryma-jobi in vertical flow constructed wetlands for municipal primary sewage effluent treatment. Int J Phytoremediation:1–16
Fassman-Beck E, Wang S, Simcock R, Liu R (2015) Assessing the effects of bioretention's engineered media composition and compaction on hydraulic conductivity and water holding capacity. J Sustain Water Built Environ 1(4):04015003
Flores L, Josa I, García J, Pena R, Garfí M (2023) Constructed wetlands for winery wastewater treatment: a review on the technical, environmental and socio-economic benefits. Sci Total Environ 163547
Goukeh MN, Alamdari N (2024) Removal of contaminants in stormwater via subsurface-flow wetlands: a review with focus on nutrients, heavy metals, and PFAS. J Environ Eng 150(3):03124001
Headley T, Tanner CC, Council, A. R (2008) Application of floating wetlands for enhanced for stormwater treatment: a review. Auckland Regional Council, Auckland
Hussain Z, Arslan M, Malik MH, Mohsin M, Iqbal S, Afzal M (2018) Integrated perspectives on the use of bacterial endophytes in horizontal flow constructed wetlands for the treatment of liquid textile effluent: phytoremediation advances in the field. J Environ Manag 224:387–395
Jagaba AH, Kutty SRM, Baloo L, Birniwa AH, Lawal IM, Aliyu MK et al (2022) Combined treatment of domestic and pulp and paper industry wastewater in a rice straw embedded activated sludge bioreactor to achieve sustainable development goals. Case Stud Chem Environ Eng 6:100261
Kadlec RH (1997) Deterministic and stochastic aspects of constructed wetland performance and design. Water Sci Technol 35(5):149–156
Kadlec RH (2009) Comparison of free water and horizontal subsurface treatment wetlands. Ecol Eng 35(2):159–174
Kataki S, Chatterjee S, Vairale MG, Dwivedi SK, Gupta DK (2021) Constructed wetland, an eco-technology for wastewater treatment: a review on types of wastewater treated and components of the technology (macrophyte, biolfilm and substrate). J Environ Manag 283:111986
Keeley J, Jarvis P, Smith AD, Judd SJ (2016) Coagulant recovery and reuse for drinking water treatment. Water Res 88:502–509
Kumar S, Dutta V (2019) Constructed wetland microcosms as sustainable technology for domestic wastewater treatment: an overview. Environ Sci Pollut Res 26(12):11662–11673
Lin SH, Liu WY (1994) Treatment of textile wastewater by ozonation in a packed-bed reactor. Environ Technol 15(4):299–311
Liu M, Li X, He Y, Li H (2020) Aquatic toxicity of heavy metal-containing wastewater effluent treated using vertical flow constructed wetlands. Sci Total Environ 727:138616
Makopondo RO, Rotich LK, Kamau CG (2020) Potential use and challenges of constructed wetlands for wastewater treatment and conservation in game lodges and resorts in Kenya. Sci World J 2020:9184192
Mannucci F, Cresci G, Maiolino R, Marconi A, Gnerucci A (2010) A fundamental relation between mass, star formation rate and metallicity in local and high-redshift galaxies. Mon Not R Astron Soc 408(4):2115–2127
Mburu N, Sanchez-Ramos D, Rousseau DP, van Bruggen JJ, Thumbi G, Stein OR et al (2012) Simulation of carbon, nitrogen and Sulphur conversion in batch-operated experimental wetland mesocosms. Ecol Eng 42:304–315
McCarty, Perry L., Jaeho Bae, and Jeonghwan Kim. (2011) Domestic wastewater treatment as a net energy producer–can this be achieved?. 7100-7106
Mitsch WJ, Gosselink JG (2015) Wetlands. Wiley

Moore R, Quarmby J, Stephenson T (2001) The effects of media size on the performance of biological aerated filters. Water Res 35(10):2514–2522

Morsy KM, Mostafa MK, Abdalla KZ, Galal MM (2020) Life cycle assessment of upgrading primary wastewater treatment plants to secondary treatment including a circular economy approach. Air Soil Water Res 13:1178622120935857

Musabandesu E, Loge F (2021) Load shifting at wastewater treatment plants: a case study for participating as an energy demand resource. J Clean Prod 282:124454

Niu L, Liu W, Juhasz A, Chen J, Ma L (2022) Emerging contaminants antibiotic resistance genes and microplastics in the environment: introduction to 21 review articles published in CREST during 2018–2022. Crit Rev Environ Sci Technol 52(23):4135–4146

Ohore OE, Qin Z, Sanganyado E, Wang Y, Jiao X, Liu W, Wang Z (2022) Ecological impact of antibiotics on bioremediation performance of constructed wetlands: microbial and plant dynamics, and potential antibiotic resistance genes hotspots. J Hazard Mater 424:127495

Olawale O, Raphael DO, Akinbile CO, Ishuwa K (2021) Comparison of heavy metal and nutrients removal in Canna indica and Oryza sativa L. based constructed wetlands for piggery effluent treatment in north-Central Nigeria. Int J Phytoremediation 23(13):1382–1390

Oodally A, Gulamhussein M, Randall DG (2019) Investigating the performance of constructed wetland microbial fuel cells using three indigenous south African wetland plants. J Water Process Eng 32:100930

Patel SP, Othus M, Chen Y, Wright GP Jr, Yost KJ, Hyngstrom JR et al (2023) Neoadjuvant–adjuvant or adjuvant-only pembrolizumab in advanced melanoma. N Engl J Med 388(9):813–823

Pedescoll A, Sidrach-Cardona R, Sánchez JC, Carretero J, Garfi M, Bécares E (2013) Design configurations affecting flow pattern and solids accumulation in horizontal free water and subsurface flow constructed wetlands. Water Res 47(3):1448–1458

Pelissari C, Sezerino PH, Decezaro ST, Wolff DB, Bento AP, de Carvalho Junior O, Philippi LS (2014) Nitrogen transformation in horizontal and vertical flow constructed wetlands applied for dairy cattle wastewater treatment in southern Brazil. Ecol Eng 73:307–310

Pinheiro HM, Touraud E, Thomas O (2004) Aromatic amines from azo dye reduction: status review with emphasis on direct UV spectrophotometric detection in textile industry wastewaters. Dyes Pigments 61(2):121–139

Pitter MC, See CW, Somekh MG (2009) Full-field heterodyne interference microscope with spatially incoherent illumination. Opt Lett 29(11):1200–1202

Poindexter CM (2014) Wetland water flows and interfacial gas exchange. Doctoral dissertation, UC Berkeley

Pokhrel D, Viraraghavan T (2004) Treatment of pulp and paper mill wastewater—a review. Sci Total Environ 333(1–3):37–58

Quagraine EK (2018) Two decades constructed wetland experience in treating municipal effluent for power plant cooling at the Shand Power Station, SaskPower part VI: multivariate (quadratic) function model (s) for o-PO4^ sup 3-^-P removal. J Water Sustain 8(1):25–58

Radelyuk I, Tussupova K, Persson M, Zhapargazinova K, Yelubay M (2021) Assessment of groundwater safety surrounding contaminated water storage sites using multivariate statistical analysis and Heckman selection model: a case study of Kazakhstan. Environ Geochem Health 43:1029–1050

Rani J, Yadav KL, Prakash S (2013) Enhanced dielectric, ferroelectric and optical properties of lead free (K0. 17Na0. 83) NbO3 ceramic with WO3 addition. Mater Sci Eng B 178(20):1469–1475

Reed PB (1988) National list of plant species that occur in wetlands: national summary, vol 88 (24). US Department of the Interior, Fish and Wildlife Service, Research and Development

Rozman U, Klun B, Kalčíková G (2023) Distribution and removal of microplastics in a horizontal sub-surface flow laboratory constructed wetland and their effects on the treatment efficiency. Chem Eng J 461:142076

Saeed T, Sun G (2012) A review on nitrogen and organics removal mechanisms in subsurface flow constructed wetlands: dependency on environmental parameters, operating conditions and supporting media. J Environ Manag 112:429–448

Sen G, Rani GU, Mishra S (2013) Microwave assisted synthesis of poly (2-hydroxyethylmethacrylate) grafted agar (Ag-gP (HEMA)) and its application as a flocculant for wastewater treatment. Front Chem Sci Eng 7:312–321
Shenoy A, Shukla BK, Bansal V (2022) Sustainable design of textile industry effluent treatment plant with constructed wetland. Mater Today Proc 61:537–542
Skrzypiecbcef K, Gajewskaad MH (2017) The use of constructed wetlands for the treatment of industrial wastewater. J Water Land Dev 34(1):233
Speight JG (2015) Handbook of petroleum product analysis. Wiley
Spiniello I, De Carluccio M, Castiglione S, Amineva E, Kostryukova N, Cicatelli A et al (2023) Landfill leachate treatment by a combination of a multiple plant hybrid constructed wetland system with a solar photoFenton process in a raceway pond reactor. J Environ Manag 331:117211
Stefanakis AI (2018) Introduction to constructed wetland technology. Wiley, Chichester, pp 1–21
Subba Rao N, Ravindra B, Wu J (2020) Geochemical and health risk evaluation of fluoride rich groundwater in Sattenapalle region, Guntur district, Andhra Pradesh, India. Hum Ecol Risk Assess Int J 26(9):2316–2348
Sumathi S, Hung YT (2004) Treatment of pulp and paper mill wastes. In: Handbook of industrial and hazardous wastes treatment. CRC Press, pp 507–555
Sumathi C, MohanaPriya D, Swarnalatha S, Dinesh MG, Sekaran G (2014) Production of prodigiosin using tannery fleshing and evaluating its pharmacological effects. Sci World J 2014
Tao M, Jing Z, Tao Z, Luo H, Zuo S, Li YY (2022) Efficient nitrogen removal in microbial fuel cell–constructed wetland with corncobs addition for secondary effluent treatment. J Clean Prod 332:130108
Thalla AK, Devatha CP, Anagh K, Sony E (2019) Performance evaluation of horizontal and vertical flow constructed wetlands as tertiary treatment option for secondary effluents. Appl Water Sci 9:1–9
Tiner RW (2016) Wetland indicators: a guide to wetland formation, identification, delineation, classification, and mapping. CRC Press
Tripathi N, Goshisht MK (2022) Recent advances and mechanistic insights into antibacterial activity, antibiofilm activity, and cytotoxicity of silver nanoparticles. ACS Appl Bio Mater 5(4):1391–1463
Valipour A, Ahn YH (2016) Constructed wetlands as sustainable ecotechnologies in decentralization practices: a review. Environ Sci Pollut Res 23:180–197
Ventura D, Barbagallo S, Consoli S, Ferrante M, Milani M, Licciardello F, Cirelli GL (2019) On the performance of a pilot hybrid constructed wetland for stormwater recovery in Mediterranean climate. Water Sci Technol 79(6):1051–1059
Verlicchi P, Zambello E (2014) How efficient are constructed wetlands in removing pharmaceuticals from untreated and treated urban wastewaters? A review. Sci Total Environ 470:1281–1306
Vymazal J (2010) Constructed wetlands for wastewater treatment. Water 2(3):530–549
Vymazal J (2013) Emergent plants used in free water surface constructed wetlands: a review. Ecol Eng 61:582–592
Vymazal J (2014) Constructed wetlands for treatment of industrial wastewaters: a review. Ecol Eng 73:724–751
Wang Y, Lin Z, Wang Y, Huang W, Wang J, Zhou J, He Q (2019) Sulfur and iron cycles promoted nitrogen and phosphorus removal in electrochemically assisted vertical flow constructed wetland treating wastewater treatment plant effluent with high S/N ratio. Water Res 151:20–30
Wu H, Wang R, Yan P, Wu S, Chen Z, Zhao Y et al (2023) Constructed wetlands for pollution control. Nat Rev Earth Environ:1–17
Xu D, Yin X, Zhou S, Jiang Y, Xi X, Sun H, Wang J (2022) A review on the remediation of microplastics using constructed wetlands: bibliometric, co-occurrence, current trends, and future directions. Chemosphere 134990
Yan C, Huang J, Lin X, Wang Y, Cao C, Qian X (2023) Performance of constructed wetlands with different water level for treating graphene oxide wastewater: characteristics of plants and microorganisms. J Environ Manag 334:117432

Yang X, Arias ME, Ergas SJ (2023) Hybrid constructed wetlands amended with zeolite/biochar for enhanced landfill leachate treatment. Ecol Eng
Zhang B, Lin N, Chen X, Fan Q, Chen X, Ren T et al (2021) Nonlinear water quality response to numerical simulation of in situ phosphorus control approaches. Water 13(5):725
Zheng F, Fang J, Guo F, Yang X, Liu T, Chen M et al (2022) Biochar based constructed wetland for secondary effluent treatment: waste resource utilization. Chem Eng J 432:134377

Chapter 3
Waste Stabilization Ponds for Decentralized Wastewater Treatment

Manoj Kumar, Kalp Bhusan Prajapati, Smita S. Kumar, and Rajesh Singh

Abstract Waste Stabilization Ponds (Waste Stabilization Ponds (WSPs)) are an effective, low-cost method for treating wastewater, particularly in small communities across Europe, including France and Germany. These man-made water bodies utilize natural processes, influenced by solar light, wind, microorganisms, and algae, to treat blackwater, gray water, or fecal sludge. The WSP systems consist of three types of ponds—anaerobic, facultative, and aerobic (maturation) each contributing uniquely to the treatment process. Waste Stabilization Ponds (WSPs) are especially relevant in regions affected by climate change where there is increasing demand for irrigation water. Despite their high efficiency in removing Biochemical Oxygen Demand (BOD) and pathogens, Waste Stabilization Ponds (WSPs) require large surface areas and expert design. The effluent produced retains nutrients such as nitrogen and phosphorus, making it suitable for agricultural reuse but unsuitable for direct discharge into surface waters. This makes Waste Stabilization Ponds (WSPs) an ideal solution for sustainable water management in agriculture while minimizing environmental impact. This chapter provides a comprehensive overview of Waste Stabilization Ponds (Waste Stabilization Ponds (WSPs)), beginning with an introduction to their design and application. It explores the different types of Waste Stabilization Ponds (WSPs)—anaerobic, facultative, and aerobic (maturation) and discusses their integration with constructed wetlands. The chapter delves into the removal mechanisms and treatment performance of Waste Stabilization

M. Kumar (✉)
Department of Hydro and Renewable Energy, Indian Institute of Technology Roorkee, Roorkee, Uttarakhand, India

K. B. Prajapati · S. S. Kumar
Department of Environmental Studies, Central University of Haryana, Mahendragarh, Haryana, India

R. Singh
Department of Environmental Science, School of Earth and Environmental Sciences, Babasaheb Bhimrao Ambedkar University, (A Central University), Lucknow, Uttar Pradesh, India

S. S. Kumar, V. Kumar (eds.), *Nature-Based Solutions for Decentralised Liquid Waste Management*, Environmental Challenges and Solutions,
https://doi.org/10.1007/978-3-032-08362-3_3

Ponds (WSPs), highlighting the factors that influence their efficiency. Additionally, it examines the advantages and disadvantages of WSP systems, their implementation, and the current challenges in their application. The chapter concludes with an analysis of future trends and research directions in the field of Waste Stabilization Ponds (WSPs), offering insights into potential advancements and innovations.

Keywords Wastewater · Waste stabilization ponds · Constructed wetlands · Decentralized wastewater treatment · Treatment performance

3.1 Introduction

Groundwater and surface water are contaminated by an increase in the use of fertilizers and chemical pesticides, as well as by raw wastewater used in irrigation. Waste from industry is still frequently dumped straight into waterways. Severe negligence is being shown toward wastewater management. Single-family dwellings' comparatively tiny amounts of wastewater are treated and disposed of using decentralized wastewater management or groups of homes that are relatively close together (approximately 3–5 km, at most) and are not linked to a regional Wastewater Treatment Plant (WWTP) by a central sewage infrastructure. Naturally, a local collecting system is still required. Still, considering that the parts of the black flow have been separated from the gray water, comparing this system to those used for traditional, centralized treatment, it will most likely be far smaller and less expensive (Parkinson and Tayler 2003; Chen and Wang 2009). Systems that cater to specific areas (clusters) in an urban region based on hydrology, geography, and local ecological factors are also classified as decentralized. The main goals of sustainable decentralized sanitation are the wastewater treatment on-site and the provincial reprocessing and recycling of the resources (mostly water) observed in domestic wastewater. Reusing and recycling water nearby is allegedly encouraged by decentralized treatment systems (Opher and Friedler 2016). Decentralized wastewater treatment is when sewage or manure is treated on-site at the point of creation. Alternatively, it might be handled nearby using a more straightforward conveyance system (NIUA 2016).

Decentralized wastewater systems transport, clean, and remove or recycle wastewater from isolated residences, small, low-density communities, and particular public or private resources (USEPA (United States Environmental Protection Agency) 2005). When there is a suitable water supply inside or near a building, wastewater flow occurs. Decentralized wastewater treatment systems recycle, treat, or dispose of wastewater relatively near to the point of generation. By significantly lowering environmental and health risks, they aim to safeguard the general public's health and the environment. The term "decentralized wastewater treatment systems" is

Table 3.1 Fundamental specifications for DEWATS

Notes	Requirements	Issue/Impact
It is advisable to offer a certain level of disinfection in case of reuse.	Reduce the chance that people will come into contact with viruses and dangerous materials that could have an impact on public health, both during disposal and after final discharge/reuse.	Healthiness and sanitation
The getting environment is important when choosing a technology. It is necessary to conduct site evaluation procedures and an estimate of it carrying capacity.	The efficiency of removing pollutants must be considered, together with the resources needed for system operation and development and the emissions that arise. The degree of nonrenewable resource protection (such as the creation of biogas) and recycling and reusing from operations (such as returning water, fertilizers, and organic material to agriculture) must also be combined.	Environment and natural resources
Sturdiness and/or susceptibility to natural disasters like floods, water shortages, and power outages are crucial.	In the case of decentralized systems, maximizing functionality and making construction, operation, and monitoring simple for local utilities are critical.	Technology
Systems ought to function without placing an undue financial strain on the community of users.	Decentralized systems have to consider the ability of nearby individuals and communities to cover all system costs, including setup, running costs, upkeep, and required reinvestments.	Financial and economic
Reuse and recycling of resources are two fundamental practices that users need to be aware of and accept.	It is necessary to consider institutional contexts, convenience, perception, impact on human dignity, legal compliance, and sociocultural acceptance.	Sociocultural and institutional
Odors are one of the most subjectively annoying things, and the public is particularly sensitive to them. Reject risk can be decreased with good practices.	Since decentralized systems are typically located closer to residential areas, the aesthetic impact should be minimized. On the other side, design should incorporate mimetism or camouflage and integrate with nature and ecological function. Odors are one example of a potential annoyance factor that needs to be considered and prevented in design and operation.	Aesthetics and nuisance

Source: Capodaglio (2017)

also used to describe them, as the primary technical obstacle is selecting an appropriate treatment and disposal facility (USEPA (United States Environmental Protection Agency) 2005) which stands for DEWATS, is a commonly used word Decentralization at the township or suburban level of an urban area can, when applied in remote villages, be regarded as a form of small-scale centralized system or minor centralized system. However, if they are limited to a few chosen huge population centers with the existing centralized system, the system as a whole may be described as hybrid, with decentralization applied to some portions of the entire drained area (Tchobanoglous and Leverenz 2013). Neighborhood decentralization refers to dividing a community into tiny districts, gated communities, residential neighborhoods, and regions with vacuum sewer service. Decentralization in "on-site" sanitation situations is when the entire system is housed on a single property and serves one or more buildings (Tchobanoglous and Leverenz 2013). Table 3.1 lists the fundamental specifications for DEWATS.

Technologies for decentralized systems that are applicable cover a broad spectrum of operations with differing levels of complexity and sophistication. More sophisticated technical alternatives quickly catch up to centralized applications in terms of cost per unit of treated pollution load (Fane and Fane 2005). Decentralized facilities can now be easily managed remotely, which makes O&M tasks easier. However, in the past, decentralization faced significant challenges due to unreliable remote monitoring technology, which frequently led to unsustainable staffing levels and unreliable treatment outcomes. Robust remote sensor technology has become widely available, significantly reducing the need for on-site monitoring and facilitating remote facility monitoring and demand-based onsite maintenance (Capodaglio 2017). Septic tanks have long been a popular on-site, low-tech method of treating wastewater; they are frequently followed by soil disposal. Low conversion of organic components results from septic tank processes, among them are settling and the anaerobic breakdown of organic materials and the development of inert particles with sludge and wastewater interacting loosely (Capodaglio 2017). These days, site-specific decentralized systems can be created to overcome site-specific issues such as limestone formations, impermeable soils, shallow bedrock, and raised

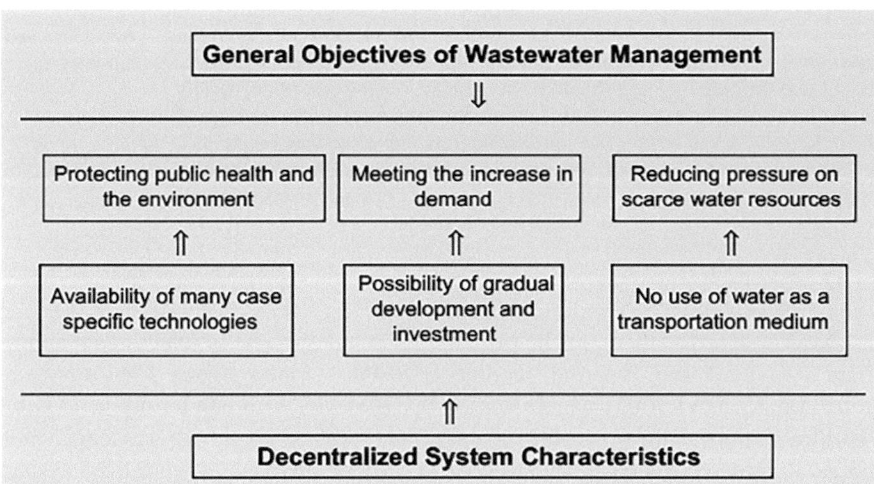

Fig. 3.1 The overall goals of managing wastewater as opposed to decentralized systems features. (Source: Massoud et al. 2009)

groundwater tables. Decentralized systems provide flexible management by combining many procedures to achieve treatment aims and fulfill environmental and community health safety mandates. Figure 3.1 illustrates the overall goals of managing wastewater as opposed to decentralized systems features.

3.1.1 Overview of Waste Stabilization Ponds

A natural device called a waste stabilization pond (WSP) can be added to centralized or semi-centralized sewage treatment systems for septage and sludge, and wastewater from residences and institutions. Waste Stabilization Ponds (WSPs) are inexpensive, easy to use, build, and efficient. It can be coupled with or utilized alone as a treatment plant's secondary or tertiary treatment unit. To identify the best strain of algae, enhance algal growth yield, and improve harvesting techniques, research is required (Mahapatra et al. 2022).

3.2 Types of Waste Stabilization Ponds

Three types of ponds: facultative, aerobic (maturation), and anaerobic. Every class has a different approach and set of structures. Waste Stabilization Ponds (WSPs) offer a high pathogen elimination rate and are inexpensive for O&M and BOD. However, extensive surface areas and expert design are required.

3.2.1 Anaerobic Ponds

In the series, anaerobic ponds are the smallest units, measuring between two and five meters (roughly) deep. Higher temperatures guarantee better digestion. Hence, temperature controls the subsequent procedure. The influent's pH is an additional consideration, as anaerobic bacteria are susceptible to pH variations (<6.2) (Mbwette et al. 2005). Acidic wastewater needs to be neutralized for anaerobic pond treatment to proceed correctly. Numerous researches have reported on odor issues. With appropriate design and control over the amount of sulfate in wastewater, this problem can be solved (Varón and Mara 2004). Wastewater is sent to anaerobic ponds untreated. They are deeper (about 3.0–5.0 m) and are smaller in surface area than ponds that are facultative. Anaerobic conditions result from the depth's reduction of photosynthesis's impact on oxygen production. Two-thirds to half of the influent BOD may be eliminated by these ponds, depending on loading and weather conditions. As a result, the facultative ponds require less space because the amount of organic stuff that enters them is much reduced (Von Sperling 2005). The drawback of anaerobic stabilizing ponds is the possibility of emitting foul-smelling gasses.

This includes, in particular, hydrogen sulfide that smells like rotten eggs if the system is experiencing malfunctions (Morris 2010). The first pond ecosystem breaks down putrescible solids suspended in the wastewater treated in a sequence of stabilization ponds. Solids can become sludge at the bottom of anaerobic ponds. A fraction of the particle organic material is removed by this settling (Morris 2010). Near the location where wastewater enters the pond, a sizable portion of the settled solids will gather. As a previous result, anaerobic ponds need to be built much more profoundly than facultative or aerobic ponds. Because there is less oxygen down depth, anaerobic bacteria can more effectively break down the waste (Morris 2010). Anaerobic microbes in anaerobic ponds can dissolve complex organic waste into straighter forward, less environmentally hazardous chemicals (Cincinnati 2010).

3.2.2 Facultative Ponds

Depending on the effluent that enters the pond, there are two kinds of ponds. Raw wastewater is sent to primary facultative ponds, and particle-free anaerobic pond effluent is sent to secondary facultative ponds (Kayombo et al. 2005). In general, facultative ponds range in depth from one to two meters. Unlike anaerobic or maturation ponds, facultative ponds have a comparatively more extended water retention period (Varón and Mara 2004). An integral part of the secondary facultative pond system is algae. Algae gives bacteria the oxygen they need to thrive and remove BOD from the environment. A facultative pond's algae population is thought to be a sign of the pond's health (Pearson et al. 1987). Algae play a crucial role in the pond system by eliminating fecal pathogens by supersaturating Dissolved Oxygen (DO) (up to 20 mg L^{-1}) and raising pH (above 9) levels. The microbial activity in the pond is controlled by temperature, pH, and light intensity (Aguiar et al. 1996). A lagoon that serves as a waste stabilization pond (WSP) is the second step of treatment. It's a shallow pond (1–2 m) with a deeper anaerobic zone and a closer-knit aerobic zone. The oxygen produced by the algae as they develop on the surface is ingested by aerobic microbes in the center of the pond, which lowers the biochemical oxygen demand (Tilley et al. 2014).

3.2.3 Aerobic (Maturation)

The final pond in the sequence of stabilization pond systems is the maturation pond. These are shallow, ranging in depth from one to two meters; their quantity is decided by the necessary specifications and the final effluent's quality. Removing pathogens, nitrogen, and phosphorus is the main goal of maturation ponds (Mbwette et al. 2005). When the maturation pond is filled with effluent from facultative ponds, it becomes apparent how the diversity of algae varies throughout ponds that are associated in the same sequence (Houghton and Mara 1992). In maturation ponds, certain excess OM and other contaminants might be removed. These ponds form a distinct part of the treatment system, designed to remove pathogens before effluent discharge into surface waters effectively. Although facultative ponds are typically utilized first, they can also come after other treatment procedures like up-flow Upflow anaerobic sludge blanket (UASB) system (Chernicharo 2007). They might also be positioned once the sludge is turned on. Maturation ponds must have a big surface area and be quite shallow (less than 1.0 m deep) to allow more O_2 to dissolve into the H_2O and provide the microbes with enough O_2 to function as intended (Morris 2010). The high photosynthetic activity brought on by sun radiation penetration benefits shallow ponds. Because of the high levels of photosynthesis and UV radiation penetration in the top layers, the pH values are high. These two elements support the elimination of harmful germs and viruses. Because the maturation ponds have a large surface area, sedimentation serves as the primary mechanism for removing helminth eggs and protozoan cysts (Verbyla et al. 2017; Von Sperling 2007). Temperature, Hydraulic Retention Time (HRT), the series' total number of ponds, the inclusion of baffles, and one of the factors that can result in very high pathogen elimination effectiveness is the depth of the ponds (Verbyla et al. 2017). Waste Stabilization Ponds (Waste Stabilization Ponds (WSPs)) are shown schematically in Fig. 3.2.

Please spell out "Hydraulic Retention Time (HRT)", if necessary.Hydraulic Retention Time (Hydraulic Retention Time (HRT))

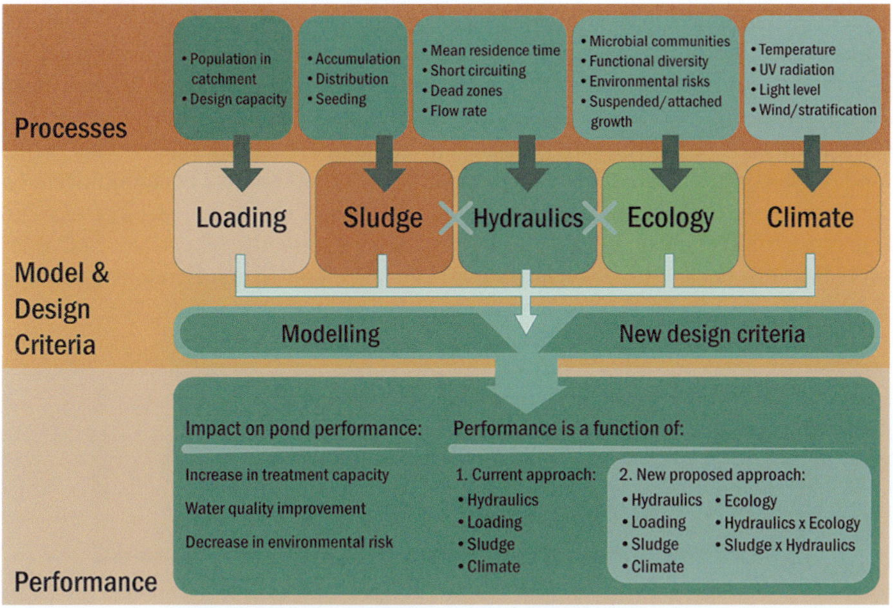

Fig. 3.2 Waste Stabilization Ponds (WSPs). (Source: Coggins et al. 2019)

3.2.4 Ponds with Constructed Wetlands (P-CWs)

In urban planning, constructed ponds and wetlands are frequently used for various purposes, such as stormwater management. Stormwater management constructed wetlands require the involvement of multiple multidisciplinary fields in their design. Appropriate management of the wetland's hydrologic regime and ideal flow hydrodynamics inside it are essential to its ongoing operation (Persson et al. 1999). CWs are another addition to pond technology that has been used more and more in temperate and tropical climates (Fig. 3.3) (Gschlößl et al. 1998; Steinmann et al. 2003). CWs can complement conventional Waste Stabilization Ponds (WSPs) and High-Rate Algal Pond (HRAP)s as a tertiary treatment, possessing the same robustness, affordability, and simplicity as natural systems for wastewater treatment. Large wetland regions can also encourage the biological removal of nutrients and plant uptake, and particular P-sorptive media can improve the removal of P even further (Tanner and Sukias 2003). As a previous result, the New Zealand, Thailand, Brazil, Spain, and France currently have many installations of this integrated system in operation (Shilton 2006). The pond systems can be integrated with all three of the traditional forms of CWs: VF-CWs, HF-CWs, and FWS-CWs (Vymazal 2010).

3.3 Removal Mechanisms and Treatment Performance

It has been found that adding a ceramic filter to sewage wastewater treatment can effectively eliminate over 93% of organics and increase the rate at which N is eliminated (0.017 kg m^3 d^{-1}) (Hasan et al. 2019). Sludge accumulation, wind patterns, and rainfall all mutually affect the effectiveness of WSP treatment (Izdori et al. 2019). The design and functioning of the stabilizing ponds are seriously considered throughout their establishment. Pond design is essential and varies depending on the kind of wastewater that needs to be treated. Several studies have been published on the impact of stabilization ponds' substandard design on treatment effectiveness. Ignorance of technical factors during building is directly related to foul odor problems and pond failure. Negligent maintenance and hydraulic short-circuiting

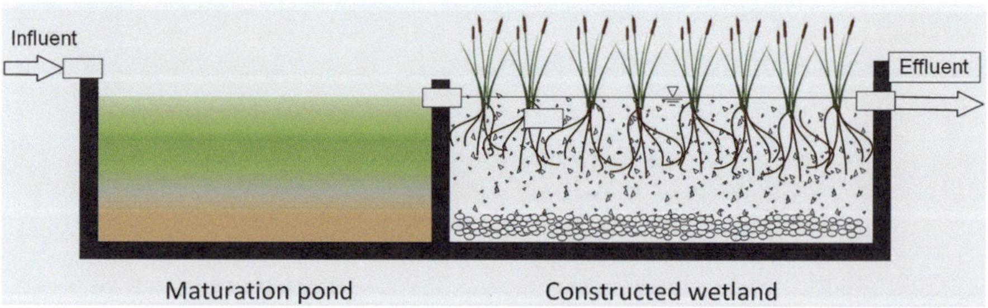

Fig. 3.3 Maturation pond with CWs-FWS. (Adapted and Modified Ho and Goethals 2020)

have a negative, damaging impact on performance. It has been stated that research on the addition of baffles, flow deflectors, and other procedure designs enhances the overall performance of stabilization ponds (Shilton and Harrison 2003). Figure 3.4 displays the causal loop diagram illustrating how sunlight affects pathogen die-off and the main elements influencing pathogen eradication in WSP.

3.3.1 *Factors Affecting Pond Performance*

Wastewater stabilization ponds (Waste Stabilization Ponds (WSPs)) use natural purifying methods for treatment, which can vary depending on the temperature and location. The prior study, conducted in a temperate climate zone of Eastern Ontario, Canada, examined the impact of three environmental parameters on decontamination efficacy in a WSP system containing three facultative cells: °C, Dissolved Oxygen (DO), and pH. The elimination of the indicator bacterium *Escherichia coli* in WSP systems is influenced by several distinct variables. Algae-related elevated pH and Dissolved Oxygen (DO) concentrations are critical to an efficient disinfection system. Thus, algae can significantly aid in disinfection in natural wastewater treatment systems. Therefore, during the course of a year, the removal efficiency of pathogenic microorganisms during wastewater treatment can vary greatly depending on algal densities, with summer and fall expected to have better removal

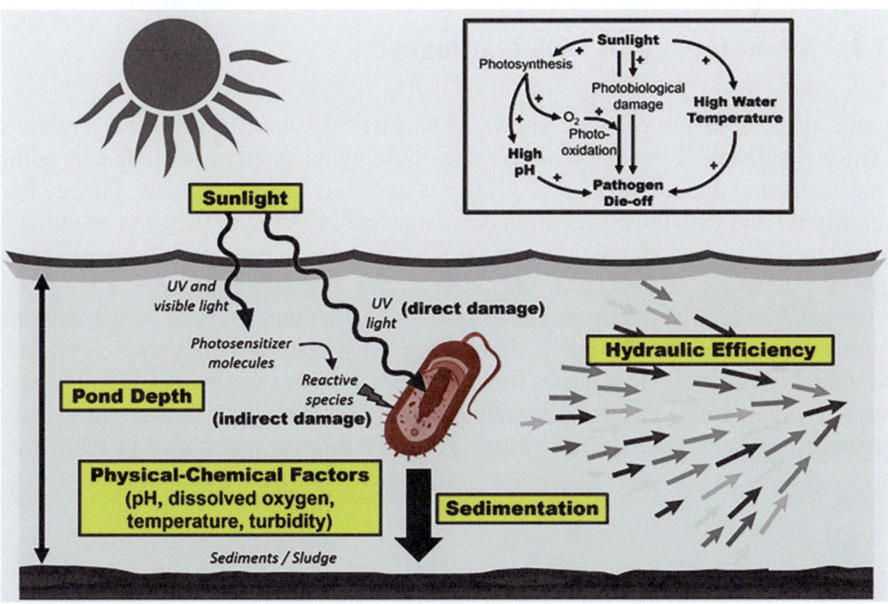

Fig. 3.4 Significant variables influencing pathogen elimination in WSP, along with a causal loop figure illustrating how sunshine affects pathogen die-off. (Source: Verbyla et al. 2017)

efficiency (Liu et al. 2015). Temperature, pH, Dissolved Oxygen (DO), and other environmental variables all affect how inactive indicator organisms become. According to reports, the ideal pH range for fecal bacteria growth is neutral to slightly acidic (Awuah et al. 2001). However, it has been suggested that pH values greater than 9 are helpful in eliminating pathogens (Ansa et al. 2011). Because they have statistically solid correlations with *E. coli* concentrations, temperature, and elevated pH brought on by the presence of algae are regarded as significant variables. In the previous investigation, pH values greater than 8 were helpful for disinfection. Algal activity appeared to be inhibited at temperatures below 5 °C, although *E. coli* could continue to grow at the same temperature. Reduced algal activity can cause the pH to drop to less than 8, which is considered ineffective for disinfection. Other factors could be at play if there isn't a strong link between Dissolved Oxygen (DO) concentration and the removal of *E. coli*, such as sunshine intensity and penetration. In Waste Stabilization Ponds (WSPs), pathogen removal efficiency can vary greatly and show seasonal trends. For example, reduced *E. coli* concentrations have been seen throughout the warmest time of year. This would indicate that location and climate have an impact on WSP performance, which increases the difficulty of treating wastewater with natural systems. It was observed that the final two WSP cells at the WPCP significantly reduced the amount of *E. coli*. This suggests that using several WSP cells in succession could yield a more dependable removal performance of the indicator microorganism (*E. coli*) than a single WSP system (Liu et al. 2015).

3.4 Advantages and Disadvantages

Large artificial bodies of water known as Waste or Wastewater Stabilization Ponds (Waste Stabilization Ponds (WSPs)) utilize light, wind, bacteria, and algae in addition to natural processes to purify blackwater, gray water, or fecal sludge. For improved treatment, the ponds might be connected in series, or they can be utilized separately. Waste Stabilization Ponds (WSPs) have a high pathogen elimination rate and are inexpensive for O&M and BOD. Large surfaces and expert design are essential, however. The effluent can be reused in agriculture because it still contains nutrients (such N and P), but it shouldn't be dumped back into surface waters. Similar benefits and drawbacks can be found here (sswm.info/factsheet/waste-stabilization-ponds). The same advantages and drawbacks of waste or wastewater stabilization ponds (Waste Stabilization Ponds (WSPs)) are presented in Table 3.2.

Table 3.2 Same advantages and disadvantages of waste or wastewater stabilization ponds (Waste Stabilization Ponds (WSPs))

S. No.	Advantages	Disadvantages
1.	Robust against hydraulic and organic shock loading	Needs much land
2.	Significant decrease in pathogens, BOD, and solids	High capital expense based on land pricing
3.	High elimination of nutrients when paired with aquaculture	Requires skilled building and design
4.	Minimal running costs	Sludge needs to be removed and treated properly
5.	No need for electrical energy	Removing sludge (usually every couple of years)
6.	If built and sustained correctly, there shouldn't be any important fly or odor problems	Mosquito management is necessary
7.	Buildable and repairable using materials readily available locally	Monitoring of salinity is necessary if the effluent is repurposed
8.	Reusing wastewater for irrigation in agriculture or aquaculture	Not usually suitable for colder regions

Source: SSWM

3.5 Successful Implementation of WSP

The previous study performance assessment was conducted on WSP, a model of a rural residential wastewater treatment facility. The facility included two series of facultative, maturation, and anaerobic ponds. The Biochemical Oxygen Demand (BOD), Chemical Oxygen Demand (COD), and Total Suspended Solids (TSS) of the WSP effluents that are discharged into the drain were removed to 50.65%, 48.95%, and 44.3% total coliform (TC), faecal coliform (FC), respectively. Additionally, the percentages of infectious rotaviruses and coliphages removed were 99.66% and 49.03%, respectively. Pollution has been observed to impact the diversity of zooplankton species; in facultative ponds with high pollution levels, there were only eight species, whereas in maturation ponds, there were 21 species. Moreover, ciliates (Protozoa), recognized as bioindicators of organic pollution, became more densely concentrated in the anaerobic pond due to corrosion. In the maturation pond, 70% of the ciliates were removed. It is advised that various design changes be made to improve WSP's efficiency (Ghazy et al. 2008).

Waste Stabilization Ponds (WSPs) are frequently home to cyanobacteria and associated toxins, such as microcystins, which threaten both ecological and human health. More research needs to be done on their removal from Waste Stabilization Ponds (WSPs), even though numerous studies have looked into their elimination from batch cultures and potable water reservoirs. It has been suggested that hydrogen peroxide (H_2O_2) is a safe substance that reduces cyanobacteria. The applicability of using H_2O_2 to remove microcystins and cyanobacteria from wastewater is reported in this study. A technique is described that provides a rapid and low-cost means of figuring out how much is appropriate for full-scale application to Waste Stabilization Ponds (WSPs). Scale-up experiment data shows that wind-induced mixing during application, when H_2O_2 is added upwind, is adequate to treat a whole WSP. Results also demonstrate a greater removal of cyanobacteria in the upper layer of waste stabilization ponds. If surface samples are the only ones considered, this could result in an overestimation of the overall reduction. We propose that H_2O_2 may be an effective algicide treatment in Waste Stabilization Ponds (WSPs) since, within days of application, it drastically reduced the cyanobacterial proportion and microcystin concentrations while increasing the development of eukaryote phytoplankton. The duration of this effect was approximately 3 weeks, suggesting that more application may be required to prevent the cyanobacteria from becoming dominant once again. Nevertheless, as there is currently little data on the impact of repeated application on other microbes in full-trial settings at Waste Stabilization Ponds (WSPs), close observation is required. For example, while recent studies in the lab indicate that the average doses used in tests may cause zooplankton to die within 24 h, this is unlikely to occur in Waste Stabilization Ponds (WSPs) since zooplankton can actively avoid unfavorable conditions. In conclusion, our work provides WSP workers with the ability to evaluate the advantages of employing H_2O_2 to suppress the amounts of cyanobacteria and microcystin quickly, thereby preventing their introduction into the environment (Barrington et al. 2013).

The previous work examines the solute transport processes and hydraulic performance in Waste Stabilization Ponds (WSPs). The advection-diffusion solute transport model and the traditional $k - \varepsilon$ turbulence closure constitute a numerical model in a three-dimensional Cartesian coordinate system and the Reynolds-averaged Navier–Stokes (RANS) flow hydrodynamic model. The impact of two deflector island designs (rotated and parallel) on improving the hydraulic performance of the WSP is examined through simulation. The numerical findings' analysis shows that deflector island retrofitting has a generally beneficial effect on the WSP's hydraulic performance. It is demonstrated that a significant factor influencing the WSP's overall performance is the slope of the side walls. By including both parallel and rotating islands, the hydraulic efficiency of the WSP was increased for the situations with side-wall inclines of 1:1, 0.5:1, and 0:1. Nevertheless, it is demonstrated that the installation of an island deflector has a detrimental effect on the waste stabilization pond's hydraulic efficiency in circumstances when the side walls have a 2:1 or 1.5:1 slope (Goodarzi et al. 2022).

To learn more about the internal workings and interactions of Waste Stabilization Ponds (WSPs), it is worthwhile to investigate the hydrodynamics simulation used at these facilities. To achieve optimal substrate usage rates and high hydrodynamic efficiency, pond design incorporates several geometrical, hydrological, physical, and dynamic factors. Computational fluid dynamic modeling allows these characteristics to be grouped in different configurations to predict pond behavior. Hydrodynamics and water quality were simulated using the two-dimensional depth-integrated model SMS. Baffling and pond geometries were included in a set of 12 modeled configurations. The model was run in a steady state utilizing raw wastewater and altering water depth, flow rate, and Hydraulic Retention Time (HRT) in order to investigate the effects of the assumed rectangular forms and dimensions with fixed area. The previous study examined the distribution of BOD, Dissolved Oxygen (DO) concentrations, and velocities for multiple rectangular forms with different numbers of baffles (Abbas et al. 2006).

Waste Stabilization Ponds (WSPs) are extensively applied in wastewater treatment because of their inexpensive cost, low skill requirement, and straightforward construction. In the prior study, the effectiveness of a WSP system in falling the amounts of pollutants in hospital wastewater in a rural South African region was examined, along with the system's current management. During several monitoring months, the bacteriological parameters (*E. coli* and *enterococci*) in the outlet were more significant than those in the inlet. Additionally, there was evidence of limited pathogen clearance effectiveness ($\leq$1 log reduction). Despite removal performance results for February, April, and June of 7.7%, 49.1%, and 31.1%, respectively, the effluent's COD (82–200 mg/L) was higher above the wastewater discharge guideline value of 75 mg/L set (Edokpayi et al. 2021).

3.6 Future Trends and Research Directions

In the future, more research on eliminating viruses and nitrogen is anticipated to improve our knowledge of stabilizing pond dynamics. It is essential to pay consideration to optimization details like the impact of various Hydraulic Retention Time (HRT). To close some information gaps regarding the complex mechanisms in a stabilization pond, a functional model of a WSP system that has been completely calibrated and validated would be necessary. More investigation is required into recent worries about greenhouse gas emissions and the withdrawal of antibiotics from Waste Stabilization Ponds (WSPs). Research to improve stabilization pond performance has demonstrated the potential for improvement in scaling up and designs, encompassing wind, pond shape, flow rates, and inlet/outlet configurations. The design processes and cultural acclimation to changing conditions will be aided by the modeling of the biotic pathways and techniques in a stabilizing pond. To address the shortcomings above, it is anticipated that as computational knowledge advances, the number of complicated integrated models will rise.

3.7 Conclusion

To treat sludge, septage, and wastewater from residences and institutions, a waste stabilization pond (WSP) is a naturally occurring device that might be added to semi-centralized sewage systems. Waste Stabilization Ponds (WSPs) are inexpensive, easy to use, easy to build, and very efficient. It might be used as a secondary unit in a treatment plant in conjunction with other units or on its own. Being one of the appropriate substitutes for traditional wastewater treatment techniques. WSP is an energy-efficient treatment technology that eliminates organic materials, heavy metals, and nutrients by the simple interplay of sunlight, bacteria, and algae. Under the right circumstances, these ponds exhibit high BOD, COD, nitrogen, and phosphorus removal efficiency; these findings are covered in the paper along with various study findings. The benefits of this treatment procedure are further enhanced by the fact that the algae produced during this process can be utilized to make vegetable oil, biofuel, animal feed, and energy. Several academics have devised several equations that consider every aspect of the WSP design. The most effective treatment method for both developed and small communities is WSP because of its low cost, simple process, and ecologically beneficial features.

References

Abbas H, Nasr R, Seif H (2006) Study of waste stabilization pond geometry for the wastewater treatment efficiency. Ecol Eng 28(1):25–34

Aguiar MDFM, Cavalcanti M, Barbosa H, Vilela SL, Mendonça JL, Horta E (1996) Síndrome de Aicardi e papiloma do plexo coróide: uma associação rara. Relato de caso Arquivos de Neuro-Psiquiatria 54:313–317

Ansa EDO, Lubberding HJ, Ampofo JA, Gijzen HJ (2011) The role of algae in removing Escherichia coli in a tropical eutrophic lake. Ecol Eng 37(2):317–324

Awuah E, Anohene F, Asante K, Lubberding H, Gijzen H (2001) Environmental conditions and pathogen removal in macrophyte-and algal-based domestic wastewater treatment systems. Water Sci Technol 44(6):11–18

Barrington DJ, Reichwaldt ES, Ghadouani A (2013) The use of hydrogen peroxide to remove cyanobacteria and microcystins from waste stabilization ponds and hypereutrophic systems. Ecol Eng 50:86–94

Capodaglio AG (2017) Integrated, decentralized wastewater management for resource recovery in rural and peri-urban areas. Resources 6(2):22

Chen R, Wang XC (2009) Cost-benefit evaluation of a decentralized water system for wastewater reuse and environmental protection. Water Sci Technol 59(8):1515–1522

Chernicharo CAL (2007) Anaerobic reactors, Biological wastewater treatment series, vol 4. IWA Publishing, London. ISBN 978-1843391623

Cincinnati O (2010) NEPIS document display. Decentralized wastewater systems: a program strategy. EPA, Washington, DC. (PDF) 2017

Coggins LX, Crosbie ND, Ghadouani A (2019) The small, the big, and the beautiful: emerging challenges and opportunities for waste stabilization ponds in Australia. Wiley Interdiscip Rev Water 6(6):1383

Edokpayi JN, Odiyo JO, Popoola OE, Msagati TA (2021) Evaluation of contaminants removal by waste stabilization ponds: a case study of Siloam Waste Stabilization Ponds (WSPs) in Vhembe District, South Africa. Heliyon 7(2)

Fane AG, Fane SA (2005) The role of membrane technology in sustainable decentralized wastewater systems. Water Sci Technol 51(10):317–325

Ghazy MM, Senousy WM, Aatty AM, Kamel M (2008) Performance evaluation of a waste stabilization pond in a rural area in Egypt. Am J Environ Sci 4(4):316

Goodarzi D, Mohammadian A, Pearson J, Abolfathi S (2022) Numerical modelling of hydraulic efficiency and pollution transport in waste stabilization ponds. Ecol Eng 182:106702

Gschlößl T, Steinmann C, Schleypen P, Melzer A (1998) Constructed wetlands for effluent polishing of lagoons. Water Res 32(9):2639–2645

Hasan MM, Saeed T, Nakajima J (2019) Integrated simple ceramic filter and waste stabilization pond for domestic wastewater treatment. Environ Technol Innov 14:100319

Ho L, Goethals PL (2020) Municipal wastewater treatment with pond technology: historical review and future outlook. Ecol Eng 148:105791

Houghton SR, Mara DD (1992) The effects of sulphide generation in waste stabilization ponds on photosynthetic populations and effluent quality. Water Sci Technol 26(7–8):1759–1768. https://sswm.info/factsheet/waste-stabilisation-ponds

Izdori F, Semiao AJ, Perona P (2019) The role of environmental variables in waste stabilization ponds' morphodynamics. Front Environ Sci 159

Kayombo STSA, Mbwette TSA, Katima JHY, Ladegaard N, Jørgensen SE (2005) Waste stabilization ponds and constructed wetlands design manual. UNEP-IETC with the Danish International Development Agency (Danida), pp 1–59

Liu L, Hall G, Champagne P (2015) Effects of environmental factors on the disinfection performance of a wastewater stabilization pond operated in a temperate climate. Water 8(1):5

Mahapatra S, Samal K, Dash RR (2022) Waste Stabilization Pond (WSP) for wastewater treatment: a review on factors, modelling, and cost analysis. J Environ Manag 308:114668

Massoud MA, Tarhini A, Nasr JA (2009) Decentralized approaches to wastewater treatment and management: applicability in developing countries. J Environ Manag 90(1):659

Mbwette TSA, Kayombo S, Katima JHY, Jorgensen SE (2005) Waste stabilization ponds and constructed wetlands design manual. IKR, Dar es Salaam

Morris J (2010) Michigan department of natural resources. State of Michigan Department of Natural Resources

NIUA (2016) Handbook on decentralized waste water treatment module, Delhi

Opher T, Friedler E (2016) Comparative LCA of decentralized wastewater treatment alternatives for non-potable urban reuse. J Environ Manag 182:464–476

Parkinson J, Tayler K (2003) Decentralized wastewater management in peri-urban areas in low-income countries. Environ Urban 15(1):75–90

Pearson HW, Mara DD, Bartone CR (1987) Guidelines for the minimum evaluation of the performance of full-scale waste stabilization pond systems. Water Res 21(9):1067–1075

Persson J, Somes NLG, Wong THF (1999) Hydraulics efficiency of constructed wetlands and ponds. Water Sci Technol 40(3):291–300

Shilton A (ed) (2006) Pond treatment technology. IWA Publishing, London

Shilton A, Harrison J (2003) Integration of coliform decay within a CFD (computational fluid dynamic) model of a waste stabilisation pond. Water Sci Technol 48(2):205–210

Steinmann CR, Weinhart S, Melzer A (2003) A combined system of lagoon and constructed wetland for an effective wastewater treatment. Water Res 37(9):2035–2042

Tanner CC, Sukias JPS (2003) Linking pond and wetland treatment: performance of domestic and farm systems in New Zealand. Water Sci Technol 48(2):331–339

Tchobanoglous G, Leverenz H (2013) The rationale for decentralization of wastewater infrastructure. In: Source separation and decentralization for wastewater management. IWA Publishing, London, pp 101–115

Tilley E, Ulrich L, Lüthi C, Reymond P, Zurbrügg C (2014) Compendium of sanitation systems and technologies, 2nd edn. Swiss Federal Institute of Aquatic Science and Technology (Eawag), Duebendorf

USEPA (United States Environmental Protection Agency) (2005) Handbook for managing onsite and clustered (decentralized) wastewater treatment systems, EPA/832-B-05-001. Office of Water, Washington, DC, p 66

Varón MP, Mara D (2004) Waste stabilisation ponds. IRC International Water and Sanitation Centre, Delft

Verbyla M, von Sperling M, Maiga Y (2017) Waste stabilization ponds. In: Rose JB, Jiménez Cisneros B (eds) Global water pathogens project. http://www.waterpathogens.org (C. Haas, J.R. Mihelcic and M.E. Verbyla) (eds) part 4 management of risk from excreta and wastewater)

http://www.waterpathogens.org/book/waste-stabilization-ponds Michigan State University, E. Lansing, MI, UNESCO
Von Sperling M (2005) Biological wastewater treatment in warm climate regions. Chernicharo, Carlos Augusto de Lemos. IWA, London. ISBN 9781843390022. OCLC 62306180
Von Sperling M (2007) Waste stabilisation ponds, London. ISBN 9781843391630. OCLC 878137182
Vymazal J (2010) Constructed wetlands for wastewater treatment. Water 2(3):530–549

Chapter 4
Access to Clean Water: Sustainable Biochar-Based Water Treatment

Shivangi Pathak, K. K. Pant, and Priyanka Kaushal

Abstract Biochar has drawn much attention as a powerful tool in reducing wastewater pollution due to its superior adsorption capability, distinctive surface chemistry, and high specific surface area. This chapter reviews the development of biochar applications in the past few years. Although biochar is derived from various agricultural and industrial wastes, its complex structure, functional groups, and surface properties provide remarkable wastewater treatment capabilities. Using biochar in wastewater treatment technologies is a promising strategy to lessen environmental pollution and support sustainable waste management practices. As a result, biochar has become a novel, affordable, and environmentally friendly material that can be utilized to raise the effectiveness of wastewater treatment technologies.

Keywords Clean water · Water treatment · Biochar · Adsorption · Surface modification · Point-of-use · Sustainability · Economic viability

S. Pathak · P. Kaushal (✉)
Centre for Rural Development and Technology, Indian Institute of Technology Delhi, New Delhi, India
e-mail: priyankak@iitd.ac.in

K. K. Pant
Department of Chemical Engineering, Indian Institute of Technology Delhi, New Delhi, India

S. S. Kumar, V. Kumar (eds.), *Nature-Based Solutions for Decentralised Liquid Waste Management*, Environmental Challenges and Solutions,
https://doi.org/10.1007/978-3-032-08362-3_4

4.1 Introduction

4.1.1 Water Crisis and the Need for Sustainable Water Treatment

The worldwide water crisis is an urgent problem threatening ecosystems, human health, and economic growth. Urbanization, industrialization, climate change, and rapid population growth have all contributed to an increased lack of clean and safe water resources. Millions lack access to proper drinking water and sanitary facilities worldwide, with severe socioeconomic ramifications and the spread of waterborne diseases. Furthermore, the water issue is worsened by pollution from industrial discharges, agricultural runoff, and improper wastewater management, which contaminate water sources with dangerous chemicals, diseases, and heavy metals.

Sustainable water treatment methods ensure everyone can access clean, safe water while addressing the water crisis. Utilizing techniques and technology that reduce environmental impact, preserve water supplies, and support long-term profitability constitutes sustainable water treatment. It covers various topics, such as effective water use, reducing pollution, conserving water, and implementing cutting-edge treatment techniques. By applying sustainable water treatment technologies, we can safeguard human health, maintain ecosystems, improve water quality, and create resilient communities less susceptible to water problems.

The pyrolysis process produces biochar, a porous material rich in carbon, which involves heating biomass, such as wood, crop leftovers, or organic waste, without oxygen. Pyrolysis releases volatile chemicals, leaving behind biochar, a stable form of carbon. Due to its numerous advantages for agriculture and the environment, this carbon-rich substance has recently attracted much interest. The manufacture of biochar helps to enhance soil quality and carbon sequestration while providing a sustainable method of handling organic waste and lowering greenhouse gas emissions. Biochar is an excellent soil additive due to its huge surface area, porosity, and capacity to hold moisture and nutrients. Biochar increases the soil's fertility, water-holding capacity, nutrient retention, and microbial activity, thus increasing crop yields.

4.1.2 Biochar as a Promising Solution for Water Treatment

Biochar has distinct characteristics and adsorption capacity, making it a viable replacement for conventional water treatment methods. Its high surface area, porosity, and availability of functional groups make it an effective adsorbent for various contaminants in water. The porous nature of biochar provides a larger surface area for adsorbing pollutants, while the functional groups on its surface encourage chemical interactions with pollutants. This makes biochar a top option for removing contaminants like heavy metals, organic pollutants, and nutrients from water.

Numerous studies have demonstrated the effectiveness of biochar in water treatment applications. For instance, research has shown that biochar can remove lead, cadmium, and copper from contaminated water sources. The ability of biochar to absorb organic contaminants, including pharmaceuticals, pesticides, and dyes, is also great. Additionally, biochar has demonstrated the ability to absorb excess nutrients from agricultural runoff and wastewater, such as nitrogen and phosphorus, reducing water pollution and eutrophication (Ahmad et al. 2014).

There are various benefits to using biochar in water treatment. First, it can be made from different organic materials, including agricultural waste. This aids in waste management and provides a greener option to conventional treatment methods. Using biochar as a soil supplement after water treatment can sequester carbon by keeping it in the soil, which is the second benefit. This has the added benefit of slowing down climate change for the environment.

Therefore, biochar has much potential as an efficient and durable water treatment method. Its unique properties, high adsorption capacity, and potential for carbon sequestration make it a flexible and practical tool for addressing issues with water contamination. Research and development will continue to enhance the efficiency of biochar-based water treatment, which will also aid in the sustainable management of water resources (Lehmann and Joseph 2009).

4.2 Biochar for Water Treatment: Adsorptive Properties and Mechanisms

4.2.1 Biochar Characteristics and Their Influence on Adsorption Capacity

The capacity of biochar to adsorb pollutants is greatly influenced by its physical and chemical characteristics. Biochar has a large surface area and many active sites for pollutant adsorption because of its porous structure (Ambaye et al. 2021). The presence of micropores, mesopores, and macropores in the biochar matrix increases its surface area and enhances accessibility to the pollutants, allowing for efficient adsorption.

The interaction of biochar with water pollutants is greatly influenced by its surface chemistry. Carboxyl (-COOH), hydroxyl (-OH), phenolic (-Ph), and amino (-NH2) groups are a few of the functional groups that cover the surface of biochar. These functional groups are significant in adsorption and aid in removing contaminants from water by forming chemical bonds or electrostatic interactions with pollutants. Biochar helps eliminate heavy metals because carboxyl groups, for instance, can draw and bind positively charged species by ion exchange or complexation. However, organic pollutants can be absorbed because phenolic and hydroxyl groups interact with polar contaminants via hydrogen bonding. Additionally, biochar can interact with some contaminants as an electron donor or acceptor due to its carbonaceous structure, which improves its capacity to adsorb pollutants (Pathak et al. 2023). Pesticides and hydrocarbons are examples of hydrophobic organic molecules that can be adsorbed on biochar surfaces.

The type and concentration of functional groups can vary depending on the feedstocks and pyrolysis temperatures, which may affect the ability of biochar to adsorb specific pollutants (Janu et al. 2021). The functional groups of biochar can be further increased through surface alterations like chemical activation or impregnation with nanoparticles, enabling more specialized surface chemistry to remove particular pollutants (Akhil et al. 2021). It is essential to comprehend how the properties of biochar affect adsorption capacity to enhance its effectiveness and create viable biochar-based water treatment systems.

4.2.2 Adsorption Mechanisms of Contaminants on Biochar

The adsorption of contaminants on biochar occurs at the biochar-water interface, where physical and chemical reactions occur (Barquilha and Braga 2021). These processes are essential for removing contaminants from water sources. Some of the main adsorption mechanisms include the following:

1. Physical Adsorption: Physical adsorption, known as physisorption or van der Waals forces, occurs when weak forces are present between the biochar surface and the contaminants. London dispersion forces and dipole-dipole interactions are two examples of van der Waals forces that can attract and bind non-polar and weakly polar pollutants to the surface of biochar. Biochar's high surface area and porosity, which allow contaminants to be trapped inside its porous structure, essentially lead to physical adsorption (Yao et al. 2016).
2. Chemical Adsorption: Chemical interactions between contaminants and functional groups on the surface of biochar result in chemical adsorption, also known as chemisorption. The complexation or formation of chemical bonds between the pollutants and the biochar surface is required for this adsorption type. Positively charged species can be eliminated from water through ion exchange reactions involving functional groups like carboxyl, hydroxyl, and phenolic and coordination complexes with metal ions (Ahmad et al. 2014).
3. Electrostatic Interactions: Electrostatic interactions are significant in the adsorption of charged pollutants on biochar. Electrostatic attraction and binding of oppositely charged species are made possible by functional groups with the presence of functional groups having ionizable characteristics, such as carboxyl and amino groups. Ion exchange, in which cations in the water phase swap out other cations linked to the functional groups on the surface of the biochar, is one method by which electrostatic interactions can take place (Chen and Zhou 2011a, b).
4. π-π Electron Donor-Acceptor Interactions: Due to its carbonaceous composition, a few contaminants can combine with biochar to create π-π electron donor-acceptor interactions. The aromatic rings in the biochar structure and the aromatic contaminants interact in these ways. Biochar's capacity to bind hydrophobic

organic compounds, such as pesticides and hydrocarbons, is boosted by the π-π interactions (Li et al. 2017a, b).

The adsorption mechanisms may be impacted by the unique characteristics of the biochar, such as surface chemistry, pore structure, and surface area. When these concepts are understood, it is simpler to construct and run biochar-based water treatment systems that successfully remove impurities.

4.2.3 Factors Influencing Adsorption Efficiency

Numerous factors can affect how well biochar works to remove contaminants from water sources. Understanding these elements is crucial to maximizing the efficacy of biochar-based water treatment systems (Dwivedi and Dey 2022). The following are some significant factors that influence the capacity of biochar for adsorption:

1. Surface Area and Porosity: The adsorption capacity of biochar is substantially impacted by its porosity and surface area. More surface area offers more active sites for adsorption, even while micropores, mesopores, and macropores increase the accessibility of contaminants to the surface. Using the optimum pyrolysis parameters and feedstocks, it is possible to increase the surface area and porosity of the biochar and thus improve its ability to adsorb substances (Chen and Zhou 2011a, b).
2. Feedstock Composition: The type of feedstock used to make biochar affects its capability to adsorb contaminants. Different feedstocks differ chemically, including in the abundance and variety of functional groups. For instance, feedstocks with more lignin may produce biochar with more aromatic structures, improving its capacity to adsorb specific pollutants (Chen and Zhou 2014).
3. Pyrolysis Conditions: Temperature, heating rate, and residence time are pyrolysis process variables that affect the biochar's physical and chemical properties. Higher pyrolysis temperatures frequently result in biochar with more carbon content, more pores, and better adsorption abilities. However, excessively high temperatures may cause the loss of functional groups, decreasing adsorption efficiency (Mohan et al. 2014).
4. pH and Solution Chemistry: Biochar adsorption efficiency depends on the solution's pH and chemistry. The pH affects the surface charge of biochar and the ionization of functional groups, affecting the electrostatic interactions with charged contaminants. Additionally, competing ions in the solution can lessen the efficacy of adsorption by interacting with the binding sites on the biochar surface (Cao et al. 2011; Ahmad et al. 2014).
5. Contaminant Characteristics: Contaminant's nature and characteristics, such as their size, charge, polarity, and concentration, have a significant impact on how well they adhere to biochar. Depending on their compatibility with the surface

functional groups and their capacity to interact with the biochar matrix, various pollutants exhibit variable affinities for biochar (Saleh 2018).

6. Contact Time and Agitation: The amount of agitation or mixing and the contact time between the biochar and water affect the adsorption efficiency. Greater interactions between pollutants and biochar are made possible by longer contact time, which improves adsorption. Agitation assists in enhancing the mass transfer of pollutants to the surface of the biochar and avoids the development of concentration gradients that could impede adsorption (Tsai et al. 2012).

Understanding and modifying these factors is the only way to maximize the adsorption efficiency of biochar in water treatment applications. When designing and operating biochar-based water treatment systems, these factors should be considered to efficiently remove contaminants and make it easier for people to access clean and safe water supplies.

4.2.4 Examples and Case Studies of Biochar Adsorption for Water Treatment

Case studies and examples demonstrate the water treatment process of biochar adsorption and emphasize its potential as a sustainable substitute. The following examples show how biochar can successfully remove various contaminants from water sources:

1. Removal of Heavy Metals: Tan et al. study focused on the removal of lead (Pb) and copper (Cu) from contaminated water using biochar made from coconut shells (2018). The outcomes demonstrated that biochar has a significant potential for adsorption, with removal efficiencies for both metals above 90%. The study found that the biochar's functional groups and substantial surface area enabled it to absorb contaminants effectively. Similarly, Khadem et al. investigated the removal of several metal ions from wastewater, including Fe (III), Cd (II), Cu (II), Pb (II), and Mn (VII), using activated biochar made from Jackfruit seed waste (JSW) at 500 °C. For Fe (III), Pb (II), Cu (II), Cd (II), and Mn (VII), activated JSW biochar has an average heavy metal absorption of 76.4 mg/g, 79.4 mg/g, 97.9 mg/g, 79.9 mg/g, and 79.8 mg/g, respectively (Khadem et al. 2022).
2. Removal of Organic Pollutants: Li et al. (2020a, b) investigated biochar generated from sewage sludge to remove organic pollutants from water, such as phenol and bisphenol A. The study found that biochar had excellent adsorption potential and removal efficiencies of over 90% for both pollutants. The study found that interactions between the hydrophobic and aromatic organic contaminants and the biochar eliminated organic pollutants (Li et al. 2020a, b). Another work by Guo et al. concentrated on using rice husk biochar (MBC) that had been magnetically modified to remove phenanthrene (PHE) from wastewater. PHE

Table 4.1 A few examples of the adsorption capacity of biochar for different pollutants

Feedstock	Pyrolysis temperature (°C)	Residence time (mins)	Pollutant	Adsorption capacity	References
Poplar wood	500–550	15	Cr (III) Cd (II) Cu (II) Pb (II)	27.9 μmol/g 2.39 μmol/g 31.7 μmol/g 17.5 μmol/g	Zhao et al. (2019)
Cattle manure	600	60	Pb (II) Cd (II) Ni (II)	40.8 mg/g 24.2 mg/g 25.1 mg/g	Ye et al. (2022)
Spent mushroom substrate	700	120	Pb(II)	398.06 mg/g	Wu et al. (2019)
Cotton shell	350	120	Sulfadiazine	86.89 mg/g	Meng et al. (2020)
CO_2-activated biochar	500	180	Phenol	50–60%	Sun et al. (2020)
Pinewood sawdust	500	120	NAP PHE ANT	~90%	Rashad et al. (2022)
Rape straw	250	240	NAP ACE PHE	525 mg/g 448 mg/g 808 mg/g	Cheng et al. (2019)

NAP Naphthalene, *PHE* Phenanthrene, *ANT* Anthracene, *ACE* Acetylene

had a maximum adsorption capacity on MBC of 97.6 mg/g, much more significant than other Biochar sources (Guo et al. 2018).

3. Removal of Nitrate: Hu et al. (2023) focused on using biochar made from rice husks to remove nitrate from contaminated water. According to the study, biochar can remove up to 90% of nitrate and has a significant capacity to do so. The study linked effective adsorption to the biochar's porous structure and the presence of functional groups (Hu et al. 2023). Similarly, a study by Revilla et al. found that wastewater treated with activated biochar made from municipal sewage waste decreased nitrate concentration from 30 mg/L to 11.27 mg/L (Revilla et al. 2020) (Table 4.1).

These case studies highlight the potential of biochar as a potent adsorbent for filtering out various contaminants from water sources. Biochar is a great option for environmentally friendly water filtration because of its wide surface area, presence of functional groups, and porous structure (Fig. 4.1).

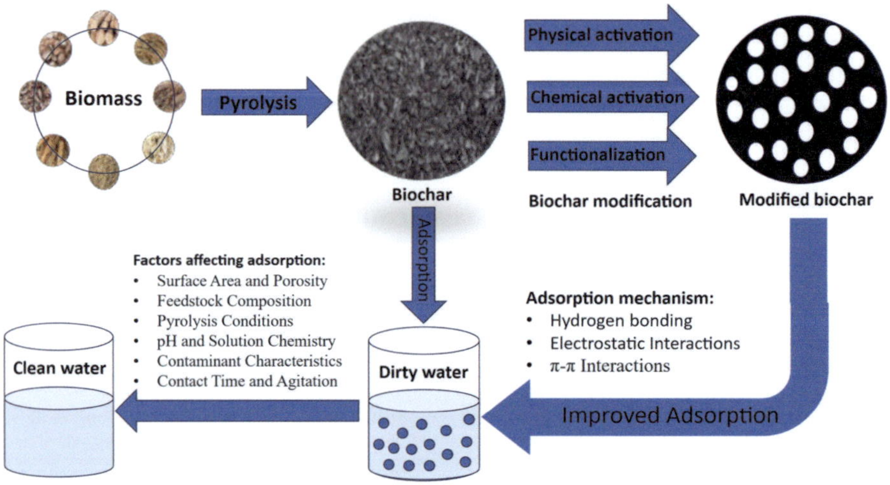

Fig. 4.1 Process of cleaning water using biochar

4.3 Surface Modification Techniques for Enhancing Biochar Performance

4.3.1 Physical, Chemical Activation, and Functionalization of Biochar

Physical and chemical activation and functionalization of biochar increase its adsorption capacity and modify its characteristics to meet the needs of water treatment applications.

Biochar must be heated to a high temperature and roasted with gases like carbon dioxide or steam to develop an activated carbon-like structure. As a result of this process, biochar has more micropores and surface area, which improves its ability to absorb contaminants. Biochar can be physically activated to alter its physical characteristics and enhance its suitability for adsorbing pollutants with smaller molecular sizes (Ahmed et al. 2018).

Biochar must be subjected to chemicals like potassium hydroxide or phosphoric acid to be chemically activated. The chemical reaction causes biochar to develop larger pores and surface functional groups. This raises the adsorption capacity because there are more surfaces for pollutants to adhere to. Chemical activation can significantly increase the efficiency of biochar's adsorption, especially for removing organic pollutants and heavy metals (Sajjadi et al. 2019).

Its surface must be modified with functional groups to improve biochar's ability to adsorb specific pollutants. Chemical processes or the impregnation of materials like nanoparticles or polymers can accomplish this. The functional groups added throughout the procedure can raise biochar's affinity for particular pollutants through chemical interactions, resulting in more effective adsorption. Biochar can be customized, and its surface chemistry can be optimized thanks to functionalization, making it a flexible and powerful adsorbent for various water treatment applications (Hamid et al. 2022).

Thus, to increase the adsorption capacity of biochar and alter its physical and chemical properties to satisfy the needs for water treatment, functionalization, physical activation, and chemical activation are available. By expanding the uses and effectiveness of biochar-based water treatment systems, these techniques aid in the efficient and long-lasting removal of contaminants.

4.3.2 Biochar Surface Modification for Water Treatment

Through surface modification, biochar can be made more adsorbent and have its surface characteristics tailored for specific pollutant removal in water treatment applications. This technique modifies the surface of the biochar utilizing various techniques, including chemical treatments, impregnation with compounds, or functionalization with particular functional groups. Making these modifications, intends to increase biochar's affinity and selectivity for pollutants and thus its overall efficacy in the treatment of water.

Li et al. (2019) looked at how coating biochar with the organic biopolymer chitosan could change the surface. The cationic and anionic pollutant adsorption capacity of the modified biochar was significantly improved. Heavy metals and organic contaminants were able to be absorbed onto the surface of the biochar due to the chitosan coating's increased availability of the amino groups (Li et al. 2019).

Kyzas et al. (2017) studied the potential impact of zero-valent iron nanoparticles on the surface of biochar. The modified biochar could be used to absorb heavy metals, colors, and medicines more efficiently. Adding zero-valent iron nanoparticles to the biochar surface formed more sites for reducing and immobilizing contaminants, increasing the removal process's efficacy (Kyzas et al. 2017).

Impregnating biochar with nanoparticles and composite materials is a promising method for enhancing its adsorption capacity and extending its value for water treatment applications. Nanoparticles like iron oxide, silver, or titanium dioxide can be added to the surface of biochar to create nanocomposites based on biochar. As they combine the distinct characteristics of nanoparticles and biochar, these nanocomposites provide synergistic advantages for removing pollutants. The nanoparticles increase the number of adsorption sites, catalytic activity, or antibacterial properties, while the charcoal matrix provides a dependable support structure. This impregnation process creates composites with enhanced adsorption capacity and additional features that effectively remove various impurities from water sources (Li et al. 2020a, b).

These examples show how surface modification approaches can improve biochar's adsorption capacity and selectivity for removing specific contaminants. Surface modification is a flexible method for enhancing the efficiency of biochar-based water treatment systems, providing customized responses to problems with water quality.

4.3.3 *Case Studies and Examples of Surface-Modified Biochar for Water Treatment*

For targeted contaminant removal in water treatment, surface modification of biochar has been intensively researched to increase its adsorption capability and selectivity. Numerous case studies and illustrations demonstrate how successfully surface-modified biochar is used in water treatment procedures.

Liu et al. (2019a, b) concentrated on the surface modification of biochar by impregnating it with nanoscale zero-valent iron to eliminate arsenic from water (nZVI). The modified biochar showed improved adsorption capacity and reactivity toward arsenic due to adding nZVI. The results showed significant removal of arsenic, with adsorption abilities exceeding those of unaltered biochar. The surface-modified biochar with nZVI remediation of arsenic-contaminated water sources demonstrated promising potential (Liu et al. 2019a, b).

Li et al. (2018) investigated grafting poly (acrylic acid) (PAA) onto the surface of biochar to remove heavy metals from water. The modified biochar showed improved heavy metal adsorption due to the addition of carboxylic acid groups by PAA grafting. In comparison to unaltered biochar, the results demonstrated higher removal efficiency for the metal ions lead (Pb), copper (Cu), and cadmium (Cd). With PAA grafting, the surface-modified biochar showed excellent potential for removing heavy metals from aqueous solutions (Li et al. 2018).

Table 4.2 shows biochar produced from different biomasses undergoes various types of modifications and its effect on adsorption capacity.

The effectiveness of surface-modified biochar in purifying water is shown in these case studies. Due to surface modification techniques like functional group grafting or nanoparticle impregnation, biochar is a versatile and effective adsorbent for removing diverse pollutants from water sources. The successful use of surface-modified biochar in these case studies demonstrates the material's potential as an environmentally acceptable water treatment process.

4.4 Biochar-Based Water Treatment Systems

One of the inventive and sustainable solutions to issues with water quality is biochar-based water treatment. These devices filter contaminants out of water sources using biochar as an adsorbent. Biochar efficiently adsorbs contaminants such as heavy metals, organic molecules, and nutrients due to its porous structure, sizable surface area, and affinity for different pollutants. Various water treatment setups, such as fixed-bed filters, batch reactors, and continuous flow systems, can use biochar. These systems not only remove pollutants but also provide other advantages like filtration, pH adjustment, and the potential for nutrient recovery. Utilizing biochar-based water

Table 4.2 Recent studies on biochar modification and pollutant adsorption from aqueous media

Feedstock	Modifications	Findings	Pollutants	Adsorption capacity	References
Banana straw	Mg-loaded biochar prepared by adding $MgCl_2$	The presence of spherical to asymmetrically shaped magnesium particles on the surface of the biochar, in the forms of $Mg_2(OH)_3Cl.4H_2O$ and MgO, offers the potential for ion exchange and other processes.	Cd (II) Cu (II) Pb (II)	333.33 mg/g 370.37 mg/g 302.58 mg/g	Li et al. (2020a, b)
Cassava straw				238.10 mg/g 294.12 mg/g 200.00 mg/g	
Chinese fir straw				75.19 mg/g 111.11 mg/g 61.73 mg/g	
Corn straw				96.15 mg/g 169.49 mg/g 90.91 mg/g	
Camellia nutshells				66.23 mg/g 84.75 mg/g 54.47 mg/g	
Taro straw				185.19 mg/g 217.39 mg/g 166.67 mg/g	
Agar	Produced N_2-mixed magnetic biochar through pyrolysis in an NH_3 environment	NH_3 atmosphere pyrolysis can effectively modify surfaces, boost magnetism, and activate N-functional groups.	Cr(VI)	142.9 mg/g	Mian et al. (2018)
Corn cob	Internally modified pyrolysis	–	MB	16.50 mg/g	Suwunwong et al. (2020)
Aegle marmelos shell	Steam activation	Biochar modification increased active adhesion sites.	IBP	95%	Chakraborty et al. (2018)
Corn Stover	Chemical activation using KOH	On the biochar surface, several acidic surface groups containing oxygen were added, increasing the specific surface areas.	MB	406.4 mg/g	Liu et al. (2020)

MB Methylene blue, *IBP* ibuprofen

treatment systems is a workable, sustainable, and cost-effective way to guarantee that people have access to clean and safe water supplies (Ahmad et al. 2014).

4.4.1 Point-of-Use Water Treatment with Biochar

An easy and practical option to provide households with clean, safe drinking water is to use biochar to treat water at the point-of-use. Biochar can be used in simple and affordable filtration technologies like biochar filters or biochar-based filter cartridges that are easy to install and maintain. These systems effectively remove contaminants such as pathogens, heavy metals, organic compounds, and turbidity, which improves the quality of drinking water (Mohan et al. 2014). The suitable material for point-of-use water treatment is biochar due to its adsorption, filtering properties, cost, and sustainability. Adopting biochar-based point-of-use water treatment systems can lower the incidence of waterborne infections and improve a community's overall health, particularly in areas with limited access to clean water sources.

4.4.1.1 Biochar Filters and Cartridges for Household Water Treatment

As an efficient and sustainable alternative for household water treatment, biochar filters and cartridges have attracted much interest. To remove impurities and enhance the purpose of removing contaminants and enhancing water quality, these filtration systems primarily use biochar as an adsorbent material. Heavy metals, organic compounds, pathogens, and turbidity are just a few of the contaminants that biochar filters catch and adsorb as water travels through layers of biochar medium (Gwenzi et al. 2017). Biochar's porous design and large surface area offer plenty of adsorption sites, enabling effective pollutant removal. The biochar's inherent antibacterial capabilities can also improve the filters' disinfecting ability.

A study by Guo et al. (2019) showed the value of biochar filters in increasing drinking water quality. The research assessed the effectiveness of a biochar filter system in eliminating different contaminants, such as heavy metals, organic pollutants, and *E. coli* bacteria. The results indicated a considerable decrease in the levels of pollutants, demonstrating the biochar filter's capacity to successfully purify water at the household level (Guo et al. 2019).

Point-of-use water treatment is made simple and portable with biochar-based filter cartridges. These cartridges can be used with individual water containers or current water treatment systems because they are small and simple to install. The cartridge's biochar material efficiently removes impurities to deliver clean, sterile drinking water. Additionally, because biochar is durable, these cartridges can be used longer, decreasing the need for frequent replacements and improving their overall sustainability.

4.4.1.2 Biochar-Based Filtration Systems for Rural and Remote Areas

Biochar-based filtration systems have become a feasible option to solve water treatment issues in rural and distant places. Deploying decentralized and sustainable technology is critical since these locations frequently lack access to centralized water treatment facilities. Biochar is a locally accessible and reasonably priced material that can be made from agricultural waste, forestry waste, or other biomass sources, making it appropriate for environments with limited resources. Additionally, biochar's high surface area and adsorptive properties allow for effectively removing pollutants such as pathogens, heavy metals, and organic compounds. Because of its adaptability, biochar can be customized to address local water quality issues and used with various water sources.

According to Liu et al. (2020), a biochar-based filtration system was effectively used in a small Chinese community. The system successfully removed impurities from the local water source, including turbidity, bacteria, and organic materials. To ensure affordability and sustainability for the neighborhood, local biochar and gravel were used to construct the biochar filters. The study concentrated on the applicability and efficiency of biochar-based filtration systems for improving water quality in rural areas (Liu et al. 2020).

Biochar-based filtration systems offer additional benefits for rural and remote areas. Community members can easily operate and maintain them without needing specialized skills or complex equipment. Moreover, the biochar media in these systems have a long lifespan, reducing replacement frequency and ensuring sustained water treatment. Biochar's potential for nutrient retention can also provide agricultural benefits by releasing nutrients into the soil when spent biochar is utilized as a soil amendment.

4.4.2 Large-Scale Water Treatment Using Biochar

Using biochar for large-scale water treatment has drawn interest as a sustainable and economical solution to water quality problems in various contexts. The efficient removal of pollutants from vast amounts of water is made possible by biochar's high surface area, porosity, and adsorptive capacity. Biochar is a dependable and long-lasting alternative for large-scale treatment systems because of its stability and durability.

One biochar application for large-scale water treatment is installing filtration systems in water treatment facilities. A multi-step treatment procedure uses biochar as a filtration medium in these systems. For instance, a biochar-based fixed-bed filtration system was put into place in a water treatment plant in a study by Mohan et al. (2014) to remove organic and inorganic impurities. The biochar filter provided treated water of higher quality by successfully reducing the concentration of pollutants, such as pesticides, heavy metals, and dissolved organic matter (Mohan et al. 2014).

Biochar can be used in activated sludge systems, anaerobic digesters, or built wetlands to improve the removal of pollutants such as organic matter, nitrogen, and phosphorus. Engineered systems (built wetlands) clean wastewater using natural processes. Biochar can be incorporated into the substrate or used as a separate layer within the wetland system. Biochar's porous structure and substantial surface area encourage microbial activity and remove organic materials, nitrogen, and phosphorus through biological and physical processes (Li et al. 2017a, b). Since studies have shown that biochar effectively boosts pollution removal in artificial wetlands, it is a valuable addition to these systems.

Surface water contaminated with heavy metals and organic pollutants was treated with a biochar filter in a study by De La Rosa et al. (2016). The biochar filter effectively decreased contaminant concentrations, demonstrating its potential for point-of-use water treatment.

Xu et al. (2017) looked specifically at the usage of biochar in a large-scale water treatment plant. The study examined biochar as a filter medium for eliminating turbidity, micropollutants, and organic matter from raw water sources. The biochar-based filtration system significantly improved the water quality, demonstrating the effectiveness of biochar in large-scale water treatment applications (Xu et al. 2017).

By incorporating biochar into these systems, treatment efficiency is improved, the load on traditional treatment facilities is reduced, and the conversion of pollutants into more stable forms is encouraged.

4.5 Environmental Sustainability and Economic Viability of Biochar-Based Water Treatment

Water treatment using biochar shows a high potential for achieving both economic viability and environmental sustainability. Biochar offers an environmentally sustainable option by using agricultural and forestry waste, avoiding waste from going to landfills, and reducing greenhouse gas emissions through carbon sequestration. Additionally, porous biochar provides a habitat for beneficial micro-organisms that promote soil health and nutrient cycling (Lehmann and Joseph 2015). Addressing waste management, reducing climate change, and boosting ecosystem resilience these factors that support environmental sustainability.

Systems for water treatment based on biochar provide several benefits in terms of economic viability. Biochar manufacturing can be economical, especially when employing biomass feedstocks that are readily available locally. Additionally, the efficient removal of pollutants from water by biochar's adsorptive characteristics eliminates the need for expensive traditional treatment techniques. The economic viability of using biochar as a filtration medium is further enhanced by its longevity, which lowers the need for frequent replacement and maintenance expenses. Additionally, the waste biochar or by-products can be used for agricultural purposes like soil amendment, increasing the overall economic worth of the biochar-based water treatment method (Mohan et al. 2014).

4.5.1 Economic Considerations and Cost-Effectiveness of Biochar-Based Water Treatment

Economic factors and cost efficiency greatly influence the acceptance and viability of biochar-based water treatment systems. The cost and long-term feasibility of such a system can be evaluated by looking at its economic features. Compared to conventional treatment techniques, biochar-based water treatment has several financial benefits. First, making biochar can be affordable, particularly when using locally accessible biomass feedstocks (Chen et al. 2015).

Biochar's stability also aids cost-effectiveness as a filtration medium. Since biochar has high durability and requires minimum replacement or regeneration, it is less expensive to operate and maintain than conventional treatment techniques. Additionally, using biochar in water treatment can lower the demand for expensive chemicals because the material's adsorptive capabilities efficiently remove impurities without the need for many chemicals (Kloss et al. 2012).

The overall financial sustainability of water treatment systems based on biochar must be determined. Therefore, cost-benefit evaluations and economic assessments are crucial. The initial investments, operational cost, and prospective savings or benefits related to enhanced water quality reduced chemical use, and less maintenance need should all be considered in these evaluations. Making educated judgments about implementation and investment is made possible by evaluating the economic viability of biochar-based water treatment systems, which enables long-lasting and affordable solutions to water treatment problems (Kuppusamy et al. 2016).

4.5.2 Social and Community Benefits of Biochar-Based Water Treatment

Systems for treating water with biochar positively impact the environment, the economy, and the community. These systems can improve access to clean and safe water while addressing issues with water quality in communities. Biochar-based water treatment can help rural and urban populations, especially those experiencing water scarcity or inadequate infrastructure, by offering a long-lasting and cost-effective alternative (Mohan et al. 2014).

Improving public health is one of the major societal advantages of biochar-based water treatment. Access to clean water is crucial for the well-being of communities and the prevention of waterborne illnesses. Biochar-based systems assist in lowering the dangers related to contaminated water sources by efficiently eliminating contaminants such as heavy metals, organic pollutants, and pathogens. This contributes to better public health outcomes, especially in areas with limited access to clean water (Liu et al. 2013).

Using biochar for water treatment can also promote community empowerment and involvement. These systems are participatory approaches to water treatment since they may be adjusted to particular community requirements and local situations. Community involvement in biochar-based systems' design, operation, and upkeep fosters ownership and a sense of responsibility, resulting in long-lasting and sustainable solutions (Atiqullah et al. 2018). Additionally, using locally accessible biomass as a feedstock to create biochar can open up business opportunities and support the growth of regional economies.

4.5.3 Policy and Regulatory Aspects of Biochar-Based Water Treatment

Promoting and regulating the use of biochar-based water treatment technologies depends heavily on policy and regulatory frameworks. To guarantee biochar's safe and efficient usage in water treatment while addressing environmental and public health issues, these frameworks establish rules, regulations, and incentives (Cimò et al. 2019). The adoption of biochar-based systems and their integration into current water management methods are made possible by policy support and laws.

Setting up rules for biochar production and quality control production and quality control of biochar is a crucial component of policy and regulatory issues. To guarantee that the biochar used in water treatment satisfies specific requirements for efficacy and safety, this includes guidelines for feedstock selection, production techniques, and quality assurance. Regulations may also cover proper biochar processing, storage, and disposal to reduce like proper biochar processing, storage, and disposal to reduce any potential environmental impacts (Lehmann and Joseph 2015).

Policy frameworks should also cover promoting research and development, information exchange, and capacity-building activities associated with biochar-based water treatment. This promotes innovation, cooperation, and sharing of best practices among stakeholders, boosting the uptake and expansion of biochar-based systems. To further encourage the cost and accessibility of biochar-based water treatment technologies, governmental support can also take the form of financial incentives, grants, and subsidies, especially in areas where water quality is a major concern (Ennouri et al. 2021).

4.6 Challenges and Future Prospects

While biochar has much potential for water treatment applications, several technical issues and knowledge gaps still need to be addressed before its full potential can be utilized. These difficulties include:

1. Standardization and optimization of biochar production: Standard protocols for biochar production are required to ensure consistent quality and performance. Biochar's adsorption capacity and stability can be greatly impacted by optimizing production variables such as feedstock source, pyrolysis temperature, and residence time (Mohan et al. 2006). The ideal manufacturing conditions for various feedstocks and pollutants improve further study.

2. Mechanisms of contaminant adsorption: Although many studies have demonstrated biochar's capacity to adsorb pollutants, the underlying mechanisms are still poorly understood. Further research is necessary to determine the precise interactions between contaminants and biochar, including surface chemistry, pore structure, and functional groups (Ahmad et al. 2014). The development of new biochar materials with enhanced adsorption capacities and specialized water treatment methods both depend on this information.
3. Long-term stability and regeneration of biochar: To ascertain biochar's long-term stability and regeneration capacity in water treatment systems, more investigation is required. The performance and durability of biochar under various working conditions, such as continuous water flow, regeneration cycles, and regeneration techniques, must be investigated (Liu et al. 2013). Understanding the mechanisms of biochar aging, fouling, and regeneration is necessary for optimizing system design and operation for long-lasting and sustainable water treatment applications.
4. Scale-up and integration of biochar-based systems: Scaling up biochar-based water treatment systems from laboratory to industrial applications is a significant challenge. Research must look at the scalability, cost, and integration of biochar-based technologies into the conventional water treatment infrastructure. This includes evaluating biochar's performance in real-world situations, taking system compatibility into account, and assessing the feasibility and practicability of large-scale applications (Mohan et al. 2014).

By addressing these technological problems and knowledge gaps, it will be feasible to understand how to use biochar for water treatment successfully. By increasing our knowledge of biochar's synthesis, adsorption mechanisms, stability, and scale-up potential, we may fully realize its potential as an efficient and long-lasting solution to water treatment issues.

4.7 Conclusion and Future Directions

A sustainable strategy based on biochar can be used to address the problem of water contamination. Since it has a high surface area, porosity, and adsorption capacity, biochar's distinct characteristics make it an efficient adsorbent for various pollutants in water. The effectiveness of employing biochar in various water treatment systems, such as point-of-use systems, large-scale treatment plants, and artificial wetlands, has been shown in several studies and case studies. Biochar has a lot of potential, but there are a lot of challenges that need to be investigated before it can be applied broadly. More study is needed to enhance biochar manufacturing methods, activation processes, and our comprehension of the fundamental ideas guiding pollutant adsorption onto biochar. Long-term study is also necessary to assess the stability, capacity for regeneration, and potential consequences of biochar on the environment.

The fundamental focus of biochar-based water treatment should be interdisciplinary cooperation involving scientists, engineers, policymakers, and industrial stakeholders. This collaboration will make it simpler to develop standardized processes, guidelines, and legal frameworks for the safe and effective use of biochar in water treatment. The economic viability and environmental sustainability of biochar-based systems also call for more detailed economic assessments.

References

Ahmad M, Rajapaksha AU, Lim JE et al (2014) Biochar as a sorbent for contaminant management in soil and water: a review. Chemosphere 99:19–33

Ahmed MB, Zhou JL, Ngo HH (2018) Insight into the impact of biochar properties on the sorption of toxic organic pollutants from aqueous solutions. Bioresour Technol 262:199–208

Akhil D, Lakshmi D, Kartik A et al (2021) Production, characterization, activation and environmental applications of engineered biochar: a review. Environ Chem Lett 19:2261–2297. https://doi.org/10.1007/s10311-020-01167-7

Ambaye TG, Vaccari M, van Hullebusch ED et al (2021) Mechanisms and adsorption capacities of biochar for the removal of organic and inorganic pollutants from industrial wastewater. Int J Environ Sci Technol 18:3273–3294. https://doi.org/10.1007/s13762-020-03060-w

Atiqullah M, Huda MK, Ahmed F et al (2018) A review of the socio-economic and environmental aspects of biochar production and utilization as an energy and nutrient recovery technology. Resour Conserv Recycl 134:339–352

Barquilha CER, Braga MCB (2021) Adsorption of organic and inorganic pollutants onto biochars: challenges, operating conditions, and mechanisms. Bioresour Technol Rep 15:100728., ISSN 2589-014X. https://doi.org/10.1016/j.biteb.2021.100728

Cao X, Ma L, Liang Y et al (2011) Effects of biochar, wood vinegar, and NPK fertilizers on nitrogen and phosphorus leaching losses in a Chinese cabbage field. Aust J Soil Res 49(2):141–147

Chakraborty P, Banerjee S, Kumar S, Sadhukhan S, Halder G (2018) Elucidation of ibuprofen uptake capability of raw and steam activated biochar of Aegle marmelos shell: isotherm, kinetics, thermodynamics and cost estimation. Process Saf Environ Prot 118:10–23

Chen B, Zhou D (2011a) Adsorption of polar and nonpolar organic chemicals to carbonaceous adsorbents: role of molecular descriptors. Environ Sci Technol 45(5):2078–2084

Chen B, Zhou D (2011b) Adsorption of organic contaminants by carbon nanomaterials in aqueous phase: Polanyi theory and its application. Chem Eng J 170(2–3):395–406

Chen B, Zhou D (2014) Advances in characterization and modeling of interactions between black carbon and microbial communities. Environ Sci Technol 48(17):9734–9748

Chen H, Zhao J, Li L (2015) Application of biochar for the removal of pollutants from aqueous solutions. Chemosphere 125:70–85

Cheng H, Bian Y, Wang F, Jiang X, Ji R, Gu C, Yang X, Song Y (2019) Green conversion of crop residues into porous carbons and their application to efficiently remove polycyclic aromatic hydrocarbons from water: sorption kinetics, isotherms and mechanism. Bioresour Technol 284:1–8

Cimò G, Liu G, Liu Y (2019) Governance of biochar: a global assessment. Glob Sustain 2:e12

De La Rosa JM, Paneque M, Miller AZ et al (2016) Using biochar for the removal of heavy metals from aqueous solutions: effect of pyrolysis temperature and biochar application rate. J Environ Manag 171:146–153

Dwivedi S, Dey S (2022) Review on biochar as an adsorbent material for removal of dyes from waterbodies. Int J Environ Sci Technol. https://doi.org/10.1007/s13762-022-04364-9

Ennouri H, Jellali S, Ghrabi A et al (2021) Biochar technology for wastewater treatment and water quality improvement: review of mechanisms, challenges, and perspectives. Environ Sci Pollut Res 28(19):23708–23722

Guo W, Wang S, Wang Y, Lu S, Gao Y (2018) Sorptive removal of phenanthrene from aqueous solutions using magnetic and non-magnetic rice husk-derived biochars. R Soc Open Sci 5:172382. https://doi.org/10.1098/rsos.172382

Guo Y, Liu H, Xu Z et al (2019) Biochar filters for household water treatment: a critical review. Sci Total Environ 648:1299–1312

Gwenzi W, Chaukura N, Noubactep C, Mukome FND (2017) Biochar-based water treatment systems as a potential low-cost and sustainable technology for clean water provision. J Environ Manage 197:732–749., ISSN 0301-4797. https://doi.org/10.1016/j.jenvman.2017.03.087

Hamid Y, Liu L, Usman M, Naidu R, Haris M, Lin Q, Zaid Ulhassan M, Hussain I, Yang X (2022) Functionalized biochars: synthesis, characterization, and applications for removing trace elements from water. J Hazardous Mater 437:129337., ISSN 0304-3894. https://doi.org/10.1016/j.jhazmat.2022.129337

Hu T, Wei J, Du L et al (2023) The effect of biochar on nitrogen availability and bacterial community in farmland. Ann Microbiol 73:4. https://doi.org/10.1186/s13213-022-01708-1

Janu R, Mrlik V, Ribitsch D, Hofman J, Sedláček P, Bielská L, Soja G (2021) Biochar surface functional groups as affected by biomass feedstock, biochar composition and pyrolysis temperature. Carbon Resour Convers 4:36–46., ISSN 2588-9133. https://doi.org/10.1016/j.crcon.2021.01.003

Khadem M, Husni Ibrahim A, Mokashi I et al (2022) Removal of heavy metals from wastewater using low-cost biochar prepared from jackfruit seed waste. Biomass Convers Biorefin. https://doi.org/10.1007/s13399-022-02748-y

Kloss S, Zehetner F, Dellantonio A et al (2012) Characterization of slow pyrolysis biochars: effects of feedstocks and pyrolysis temperature on biochar properties. J Environ Qual 41(4):990–1000

Kuppusamy S, Thavamani P, Megharaj M et al (2016) Agronomic and remedial benefits and risks of applying biochar to soil: current knowledge and future research directions. Environ Int 87:1–12

Kyzas GZ, Kostoglou M, Deliyanni EA (2017) Zero valent iron nanoparticles decorated biochar for the simultaneous removal of heavy metals and organic pollutants: a multi-response optimization approach. J Clean Prod 165:793–803

Lehmann J (2019) Biochar for mitigating climate change: carbon sequestration, properties, and health implications. In: Biochar: a guide to analytical methods. Springer, pp 229–244

Lehmann J, Joseph S (2009) Biochar for environmental management: an introduction. In: Biochar for environmental management: science, technology, and implementation. Routledge, pp 1–12

Lehmann J, Joseph S (2015) Biochar for environmental management: science, technology, and implementation. Routledge

Li H, Dong X, da Silva EB et al (2017a) Engineered biochar system for simultaneous pesticide adsorption and dechlorination of chlorinated organic compounds. Environ Sci Technol 51(19):11348–11356

Li X, Wang X, Zhao Y et al (2017b) Roles of biochar in constructed wetlands and its applications. Sci Total Environ 584–585:1088–1097

Li X, Liu H, Wu D et al (2018) Grafting poly(acrylic acid) onto biochar surface for enhancing the removal of heavy metal ions from aqueous solutions. Chem Eng J 332:766–774

Li Q, Huang Q, Zhu L et al (2019) Surface modification of biochar by chitosan for enhanced adsorption of contaminants: a review. Bioresour Technol 294:122166

Li Y, Yu S, Wu Y et al (2020a) Impregnation of biochar derived from waste biomass with magnetite nanoparticles for enhanced Pb(II) removal from aqueous solution. J Hazard Mater 398:122876

Li AY, Deng H, Jiang YH, Ye CH, Yu BG, Zhou XL, Ma AY (2020b) Superefficient removal of heavy metals from wastewater by Mg-loaded biochars: adsorption characteristics and removal mechanisms. Am Chem Soc 36(31):9160–9174. https://doi.org/10.1021/acs.langmuir.0c01454

Li Y, He Z, Han S et al (2021) Biochar-based composites for water purification: a review. J Clean Prod 296:126618

Liu Z, Quek A, Kentish S (2013) Life cycle assessment of biochar for land application. J Clean Prod 52:210–218

Liu Y, Deng Z, Zhou S et al (2019a) Surface modification of biochar derived from waste biomass for the removal of arsenic from aqueous solutions. Bioresour Technol 288:121522

Liu H, Li X, Liu G (2019b) Biochar application in wastewater treatment: a critical review. Sci Total Environ 657:1243–1253

Liu Y, Li M, Zhang X et al (2020) Application of biochar-based filter for point-of-use water treatment in rural China. J Water Supply Res Technol AQUA 69(5):465–476

Meng Q, Zhang Y, Meng D, Liu X, Zhang Z, Gao P, Lin A, Hou L (2020) Removal of sulfadiazine from aqueous solution by in-situ activated biochar derived from cotton shell. Environ Res 191:110104. https://doi.org/10.1016/j.envres.2020.110104

Mian MM, Liu G, Yousaf B, Fu B, Ullah H, Ali MU, Abbas Q, Mujtaba Munir MA, Ruijia L (2018) Simultaneous functionalization and magnetization of biochar via NH3 ambiance pyrolysis for efficient removal of Cr (VI). Chemosphere 208:712–721

Mohan D, Pittman CU Jr, Steele PH (2006) Pyrolysis of wood/biomass for bio-oil: a critical review. Energy Fuel 20(3):848–889

Mohan D, Sarswat A, Ok YS et al (2014) Organic and inorganic contaminants removal from water with biochar, a renewable, low-cost and sustainable adsorbent—a critical review. Bioresour Technol 160:191–202

Pathak S, Pant KK, Kaushal P (2023) Analysis of naphthalene adsorption from wastewater using activated and non-activated biochar produced from bagasse. Biomass Convers Biorefinery 14(16):19661–19674

Rashad E, Saleh HN, Eltaweil AS et al (2022) Pinewood sawdust biochar as an effective biosorbent for PAHs removal from wastewater. Biomass Convers Biorefin. https://doi.org/10.1007/s13399-021-02181-7

Revilla PND et al (2020) IOP conference series: materials science and engineering. 778:012135

Sajjadi B, Zubatiuk T, Leszczynska D, Leszczynski J, Chen WY (2019) Chemical activation of biochar for energy and environmental applications: a comprehensive review. Rev Chem Eng 35(7):777–815. https://doi.org/10.1515/revce-2018-0003

Saleh TA (2018) Adsorption of heavy metals on biochar: a comprehensive review. Adv Colloid Interf Sci 248:235–246

Sun C, Chen T, Huang Q, Zhan M, Li X, Yan J (2020) Activation of persulfate by CO2-activated biochar for improved phenolic pollutant degradation: performance and mechanism. Chem Eng J 380:122519. https://doi.org/10.1016/j.cej.2019.122519

Suwunwong T, Hussain N, Chantrapromma S, Phoungthong K (2020) Facile synthesis of corncob biochar via in-house modified pyrolysis for removal of methylene blue in wastewater. Mater Res Express 7(1):015518

Tsai WT, Liu SC, Juang KW (2012) Adsorption of acid dyes from water onto aminated and protonated mesoporous silicate. J Hazard Mater 209–210:57–64

Wu Q, Xian Y, He Z, Zhang Q, Wu J Yang G, Zhang X, Qi H, Ma J, Xiao Y, Long L, (2019) Adsorption characteristics of Pb(II) using biochar derived from spent mushroom substrate. Sci. Rep. https://doi.org/10.1038/s41598-019-52554-2

Xu X, Cao X, Zhao L et al (2017) Biochar-based water treatment systems as a sustainable and renewable solution for clean water production. Environ Sci Pollut Res 24(13):11884–11892

Yao Y, Gao B, Zhang M et al (2016) Mechanisms and evaluation indicators for biochar physical activation: a review. Environ Sci Pollut Res 23(3):2111–2125

Ye Q, Li Q, Li X (2022) Removal of heavy metals from wastewater using biochars: adsorption and mechanisms. Environ Pollut Bioavailab 34:385–394. https://doi.org/10.1080/26395940.2022.2120542

Zhao J, Shen XJ, Domene X et al (2019) Comparison of biochars derived from different types of feedstock and their potential for heavy metal removal in multiple-metal solutions. Sci Rep 9:9869. https://doi.org/10.1038/s41598-019-46234-4

Chapter 5
Recent Advances in Anaerobic Baffled Reactor Technology for Decentralized Wastewater Management

Dinesh Bhardwaj, Richa Sharma, Lavi Dhiman, Vikas Kumar Gahlot, Shalini Anand, and Pramod Kumar Rai

Abstract Anaerobic baffled reactor (ABR) technology has emerged as a promising approach for decentralized wastewater treatment due to its simplicity and effectiveness. This book chapter comprehensively reviews recent advancements in ABR technology, emphasizing its application in decentralized wastewater management. It covers the fundamental principles and design features of ABRs, demonstrating their suitability for such settings. The chapter discusses recent developments in reactor design, operational parameters, material selection, and the optimization of ABR performance, particularly in terms of COD (chemical oxygen demand), Biological Oxygen Demand (BOD), and Total Suspended Solids(TSS) removal efficiency with varying hydraulic retention times (HRTs). Additionally, it explores the integration of ABRs with other wastewater treatment technologies, such as pretreatment and posttreatment processes, to enhance overall treatment efficacy. Studies highlighted in the chapter show that ABRs can achieve up to 98% COD removal, with longer HRTs generally resulting in improved treatment outcomes, especially for domestic wastewater.

Keywords Anaerobic baffle reactor · Wastewater treatment · Decentralized · Biological Oxygen Demand (BOD) · COD · Hydraulic retention time

D. Bhardwaj · L. Dhiman · V. K. Gahlot · S. Anand (✉) · P. K. Rai
Centre for Fire, Explosive, and Environment Safety (CFEES), Defence Research & Development Organisation (DRDO), Delhi, India
e-mail: shalini.cfees@gov.in

R. Sharma
Maharaja Agrasen Institute of Technology, GGSIPU, New Delhi, India

S. S. Kumar, V. Kumar (eds.), *Nature-Based Solutions for Decentralised Liquid Waste Management*, Environmental Challenges and Solutions,
https://doi.org/10.1007/978-3-032-08362-3_5

Graphical Abstract

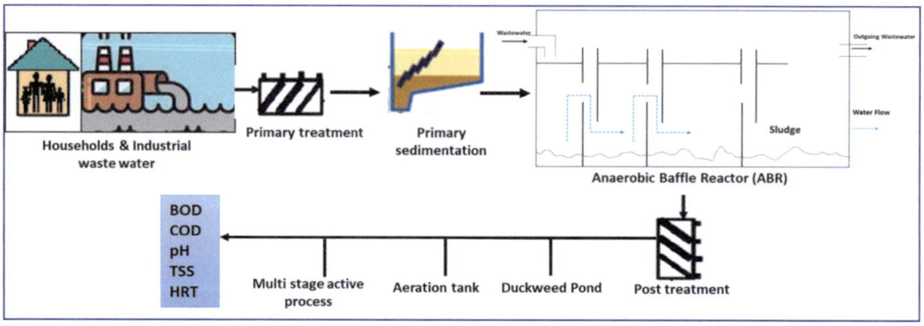

5.1 Introduction

In the current era, the release of decentralized wastewater is on the rise due to the population's constant growth and the swift progress of rural parts. It contains diverse contaminants such as nutrients, organic matter, suspended solids, and pathogens. The source of the contamination—such as municipal, industrial, or agricultural wastewater—determines the concentration of the contaminant. Clearing of decentralized wastewater in the adjoining areas without appropriate handling is adding to the decline in the water quality of rivers, lakes, and groundwater. This may have thoughtful harmful effects on natural resources and human health, necessitating effective and sustainable decentralized wastewater treatment. Also, India's water obligation is projected to be 1.5 trillion m^3 by 2030. However, the nation only manages 16.8% of the sewage produced. There is a huge gap between the water required and available for use that could only be overcome by reprocessing and utilizing treated wastewater (Pal and Bhanot 2023). Wastewater purification and recycling are therefore crucial. Owing to the diverse nature of wastewater and its likely harmful effects on the environment, various treatment technologies viz., physical, chemical, biological, and advanced treatment are in use to remove the different types of contaminants (Metcalf, and Eddy, Inc 2003; Metcalf and Eddy Inc et al. 2013). Physical treatment viz., sedimentation, screening, and sand removal involves exclusion based on their physical properties such as size and density. Chemical treatment viz., precipitation, coagulation-flocculation, and ion exchange are based on contaminant separation by forming the precipitate, coagulating, or flocculating. Biological treatment includes activated sludge, trickling filters, and rotating biological contractors employing microorganisms for contaminant removal from the wastewater. Advanced treatment technologies include membrane filtration, adsorption, and advanced oxidation process. Among the above-discussed treatment technologies, membrane-based technologies showed numerous advantages such as complete sludge retention, produced clean water, efficient energy recovery, and high removal rate (Zijing An et al. 2023).

The anaerobic biotic treatment technology has been acknowledged the most due to advancements in terms of less discharge of carbon emissions and lesser requirement of energy, minimum production of left-over sludge, and the opportunity of resource generation by biogas production (Zhu et al. 2015). In the realm of anaerobic biotic wastewater treatment technologies, the anaerobic baffled reactor (ABR) stands as a notable innovation, representing a promising solution for decentralized wastewater management with its steady performance. Since its introduction by Lettinga and associates in the late 1970s, this reactor design has changed due to developments in microbiology, engineering, and our understanding of wastewater treatment procedures (Lettinga et al. 1980). The original design calls for a series of up-flow anaerobic sludge bed (UASB) reactors with hanging baffles and multiple partitions. A variety of anaerobic microbes comprising acidogenic, acetogenic bacteria, and methanogenic bacteria in separate sections carry out the steady decline in the contaminants by supportive effort (She et al. 2006). The ABR with its creative structure has advantages in terms of relatively low construction and operation costs, good stability, and greater efficacy when subjected to highly polluted water, and greater production of biogas with the simpler process of collection (Hassan and Dahlan 2013; Zhu et al. 2015). Although a comprehensive review of research work carried out on ABR has been summarized by researchers (Barber and Stuckey 1999; Zhu et al. 2015), this present chapter provides newer insights about the scientific enhancements and applications of ABR specifically for decentralized wastewater management. The information gathered here will aid fellow researchers in working in the direction of further increasing ABR treatment competence and process steadiness.

5.2 Decentralized Wastewater Treatment Systems

Several decontamination processes for wastewater remediation have been in place for years comprising physical, chemical, and biological treatment possibilities. However, recently emphasis has been given to decentralized treatment systems over centralized wastewater treatment systems due to simplicity of operation, cost-effectiveness, and upkeep. In an evaluation of the treatment proficiency of decentralized wastewater treatment systems (DEWATS) plants (eight numbers) for their performance in terms of designated physicochemical parameters of the wastewater it was concluded that DEWATS can be a decent substitute for treating wastewater with insignificant energy and chemical consumption (Singh et al. 2019). Due to its ability to address issues with centralized systems, such as large infrastructure, energy consumption, and expensive construction and maintenance costs, decentralized wastewater management has drawn more and more attention. Within the realm of DEWATS, anaerobic baffled reactors stand out as a specific technology leveraging anaerobic digestion to treat wastewater (Tran et al. 2022). These compact systems utilize a series of baffles within a tank, forcing wastewater to flow upwards in a meandering path. Under DEWATS systems, ABR design promotes the growth of anaerobic microorganisms in the absence of oxygen (Lau and Trzcinski 2022). These beneficial microbes then break down organic matter in the wastewater, producing biogas as a by-product and significantly reducing organic pollutants (Elreedy et al. 2016).

5.3 Anaerobic Baffled Reactor

Anaerobic baffled reactors are a type of DEWATS that uses anaerobic digestion to remove organic matter and pollutants from wastewater. The ABR technology, in particular, has garnered interest for its ability to treat wastewater at a smaller scale, ensuring localized treatment and resource recovery (Singh et al. 2009). ABRs are an excellent choice for decentralized wastewater treatment in underdeveloped nations and rural communities since they are simple to construct, maintain, and reasonably priced (Varma et al. 2022). The ABR's unique design incorporates multiple chambers with alternating zones, each facilitating different stages of anaerobic digestion and organic matter degradation, making it an efficient and sustainable option for decentralized wastewater treatment (Fig. 5.1).

5.3.1 Key Components of Anaerobic Baffled Reactor

5.3.1.1 Reactor Structure and Compartmentalization

The ABR is divided into several compartments using baffles, promoting a gradual transition of wastewater from one compartment to another. Recent advances in reactor structure design have focused on optimizing the number and arrangement of

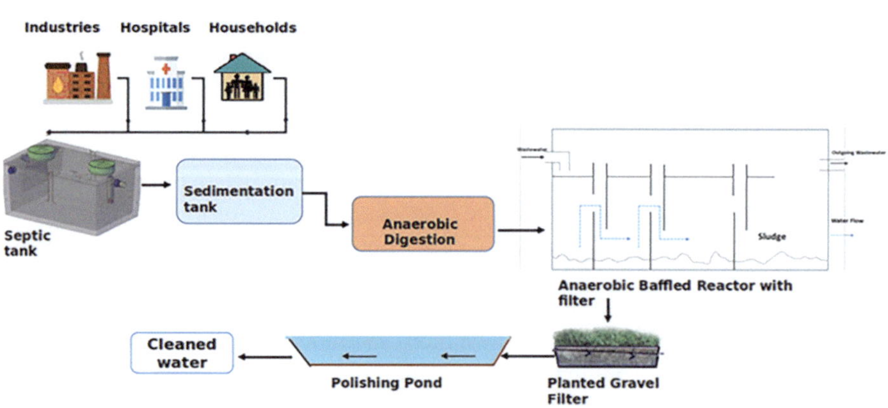

Fig. 5.1 Schematic representation of the anaerobic baffled reactor system for treatment of decentralized wastewater

compartments to enhance treatment efficiency and minimize reactor footprint. ABRs offer a simple yet effective solution for DEWAT. Their key feature lies in their compartmentalized structure, differentiating them from traditional septic tanks, which are baffles, inlets, and outlets (Bruno et al. 2020). Usually rectangular, tanks can be constructed from steel, plastic, or concrete. The intended hydraulic retention time (HRT) and the design flow rate determine the tank's size (Pererva et al. 2020). However, the distinguishing characteristic of ABR is its use of baffles, which are vertical panels set alternately to divide the tank into various compartments.

5.3.1.2 Baffle Design and Hydraulic Characteristics

The arrangement and design of baffles are essential for attaining the ideal ratio of mixing and flow (Yao et al. 2023). ABR consists of multiple baffles arranged in a series within a tank. These baffles create compartments that channel and decrease the wastewater flow, promoting anaerobic treatment processes (Yao et al. 2023). Improvements in baffle design aim to optimize hydraulic characteristics within the reactor, ensuring efficient wastewater flow, enhanced mixing, and better separation of solid particles. Novel baffle designs and materials have been explored to improve performance and longevity (Rahman 2023). The hydraulic characteristics of ABRs are HRT, flow regime, and mixing. HRT being the average time wastewater spends inside the ABR significantly impacts treatment performance, with longer HRTs generally leading to better COD and Biological Oxygen Demand (BOD) removal (Reynaud and Buckley 2016). Concerning the flow regime, plug flow is preferred, allowing wastewater to move as a uniform plug through the compartments. Short-circuiting (uneven flow distribution) should be minimized to ensure effective treatment throughout the reactor. While more mixing inside the compartments is advantageous, too much turbulence might interfere with sludge settling and impede anaerobic processes.

5.4 Advances in ABR Performance

Recent years have added to several advancements in ABR technology leading to improvement in performance and efficiency. These advances include improved reactor design improving the wastewater flow through the reactor and reducing the risk of short-circuiting. This has led to increased removal of organic matter and pollutants. Modern materials such as geomembranes and polymers, increase the durability and are easier to maintain than traditional materials, such as concrete and brick. The development of new operating strategies further improves the performance of ABRs in different climates and under varying loading conditions. This has made ABRs more reliable and versatile. These advances have made ABRs an even more attractive option for decentralized wastewater treatment.

5.4.1 Enhanced Organic Matter Removal

Recent research has focused on strategies to enhance organic matter degradation within the ABR, such as employing pretreatment techniques and optimizing operating conditions to increase organic load removal and biogas production (Amoo et al. 2023). Also, the unique baffled design of the ABR creates a multichambered system. Wastewater flows upwards through each chamber, encountering baffles that force the flow down and back up again (Tillman 1996). This extended flow path increases the contact time between the wastewater and the active biomass, which consists of various anaerobic microorganisms. Extended interaction enables these microorganisms to decompose organic materials found in the wastewater. Additionally, the baffling promotes the settling of heavier organic solids, leading to their removal from the treated effluent. ABR provides a favorable environment for anaerobic digestion meaning there's no free oxygen present during digestion (Giordani et al. 2021). This environment fosters the growth of specific bacteria adept at breaking down complex organic molecules into simpler compounds like methane, carbon dioxide, and new microbial cells through anaerobic digestion (Anderson et al. 2003). Figure 5.2 shows the breakdown process significantly reduces the overall organic content in the wastewater.

5.4.2 Nutrient Removal

Efforts have been made to improve the nutrient removal capabilities of ABR, including nitrogen and phosphorus removal, through innovative reactor configurations and integrated treatment processes (Rout et al. 2021). ABR creates an environment conducive to specific bacteria that convert organic nitrogen compounds (like proteins) into ammonia through ammonification (Barton and Fauque 2009). This ammonia can be further utilized in subsequent aerobic treatment stages, where other bacteria convert it to harmless nitrogen gas through nitrification and denitrification. Besides this ABRs can be designed to encourage specific bacteria known as phosphorus-accumulating organisms (PAOs) to take up phosphorus under anaerobic conditions and store it within their cells (Barton and Fauque 2009; Islam et al. 2017). When the wastewater flows through anoxic zones within the ABR (areas with limited oxygen), these PAOs release the stored phosphorus, for its subsequent removal through settling or biological processes in later stages.

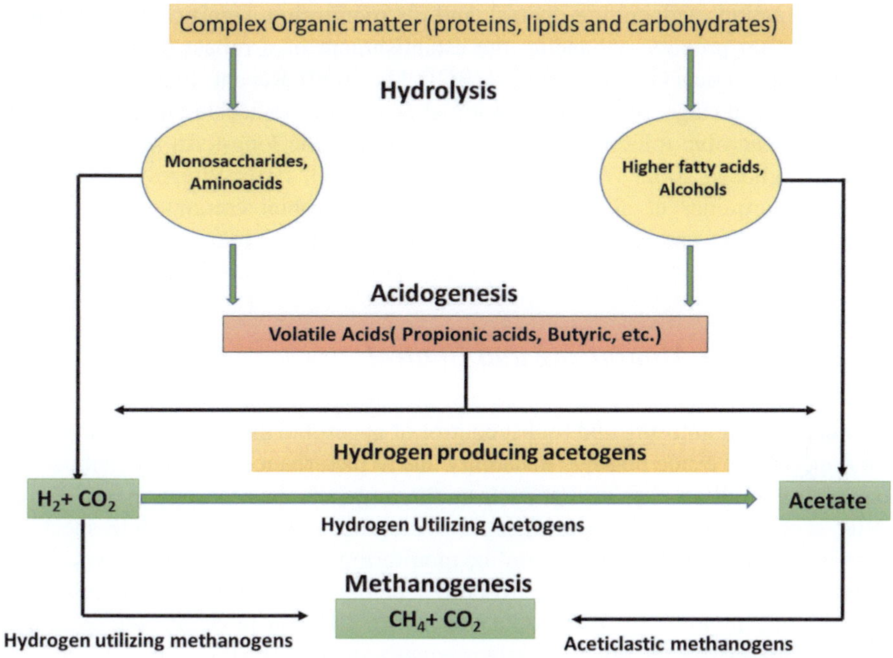

Fig. 5.2 Metabolic pathway in anaerobic baffle reactor for organic matter digestions (Lettinga et al. 1996)

5.5 Operational Considerations and Monitoring of ABR

Although the ABRs have a sturdy structure and cushion capacity to manage fluctuations in organic and hydraulic payload rates, the deficiency of unswerving procedures and maintenance practices has negatively impacted decentralized sewage treatment systems. In a study of functional ABRs in Indonesia, the final effluent did not meet the discharge standards necessitating reinvestigation in ABR design standards, procedure, and upkeep to accomplish sustainability in the functioning of cleanliness amenities (Yulistyorini et al. 2019). Implementing particular start-up, acclimatization, process monitoring, and control procedures is necessary to increase the performance efficiency for increased output of ABR.

5.5.1 Start-Up and Acclimatization

While ABRs present an attractive option for DEWAT, their initial start-up and acclimatization phase can be crucial for establishing efficient and stable performance. By implementing research-driven strategies such as seeding with pre-acclimated sludge, gradual organic loading by introducing wastewater gradually, nutrient supplementation, temperature control, and pH monitoring, significant acceleration in the start-up process, enhancing the establishment of a robust process may be achieved (Biological Oxygen Demand (BOD)khe 2009). Recent studies have delved into this critical stage, investigating optimal procedures to accelerate the formation of robust anaerobic microbial consortia and ensure the long-term stability of the reactor (Ingildsen 2002; Wu et al. 2014; Elreedy et al. 2016; Vítězová et al. 2020; Guamán-Marquines et al. 2023). The diverse microbial community ultimately ensures the stable and efficient performance of the ABR system.

5.5.2 *Process Monitoring and Control*

The real-time monitoring and robust control of anaerobic digestion (AD) processes are important to evade possible instability due to disturbances. Various methods are available for online monitoring, such as fluorescence spectroscopy, near-infrared (NIR) spectroscopy, titration, and gas chromatography. Among these, NIR spectroscopy is gaining popularity for the online monitoring of gas-phase parameters. Thus, integrating titration and NIR spectroscopy techniques can provide comprehensive and accurate monitoring of the AD process. In addition to parameter detection, effective analysis of the obtained data is equally important for reactor status evaluation and diagnosis (Dong et al. 2011). The incorporation of advanced monitoring and control systems such as online sensors to continuously measure parameters like pH, temperature, and dissolved oxygen (Safeer et al. 2022) and samplers to collect periodic samples for offline analysis of specific parameters, data acquisition systems to visualize the data for real-time monitoring, and enable automated control actions based on predefined thresholds, allowing real-time tracking of ABR performance, facilitating prompt adjustments, and optimizing operational parameters for better efficiency and stability.

5.6 Hybrid ABR Systems Integrating with Other Treatment Technologies

Integrating ABRs with other treatment technologies has shown aptitude in improving overall treatment efficiency and achieving higher levels of wastewater treatment and resource recovery.

5.6.1 ABR Coupled with Anaerobic Filters and Membrane Filtration Segment

The potentials of ABR coupled with anaerobic filters (AFs) for biological degradation have been evaluated for the extended treatment performance by further integrating fiber microfiltration membranes for water reuse. A 21 L capacity lab-scale ABR was made from an acrylic sheet. The reactor contained six chambers, a gas outlet, and a sampling port in each chamber. This was complemented by the presence of an anaerobic filter (AF). A tank (5 L) made of high-density polyethylene (HDPE) material was used for membrane filtration. A highly efficient COD removal (79%) was reported when the domestic wastewater was treated by an ABR integrated with an AF and membrane filtration (MF) segments running at an HRT of 24 hours. The released water was within the permissible limits of COD discharge (113 mg/L). However, integration of membrane filtration further reduced COD (50 mg/L) making the water fit for agricultural reuse (Khan et al. 2023).

5.6.2 Modified Packed Anaerobic Baffled Reactor

In a recent study, approximately 76.7% of chemical oxygen demand (COD) was removed from decentralized wastewater employing a modified packed ABR (MABR) at an HRT of 12 h (Li et al. 2022). MABR with an up-flow and a down-flow sludge bed alternating in series was evaluated. Anaerobic bacteria consume organics containing carbon without supplementary carbon feed through breakdown, treating organics resulting in decreased carbon dioxide discharges. The highest COD elimination rate (10.0 g/l/d) was attained at an HRT of 17 h. MABR might enhance COD treatment performance by offering a firm COD removal with lesser carbon dioxide releases (Yao et al. 2023).

5.6.3 Integrated Anaerobic Baffled Reactor and Bio-rack Wetland Planted with Phragmites sp. and Typha sp.

The performance of a novel integrated method using advanced ABR combined with bio-rack wetland was examined for effective treatment of domestic wastewater for its future implementation in decentralization practices (Jamshidi et al. 2014). The *Phragmites* sp. plantation scheme showed much greater decontamination (93–79%) concerning Total Suspended Solids(TSS), COD, Biological Oxygen Demand (BOD), and TN. The cohesive procedures could be a model approach for encouraging maintainable decentralization.

5.6.4 *Insulated Hybrid ABR*

ABRs are efficient, low-maintenance reactors that utilize anaerobic bacteria to break down organic matter in wastewater without requiring constant aeration. This translates to reduced energy consumption compared to traditional aerobic treatment methods. However, ABRs can be susceptible to temperature fluctuations in colder climates (Qi et al. 2019), impacting their performance. Insulated hybrid ABR showed improved performance in cold climates (Khalekuzzaman et al. 2018). Insulation in HABRs addresses this limitation. By adding insulation material to the reactor walls, the internal temperature is maintained within an optimal range for the anaerobic bacteria, even in fluctuating ambient temperatures. Insulated HABRs offer a sustainable and efficient solution for wastewater treatment, particularly in warm climates and resource-constrained regions. Their combination of low energy requirements, improved temperature stability, and potential for integration with other treatment processes makes them a valuable technology for addressing the global challenge of wastewater management (Zieliński et al. 2022) (Table 5.1).

5.7 Future Prospects and Challenges

While ABRs are effective for smaller-scale applications, scaling them up for larger communities or industrial settings is still difficult. Challenges related to scaling, cost-effectiveness, and public acceptance must be addressed to encourage widespread adoption and implementation of ABR technology for decentralized wastewater management. The challenges and forthcoming prospects of ABR are depicted in Fig. 5.3.

5.8 Conclusion

A decentralized wastewater treatment system offers an economically achievable and competent wastewater treatment technology for evolving a sustainable environmental cleanliness system. Anaerobic baffled reactor processes play a progressively significant part in DEWAT and resource recycling through their inventive structure and outstanding performance. Recent advances in ABR design, performance enhancement, operational considerations, and integration with other technologies have significantly improved its efficiency and applicability. Treatment of wastewater from different sources via COD removal would be more closely associated with flow rate, organic loading rate, water temperature, and influent suspended solids. Ongoing research and development are essential to overcome challenges related to these factors, enabling ABRs to reach their full potential in offering sustainable solutions for wastewater treatment and resource recovery. The potentials of advanced

Table 5.1 Types of ABR system with removal percentage over the HRT

Type of ABR system	Type of wastewater	Waste removal percent				HRT in hours	References
		Total Suspended Solids(TSS)	COD	Biological Oxygen Demand (BOD)	Others		
A laboratory-scale anaerobic baffled reactor	Domestic	96	84	–	TN-68	24	Vijay (2024)
Carrier ABR	Municipal	3	76	81	–	24–08	Ullah et al. (2023)
Modified packed ABR with fillers	Low-strength decentralized wastewater	–	76.7	–	–	12	Li et al. (2022)
Three-chamber ABR	Domestic wastewater in rural areas of Iraq	91	98	–	–	36	Mahdi et al. (2022)
Anaerobic membrane bioreactor (AnMBR)	Municipal	–	93	–	–	–	Cheng et al. (2021)
Anaerobic fluidized bed membrane bioreactor (AFMBR)	Municipal	100	95	–	Total nitrogen- 9	–	Li et al. (2020)
Anaerobic baffled reactor	Domestic	66	–	74	TP -21 NH4+- 43	20	Yulistyorini et al. (2019)
HABR	Dyeing wastewater	39.9	33.7	–	SO_4^{2-} 22.5	12	Qi et al. (2019)
Insulated hybrid ABR	Industrial	95	88	–	Turbidity-97, VSS-93	20	Khalekuzzaman et al. (2018)
Anaerobic bio-entrapped membrane reactor (AnBEMR)	Pharmaceutical	–	75	–	–	24	Shi et al. (2017)
Panelled anaerobic baffle-cum-filter reactor (PABFR)	Municipal	95	90	91	–	8	Renuka et al. (2016)
A four-cell anaerobic baffled reactor	Municipal	83	75	47	–	12	Hahn and Figueroa (2015)
Integrated ABR with bio-rack wetland containing *Phragmites* sp.	Domestic	–	87	93	SCOD- 90	21	Jamshidi et al. (2014)

(continued)

Table 5.1 (continued)

Type of ABR system	Type of wastewater	Waste removal percent				HRT in hours	References
		Total Suspended Solids(TSS)	COD	Biological Oxygen Demand (BOD)	Others		
A two-stage anaerobic fluidized-bed membrane bioreactor (SAF-MBR)	Domestic	–	91	94	–	2.3	Bae et al. (2013)
Modified anaerobic baffled reactor	Industrial	–	–	71	–	360	Zwain et al. (2013)
Settler anaerobic baffled reactor (ABR) with anaerobic filter (AF)	Domestic	24.5	21.9	49.7	–	–	Kerstens et al. (2012)

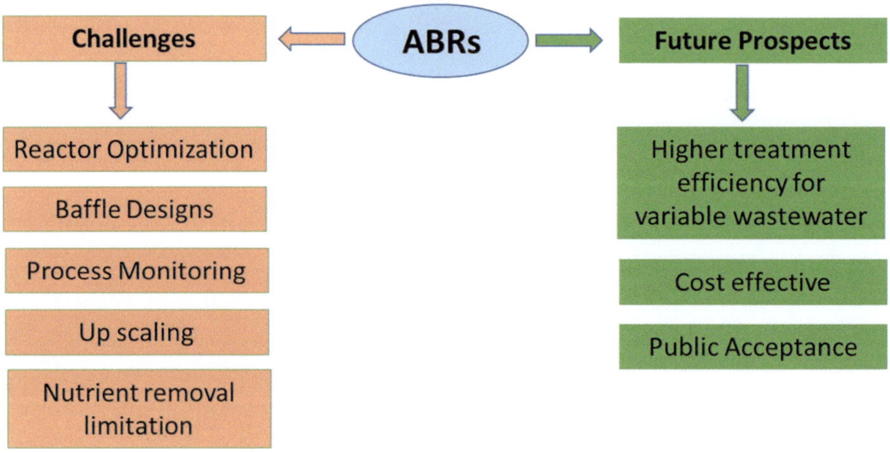

Fig. 5.3 Challenges and future prospects of ABR

configurations like hybrid ABR, carrier ABR, anaerobic fluidized bed membrane bioreactor, insulated hybrid ABR, anaerobic bio-entrapped membrane reactor, integrated ABR with bio-rack wetland containing *Phragmites* sp., two-stage anaerobic fluidized-bed membrane bioreactor, and settler ABR with anaerobic filter to further enhance performance may be exploited. However, consistent operation and maintenance remain key factors for sustained efficiency. Overall, ABR technology holds significant potential for efficient and sustainable decentralized wastewater treatment, with ongoing advancements promising even greater improvements in the future.

References

Amoo AO, Ijanu EM, Haruna A, Adeleye AO, Sabo A (2023) Review on the pre-treatment advancements of biogas production barriers. UMYU J Microbiol Res (UJMR) 8(1):6–30. https://doi.org/10.47430/ujmr.2381.002

Anderson K, Sallis P, Uyanik S (2003) Anaerobic treatment processes. In: Handbook of water and wastewater microbiology. Elsevier, Amsterdam, pp 391–426

Bae J, Yoo R, Lee E, McCarty PL (2013) Two-stage anaerobic fluidized-bed membrane bioreactor treatment of settled domestic wastewater. Water Sci Technol 68(2):394–399. https://doi.org/10.2166/wst.2013.191

Barton LL, Fauque GD (2009) Biochemistry, physiology and biotechnology of sulfate-reducing bacteria. Adv Appl Microbiol 68:41–98. https://doi.org/10.1016/S0065-2164(09)01202-7

Barber and Stucky (1999). The Use of the Anaerobic Baffled Reactor (ABR) for Waste Water Treatment: A Review. Water Research, 33, 1559–1578.

Biological Oxygen Demand (BOD)khe SY (2009) A modified anaerobic baffled reactor for municipal wastewater treatment. J Environ Manag 90(8):2488–2493. https://doi.org/10.1016/j.jenvman.2009.01.007

Bruno P, Di Bella G, De Marchis M (2020) Perforated baffles for the optimization of disinfection treatment. Water 12(12):3462. https://doi.org/10.3390/w12123462

Cheng H, Li Y, Hu Y, Guo G, Cong M, Xiao B, Li YY (2021) Bioenergy recovery from methanogenic co-digestion of food waste and sewage sludge by a high-solid anaerobic membrane bioreactor (AnMBR): mass balance and energy potential. Bioresour Technol 326:124754. https://doi.org/10.1016/j.biortech.2021.124754

Dong F, Zhao QB, Li WW, Sheng GP, Zhao JB, Tang Y, Yu HQ, Kubota K, Li YY, Harada H (2011) Novel online monitoring and alert system for anaerobic digestion reactors. Environ Sci Technol 45(20):9093–9100. https://doi.org/10.1021/es202245f

Elreedy A, Tawfik A, Enitan A, Kumari S, Bux F (2016) Pathways of 3-biofules (hydrogen, ethanol and methane) production from petrochemical industry wastewater via anaerobic packed bed baffled reactor inoculated with mixed culture bacteria. Energy Convers Manag 122:119–130. https://doi.org/10.1016/j.enconman.2016.05.067

Giordani A, Brucha G, Santos KA, Rojas K, Hayashi E, Alves MMS, Tommaso G (2021) Performance and microbial community analysis in an anaerobic hybrid baffled reactor treating dairy wastewater. Water Air Soil Pollut 232:1–16. https://doi.org/10.1007/s11270-021-05348-0

Guamán-Marquines CW, Mendoza-Loor RJ, Gómez-Salcedo Y, Baquerizo-Crespo RJ (2023) Assessment of the start-up of tubular reactors on a laboratory scale for the anaerobic digestion of slaughterhouse wastewater. Int J Thermofluids 19:100378. https://doi.org/10.1016/j.ijft.2023.100378

Hahn MJ, Figueroa LA (2015) Pilot scale application of anaerobic baffled reactor for biologically enhanced primary treatment of raw municipal wastewater. Water Res 87:494–502. https://doi.org/10.1016/j.watres.2015.09.027

Hassan SR, Dahlan I (2013) Anaerobic wastewater treatment using anaerobic baffled bioreactor: a review. Open Eng 3(3):389–399. https://doi.org/10.2478/s13531-013-0107-8

Ingildsen P (2002) Realising full-scale control in wastewater treatment systems using in situ nutrient sensors. [Doctoral thesis (monograph), industrial electrical engineering and automation]. ISSN no: 91-88934-22-5, IEA, LTH, Box 118, SE-221 00 Lund, Sweden, https://portal.research.lu.se/en/publications/realising-full-scale-control-in-wastewater-treatment-systems-usin

Islam MS, Zhang Y, Dong S, McPhedran KN, Rashed EM, El-Shafei MM, Noureldin AM, El-Din MG (2017) Dynamics of microbial community structure and nutrient removal from an innovative side-stream enhanced biological phosphorus removal process. J Environ Manag 198:300–307. https://doi.org/10.1016/j.jenvman.2017.04.074

Jamshidi S, Akbarzadeh A, Woo KS (2014) Wastewater treatment using integrated anaerobic baffled reactor and bio-rack wetland planted with Phragmites sp. and Typha sp. J Environ Health Sci Eng 12:131. https://doi.org/10.1186/s40201-014-0131-5

Kerstens S, van der Geest C, van der Hoek J, de Haan F (2012) Decentralized wastewater treatment systems (DEWATS) for sustainable sanitation in Indonesia. Water Sci Technol 65(4):584–595
Khalekuzzaman M, Alamgir M, Hasan M, Hasan MN (2018) Performance comparison of uninsulated and insulated hybrid anaerobic baffled reactor (HABR) operating at warm temperature. Water Sci Technol 78(9):1879–1892. https://doi.org/10.2166/wst.2018.401
Khan A, Khan SJ, Miran W, Zaman WQ, Aslam A, Shahzad HMA (2023) Feasibility study of anaerobic baffled reactor coupled with anaerobic filter followed by membrane filtration for wastewater treatment. Membranes 13(1):79. https://doi.org/10.3390/membranes13010079
Lau PL, Trzcinski AP (2022) A review of modified and hybrid anaerobic baffled reactors for industrial wastewater treatment. Water Sci Eng 15(3):247–256. https://doi.org/10.1016/j.wse.2022.06.004
Lettinga G, Van Velsen AF, Hobma SW, de Zeeuw W, Klapwijk A (1980) Use of the upflow sludge blanket (USB) reactor concept for biological wastewater treatment, especially for anaerobic treatment. Biotechnol Bioeng 22(4):699–734. https://doi.org/10.1002/bit.260220402
Lettinga G, Hulshoff Pol LW, Zeeman G (1996) Biological wastewater treatment. Part I: anaerobic wastewater treatment, Lectures notes. Wageningen Agricultural University
Li Y, Sim LN, Ho JS, Chong TH, Wu B, Liu Y (2020) Integration of an anaerobic fluidized-bed membrane bioreactor (MBR) with zeolite adsorption and reverse osmosis (RO) for municipal wastewater reclamation: comparison with an anoxic-aerobic MBR coupled with RO. Chemosphere 245:125569. https://doi.org/10.1016/j.chemosphere.2019.125569
Li J, Nana Z, Liutang G, Li X, Hongyu T, Peng Y, Li T (2022) A modified packed anaerobic baffled reactor based on phase separation for the treatment of decentralized wastewater: performance and microbial communities. Biochem Eng J 183:108455. https://doi.org/10.1016/j.bej.2022.108455
Mahdi FK, Abu-Alhail S, Dawood AS (2022) Performance of the anaerobic baffled reactor for primary treatment of rural domestic wastewater in Iraq. Open Eng 12(1):859–865. https://doi.org/10.1515/eng-2022-0346
Metcalf & Eddy, Inc (2003) Wastewater engineering: treatment and reuse, 4th edn. McGraw-Hill Education
Metcalf & Eddy Inc, Tchobanoglous G, Burton FL, Tsuchihashi R, Stensel HD (2013) Wastewater engineering: treatment and resource recovery, 5th edn. McGraw-Hill Professional
Pal S, Bhanot C (2023). https://india.mongabay.com/2023/01/commentary-decentralised-wastewater-treatment-systems-to-improve-water-security/. Accessed on 29.10.23
Pererva Y, Miller CD, Sims RC (2020) Approaches in design of laboratory-scale UASB reactors. PRO 8(6):734. https://doi.org/10.3390/pr8060734
Qi Z, Xiang G, Xiong D (2019) Performance evaluation of pilot-scale hybrid anaerobic baffled reactor (HABR) to process dyeing wastewater based on grey relational analysis. Appl Sci 9(10):1974. https://doi.org/10.3390/app9101974
Rahman MA (2023) Effectiveness of a tubular heat exchanger and a novel perforated rectangular flow-deflector type baffle plate with opposing orientation. World J Eng 21(6):1086–1094
Renuka R, Mohan SM, Sowmiya B, Raj SA (2016) Performance evaluation of panelled anaerobic baffle-cum-filter reactor in treating municipal wastewater. Ecol Eng 97:1–12. https://doi.org/10.1016/j.ecoleng.2016.07.020
Reynaud N, Buckley CA (2016) The anaerobic baffled reactor (ABR) treating communal wastewater under mesophilic conditions: a review. Water Sci Technol 73(3):463–478. https://doi.org/10.2166/wst.2015.539
Rout PR, Shahid MK, Dash RR, Bhunia P, Liu D, Varjani S, Zhang TC, Surampalli RY (2021) Nutrient removal from domestic wastewater: a comprehensive review on conventional and advanced technologies. J Environ Manag 296:113246. https://doi.org/10.1016/j.jenvman.2021.113246
Safeer S, Pandey RP, Rehman B, Safdar T, Ahmad I, Hasan SW, Ullah A (2022) A review of artificial intelligence in water purification and wastewater treatment: recent advancements. J Water Process Eng 49:102974. https://doi.org/10.1016/j.jwpe.2022.102974

She Z, Zheng X, Yang B, Jin C, Gao M (2006) Granule development and performance in sucrose fed anaerobic baffled reactors. J Biotech 122:198–208. https://doi.org/10.1016/j.jbiotec.2005.09.004

Shi X, Leong KY, Ng HY (2017) Anaerobic treatment of pharmaceutical wastewater: a critical review. Bioresour Technol 245:1238–1244. https://doi.org/10.1016/j.biortech.2017.08.150

Singh S, Haberl R, Moog O, Shrestha RR, Shrestha P, Shrestha R (2009) Performance of an anaerobic baffled reactor and hybrid constructed wetland treating high-strength wastewater in Nepal—a model for DEWATS. Ecol Eng 35(5):654–660. https://doi.org/10.1016/j.ecoleng.2008.10.019

Singh A, Sawant M, Kamble SJ, Herlekar M, Starkl M, Aymerich E, Kazmi A (2019) Performance evaluation of a decentralized wastewater treatment system in India. Environ Sci Pollut Res 26:21172–21188. https://doi.org/10.1007/s11356-019-05444-z

Tillman GM (1996) Water treatment: troubleshooting and problem solving. CRC Press

Tran CS, Dang TB, Nguyen PT, Nguyen KQ, Hoang QH, Le LT, Nguyen TT, Bui XT (2022) Anaerobic baffled reactor coupled with membrane bioreactor treating tannery wastewater. Case Stud Chem Environ Eng 5:100185. https://doi.org/10.1016/j.cscee.2022.100185

Ullah N, Sheikh Z, Badshah M (2023) Municipal wastewater treatment with corrugated PVC carrier anaerobic baffled reactor. Water Sci Technol 87(3):660–671. https://doi.org/10.2166/wst.2022.414

Varma VG, Jha S, Raju LHK, Kishore RL, Ranjith V (2022) A review on decentralized wastewater treatment systems in India. Chemosphere 300:134462. https://doi.org/10.1016/j.chemosphere.2022.134462

Vijay R (2024) Anaerobic baffled reactor followed by up-flow constructed wetland for treatment of municipal wastewater. J Water Process Eng 58:104810. https://doi.org/10.1016/j.jwpe.2024.104810

Vítězová M, Kohoutová A, Vítěz T, Hanišáková N, Kushkevych I (2020) Methanogenic microorganisms in industrial wastewater anaerobic treatment. PRO 8(12):1546. https://doi.org/10.3390/pr8121546

Wu S, Kuschk P, Brix H, Vymazal J, Dong R (2014) Development of constructed wetlands in performance intensifications for wastewater treatment: a nitrogen and organic matter targeted review. Water Res 57:40–55. https://doi.org/10.1016/j.watres.2014.03.020

Yao S, Zhu T, Liang B, Zou Y, Wang Y, Han J, Li Z, Li F (2023) Performance evaluation of modified anaerobic baffled reactor (MABR) treating high strength wastewater. Waste Biomass Valorization 14:2557–2568. https://doi.org/10.1007/s12649-022-02015-1

Yulistyorini A, Camargo-Valero MA, Sukarni S, Suryoputro N, Mujiyono M, Santoso H, Tri Rahayu E (2019) Performance of anaerobic baffled reactor for decentralized wastewater treatment in urban Malang, Indonesia. Processes 7:184. https://doi.org/10.3390/pr7040184

Zhu G, Zou R, Jha AK, Huang X, Li L, Liu C (2015) Recent developments and future perspectives of anaerobic baffled bioreactor for wastewater treatment and energy recovery. Crit Rev Environ Sci Technol 45(12):1243–1276. https://doi.org/10.1080/10643389.2014.924182

Zieliński M, Kazimierowicz J, Dębowski M (2022) Advantages and limitations of anaerobic wastewater treatment—technological basics, development directions, and technological innovations. Energies 16(1):83. https://doi.org/10.3390/en16010083

Zijing An, Junjie Z, Min Z, Yan Z, Xiaomei S, Hongjun L, Faqian S (2023) Anaerobic membrane bioreactor for the treatment of high-strength waste/wastewater: A critical review and update, Chemical Engineering Journal 70: 44322 https://doi.org/10.1016/j.cej.2023.144322

Zwain HM, Hassan SR, Zaman NQ, Aziz HA, Dahlan I (2013) The start-up performance of modified anaerobic baffled reactor (MABR) for the treatment of recycled paper mill wastewater. J Environ Chem Eng 1(1–2):61–64. https://doi.org/10.1016/j.jece.2013.03.007

Chapter 6
Prospects and Advances in Soil Biotechnology as a Sustainable Strategy for Decentralized Liquid Waste Management

Priyanka, Sonia Bansal, and Preet Kaur

Abstract A DEWATS (Decentralized Wastewater Treatment System) is an inexpensive and effective wastewater treatment technology, particularly in developing nations. It has significant potential for creating a long-term environmental sanitation system. The treatment of water and wastewater, as well as its reuse for household, industrial, or agricultural purposes, serves to conserve and improve water quantity and quality. Green technologies currently supply high-quality water without incorporating global warming. Several conventional treatment methods successfully eliminate various pollutants and contaminants from wastewater. Still, they require many external sources, labor, and pumping systems, as well as higher maintenance and operation costs. Due to several issues with traditional treatment procedures, soil biotechnology (SBT) was implemented to establish a pollution-free environment. Soil biotechnology (SBT) is a water regeneration technique that uses specific media containing soil micro- and macroorganism culture to process water and wastewater.

Keywords Wastewater · Soil biotechnology · Green technologies · Decentralized wastewater treatment system · Water contaminants

---

Priyanka · S. Bansal (✉)
Department of Physics, J.C. Bose University of Science and Technology, YMCA, Faridabad, India

P. Kaur
Department of Electronics Engineering, J.C. Bose University of Science and Technology, YMCA, Faridabad, Haryana, India

S. S. Kumar, V. Kumar (eds.), *Nature-Based Solutions for Decentralised Liquid Waste Management*, Environmental Challenges and Solutions,
https://doi.org/10.1007/978-3-032-08362-3_6

6.1 Introduction

The utilization of living organisms for the treatment of wastewater is known as soil biotechnology. The water that is filtered after treatment is utilized for purposes like agriculture, gardening, and plant growth. After the process of decomposition, it is used as a solid material and then turned into manure. The disposal process is eliminated as a result (Upadhyay et al. 2016). Water is composed mostly of one molecule of oxygen and two molecules of hydrogen. Because these molecules are not often present in clean water, they might be regarded as contaminants. There are numerous such contaminants in drinkable water, which we consume daily. Water covers over 71% of the earth's surface, with just 2.5% of fresh water available on the surface. The rapid growth of population and industrialization expansion results in the discharge of significant quantities of wastewater, which are utilized as an important source of irrigation for urban and pre-urban agriculture and have endemic impacts on natural resources worldwide (Mohammadreza Kamali et al. 2019).

The wastewater crisis is primarily caused by the absence of management systems and infrastructure that can handle the growing amount of wastewater (Corcoran et al. 2010). Sustainable development is centered on the treatment of wastewater, which ensures that the flow of pollutants does not harm aquatic life and relieves pressure on sources of fresh water through reuse. Several conventional technological solutions for wastewater treatment are also present. Maintenance costs, high capital, and operation, along with skilled manpower for sustainability purposes, are necessary for them. The limitations that exist in nature-based wastewater treatment systems are overcome by wetland and land management techniques. The utilization of soil-based treatment systems, including biofilters, sand filters, soil infiltration systems, and constructed wetlands for wastewater treatment has been proven to be promising and cost-effective (Njau et al. 2003; Zhang et al. 2004, 2008; Koottatep et al. 2005; Tietz et al. 2008; Torrens et al. 2009; Bester and Schäfer 2009).

SBT technology utilizes earthworm, bacteria, and mineral additive processing to create green infrastructure for wastewater treatment plants that resemble gardens. It is designed to reduce the impact of pesticides by using biological processes (Shrivithal et al. 2021). The SBT process eliminates pathogens and chemical contaminants in a single evergreen facility that is open to the atmosphere. Treated water can be utilized for gardening, car washing, and toilet flushing instead of being used as pure drinking water. The recharge of groundwater through rainwater-harvesting recharge wells can be achieved by using the processed water (Kamble et al. 2017a).

6.2 Wastewater Contamination and Its Effects

Wastewater can be defined in a variety of ways, with different interpretations for each individual. Despite this, the report has taken a comprehensive approach and defined wastewater as encompassing dissolved or suspended matter from domestic effluent, blackwater (urine, excreta, and fecal sludge), grey water from kitchens and baths, commercial water from institutions and facilities, industrial discharge, stormwater, urban runoff, and horticultural, agricultural and aquaculture waste (Raschid-Sally and Jayakody 2008). The presence of numerous contaminants can be observed in wastewater, as illustrated in Fig. 6.1. The list includes organic compounds, pathogens, synthetic chemicals, heavy metals, nutrients, and organic matter. Water quality is affected by particles or particulate matter that are transported from one source to another. These substances affect the quality of water.

These components can exhibit bioaccumulative, synergistic, and persistent characteristics that impact ecosystem function and health, human health & well-being, food production as well as human security. Water pollution is primarily caused by various sources, such as mining activities industries, sewage, radioactive waste, and chemical fertilizers; the rapid development of urbanization has adversely affected the environment and resources, while also negatively affecting the life of living things (Crini and Lichtfouse 2018). In untreated wastewater, there may be a range of pathogens present, including parasites, bacteria or virus infections, toxic chemicals such as organic chemicals, and heavy metals from industry, agriculture, and domestic sources (Drechsel et al. 2010). Before entering cities, more than 70% of the water has already been utilized for other productive purposes (Appelgren 2004; Pimentel and Pimentel 2008). The management of wastewater must address both urban and rural ecosystems through integrated ecosystem approaches, such as agriculture, fishing, and forestry.

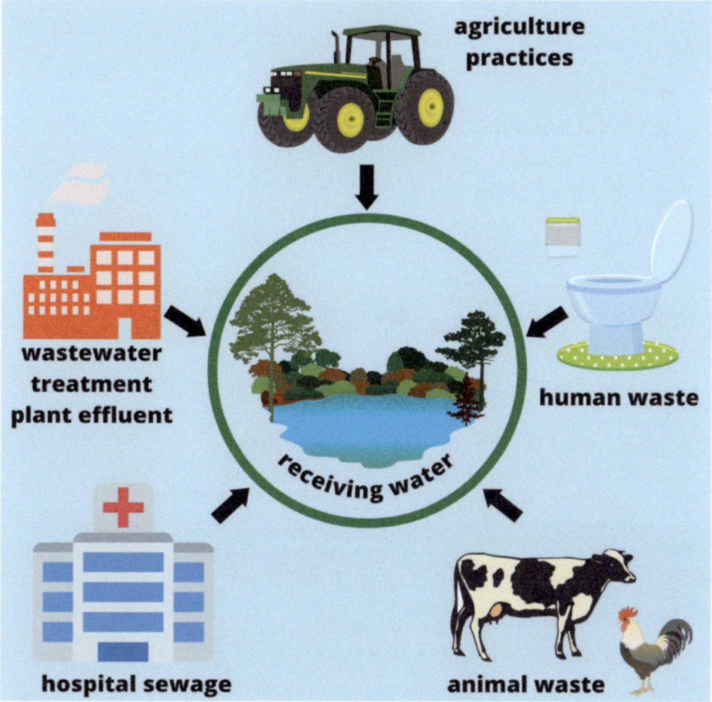

Fig. 6.1 Various contaminants in wastewater from different sources. (Alawi et al. 2022)

6.3 Types of Wastewater Management Systems

Safe wastewater treatment systems and transportation are available in most cities through two approaches: conventional centralized sewer systems also known as sewered sanitation, nonconventional systems with on-site sanitation also known as non-sewered sanitation, and decentralized systems as illustrated in Fig. 6.2. Government agencies typically operate centralized systems that collect and treat wastewater for large communities. Conversely, decentralized wastewater management (DWWM) systems handle wastewater from individual homes, apartment buildings, or small communities near their origin. The decentralized system is commonly made up of multiple technologies, such as onsite systems, low-cost collection systems (such as energy efficiency), and dispersed treatment facilities, within a specific geographic region. Treatment facilities that manage wastewater, including septic tanks, pit latrines, and bio-toilet installations are termed "On-Site Sanitation Systems (OSS)" and are utilized for partially treating wastewater in individual homes or small groups of houses. While an OSS does not necessarily need a wastewater collection system, DWWM can have one if desired—for example, it may also have some type of sewerage system. It should be emphasized that any city or town can have a combination of decentralized, on-site, and centralized wastewater management systems to fulfill the worldwide city sanitation requirements (Guidelines for Decentralised Wastewater Management 2012).

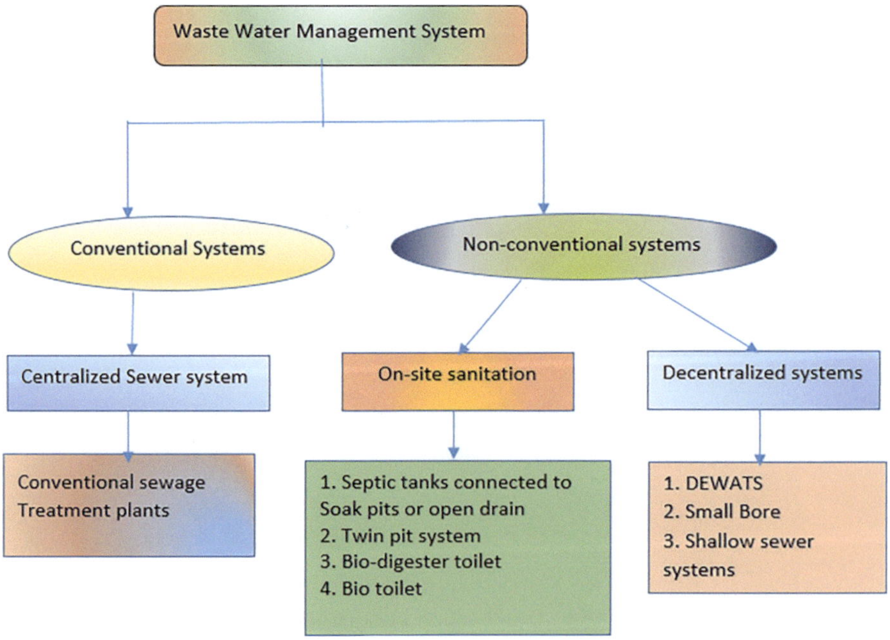

Fig. 6.2 Types of wastewater management systems

6.4 Decentralized Liquid Waste Management

Decentralized Wastewater Treatment Systems (DEWATS) is a technical approach rather than just essentially "a technology package". DEWATS are commonly structured and operated by residents and typically consist of a settler, gravel/sand filter beds, an open pond, and anaerobic baffled tanks. The system is not designed to be a large-scale installations. Naturally, the polishing tank or open pond creates a living environment for the wastewater to clean itself. Urbanization, inadequate wastewater management, and poor disposal all play a role in deteriorating the situation in many developing countries regarding wastewater treatment (CSE 2017).

Thus, sanitation solutions based on centralized systems are not feasible in many areas of the world. The collection and treatment of vast amounts of wastewater for entire towns and industrial/residential platforms is facilitated by centralized systems that use large pumping systems, pipelines, multiple access routes, and equipment. Management flexibility is available due to the independent collection, treatment, and reuse of treated wastewater by decentralized systems at or near the producing facility (Sharma et al. 2013). Environmentally and economically, wastewater should be treated near the source to the point where it is safely treated rather than having to build large and sometimes expensive sewage lines. DEWATS has become a popular choice for wastewater treatment in areas without sanitation services. Decentralized approaches can be implemented on any scale, from completely decentralized individual systems to semi-centralized facilities managing the effluents of distant neighborhoods (Iribarnegaray et al. 2018).

The term decentralized is used to describe systems that serve only small portions (clusters) of an urban area, taking into account landscape and other factors such as hydrology and local ecology. Sustainable decentralized sanitation prioritizes on-site wastewater treatment and local recycling and reuse of resources from domestic wastewater. The assertion has been made that decentralized treatment systems favor water recycling and reuse in areas near their designated sites (Opher and Friedler 2016). The primary objectives of wastewater management systems are: to safeguard human health by creating a clean environment and preventing harmful diseases and to enhance water quality and ecosystems by limiting the release of excessive pollutants (Novotny et al. 1989). Generally, the most appropriate technology for any given situation is one that is economically affordable, socially acceptable for the specific application, environmentally sustainable, institutionally and technically consistent or sustainable by all perspectives. Decentralized systems are typically compact, have greater flexibility in operation, and have a less pleasing appearance (Capodaglio et al. 2002; Torretta et al. 2016).

6.4.1 Types of DEWATS

DEWATS is classified into three groups:

1. Natural treatment systems.
2. Aerobic systems.
3. Anaerobic systems.

6.4.1.1 Natural Treatment Systems

The treatment of wastewater is typically carried out using natural wastewater treatment systems, which are also cost-effective. The easiest approach involves the interaction between bacteria and algae in a water-based pond where wastewater is collected. In other widespread systems, soil and plants serve as natural catalysts and filters for biological processes. These treatments can be used as a secondary treatment or in combination with the primary treatment. Certain techniques have been utilized in the tertiary level of treatment, such as floating wetlands with duckweed, constructed wetland areas or water hyacinth plants, and facultative ponds.

6.4.1.2 Aerobic Systems

The process of aerobic treatment involves the utilization of oxygen by microorganisms to break down organic materials into carbon dioxide and water. The environmental impact of aerobic treatment procedures is lower than nature-based systems and they can be easily replaced to produce effluents of high quality that meet the requirements for effluent disposal but require more energy. In addition, these systems only require semiskilled human power for operation and maintenance, making them an attractive option for wastewater treatment technology in both developing and underdeveloped countries. Among the most common types of aerobic systems are, extended aeration processes, oxidation ditches, membrane bioreactors, moving bed biofilm reactors, and sequential bioreactors.

6.4.1.3 Anaerobic Systems

These systems are cost-effective, straightforward, and energy-prohibitive biological treatment methods that can be utilized to produce energy. In contrast to aerobic systems, anaerobic systems take longer to start up to 4 months on average, while aerobic system start-up takes only 4 weeks. Moreover, the low effluent quality of anaerobic systems makes them incompatible with standard discharge requirements, so they are often combined with natural or aerobic processes. Upflow anaerobic systems, anabolic baffled reactors, and septic tanks are some of the systems that are frequently used (Aayog 2022).

6.4.2 Steps Involved in DEWATS

As shown in Fig. 6.3, mainly DEWATS incorporates the below-mentioned treatment processes in a modular fashion:

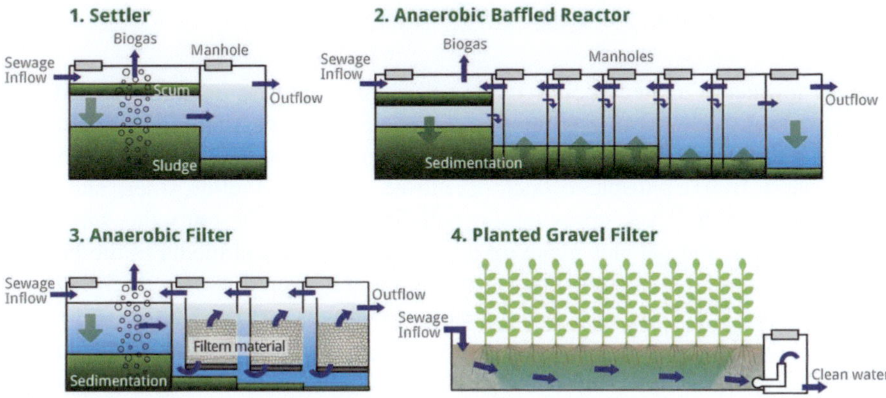

Fig. 6.3 DEWATS configuration scheme. (Geetha Varma et al. 2022)

- Primary treatment process: It involves sedimentation ponds, biodigester, settlers, and septic tanks.
- Secondary treatment process: It takes place in anaerobic filters, anaerobic baffled reactors, or facultative and anaerobic pond systems.
- Secondary aerobic or facultative treatment process: This type of treatment needs horizontal gravel filters.
- Posttreatment process: Aerobic polishing ponds are required for posttreatment.

6.5 Soil Biotechnology

The bioconversion process that enables the purification of wastewater involves the fundamental reactions of nature, including photosynthesis, mineral weathering and respiration, occurring in a medium with small and macroorganisms is known as SBT. SBT is a biological engine that can provide oxygen to water in all types, including municipal, domestic, and industrial applications. Soil biotechnology (SBT) can be utilized to process both inorganic and organic components (Arghyam 2013).

Soil biotechnology uses soil as a medium for wastewater treatment. SBT is a synthesis process that transforms carbon, waste energy, and other components into valuable “bioenergy” products such as soil with high nutrients, vegetation, full biofertilizer, and water (Zhang et al. 2014; Ramprasad and Philip 2016; Abidi et al. 2009; Arvindbhai and Vyas 2016). SBT comprises eliminating organic waste by adsorption, followed by biological breakdown and oxygen delivery to the treatment system via natural aeration. Green cover photosynthetic activity acts as a bioindicator for the kind of microhabitat in SBT (Juwarkar et al. 1995; Jeevitha et al. 2007; Gupta 2015).

6.5.1 Methodology

The wastewater is treated in SBT using soil as the treatment medium. The SBT is designed to provide necessary aeration, biochemical processing, and filtration for removing the toxicity, including chemical oxygen demand (COD), biological oxygen demand (BOD), phosphate, bacteria, nitrate, suspended solids, smell, and color.

The process involves spraying or pumping wastewater onto the sand bed. Boulders, sand, and pebbles make up a layer of cultured soil media in the bed. To prevent wastewater loss, the filtering materials are layered over a thick layer of plastic sheets. The wastewater is pumped repeatedly over the top of the soil media through a pipeline network. The wastewater that has been treated is collected in the channels located between the soil bund and then redirected to a collection well. The collection well can also function as an aeration tank. Ultimately, the water is pumped out and utilized for irrigation. To facilitate the treatment process, wild plants that are available locally are cultivated on top of the soil.

The use of SBT applies to a range of operational scales. The wastewater can be treated in various stages, such as batches, semicontinuous, or continuous mode, depending on the user's preference. This method eliminates the need for sludge production. About 99.99% of bacteria is eliminated by SBT thus resulting in a healthier environment with no adverse effects. Its energy efficiency and cost-effectiveness make it suitable for use at regular temperatures. Both urban and rural areas can utilize SBT due to its flexibility in terms of total size. SBT designs are possible within the home and may be specific to particular areas. The cost of operating and maintaining SBT is low because it requires only energy for pumping purposes and does not require chemicals (National Institute of Urban Affairs 2019). Figure 6.4 illustrates the schematic diagram of SBT.

6.5.2 Features of SBT

- The method can lower chemical oxygen demand (COD), total suspended solids (TSS), biological oxygen demand (BOD), turbidity, phosphorus, and nitrogen levels.
- Except for pumps, it is an optimized evergreen process that is natural with no moving components.
- There is no odor in the technique.
- Because of the ecosystem at work, there is no sludge.
- Due to the great natural oxygen transfer in the process, the energy usage intensity is quite low, i.e., for sewage (0.06 kWh/kL).
- It has adaptable output quality.
- Almost no maintenance is required for SBT.
- The sewage has a very low spatial intensity of 0.8–1.0 sq.m/kL per day.

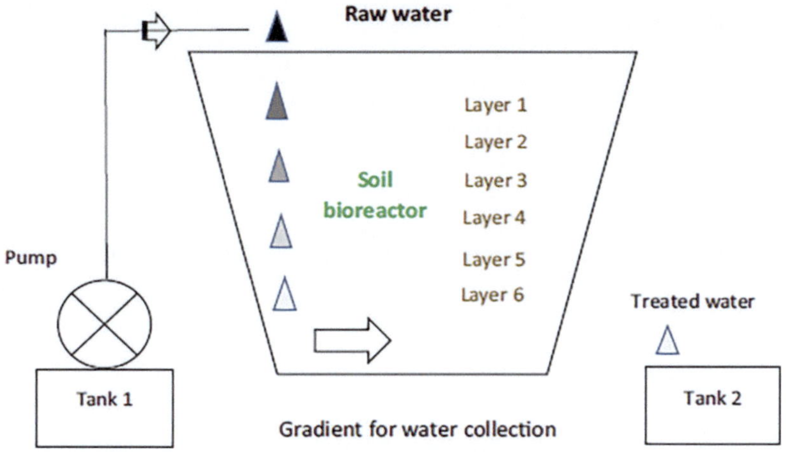

Fig. 6.4 Schematic diagram of SBT. (Kamble et al. 2017b)

- BOD, Bacteria, COD, color, suspended solids, ammonia, and odor removal are all quite high.

6.5.3 Advantages of SBT

- Soil biotechnology is a new green technology for purifying water and air as well as digesting organic materials.
- Organics can be removed from industrial or home effluent using soil biotechnology.
- There is no flooding or negative impact during the monsoon on soil biotechnology.
- SBT is less expensive than other alternative technologies such as SBR, activated sludge process membrane, constructed wetlands, and bioreactor.
- Other technologies, such as soil biotechnology, cannot be created with capacities less than 10 MLD.
- Soil biotechnology requires little energy.
- Constructed soil filters can be used to recharge groundwater. Treated wastewater is sent to the ground by ground water recharge standards, where it recharges the earth and replenishes the wells.
- SBT works in high salinity but only up to 4000 ppm. Anything over that requires additional precautions and a salinity control system.

6.5.4 *Applications of SBT*

- It helps in the treatment of industrial wastewater.
- SBT helps in the purification of industrial air.
- It helps in sewage treatment for the reuse of wastewater in cleaning, construction, groundwater recharging, gardening, water make-up for swimming pools, irrigation, industries, and so on.
- Applicable to buildings, societies, commercial structures, hospitals, and so forth.

6.6 Conclusion

Toxic compounds, pollutants, and other chemical impurities in water must be controlled to protect human health and the environment. Waste treatment in water can enhance the reuse of treated water in a variety of applications and promote water recycling for human and natural uses. Soil biotechnology not only treats wastewater but also recycles it to the point where it may be reused. Almost half of all water usage is for horticulture, flushing cleaning, agriculture, and washing, all of which may be serviced by recycled water from an SBT plant.

On the other hand, establishing an SBT facility to treat wastewater would significantly help to reduce river damage. Because the applied technology is minimally demanding, i.e., no expert personnel or expensive operating and maintenance costs are required, it can withstand unfavorable circumstances without breaking down. Processed water waste may be used in several applications, including agricultural, urban areas, and environmental objectives. As a result, the trash collected from the water may be recovered and securely disposed of in the environment.

Thus, SBT is a unique green, and cost-effective method for purifying water and air as well as digesting organic materials. SBT has a high potential with a low power need; electricity is only required when the pumping system needs to pump water into and out of the reactor. This demand can be further lowered if the water is allowed to flow naturally. Furthermore, the plant does not require continuous operation and is only intended to function for a few hours each day. As a result, this new soil biotechnology would be a boon to water treatment and a significant step toward a brighter future.

References

Aayog N (2022) Urban wastewater scenario in India. Atal Innovation Mission

Abidi S, Kallali H, Jedidi N, Bouzaiane O, Hassen A (2009) Comparative pilot study of the performances of two constructed wetland wastewater treatment hybrid systems. Desalination 246:370–377

Alawi M, Torrijos TV, Walsh F (2022) Plasmid-mediated antimicrobial resistance in drinking water. Environ Adv 8:100191., ISSN 2666-7657. https://doi.org/10.1016/j.envadv.2022.100191
Appelgren B (2004) Water in agriculture. UNESCO, New Delhi
Arghyam (2013) Piece of the puzzle: evaluating the soil biotechnology wastewater treatment plant at ACCEPT society, Bengaluru
Arvindbhai CE, Vyas DS (2016) Treatment of petrochemical wastewater by soil biotechnology. IJARIIE 2(3) ISSN (O)-2395-4396
Bester K, Schäfer D (2009) Activated soil filters (bio filters) for the elimination of xenobiotics (micro-pollutants) from storm-and waste waters. Water Res 43(10):2639–2646
Capodaglio AG, Conti F, Fortina L, Pelosi G, Urbini G (2002) Assessing the environmental impact of WWTP expansion: odour nuisance and its minimization. Water Sci Technol 46:339–346. [PubMed]
Corcoran E, Nellemann C, Baker E, Bos R, Osborn D, Savelli H (2010) Sick water? The central role of wastewater management in sustainable development. A rapid response assessment. United Nations Environment Programme, UN-HABITAT, GRID Arendal. www.grida.no ISBN: 978–82–7701-075-5 Printed by BirkelandTrykkeri AS, Norway
Crini G, Lichtfouse E (2018) Wastewater treatment: an overview. In: Green adsorbents for pollutant removal: fundamentals and design, pp 1–21. https://doi.org/10.1007/978-3-319-92111-2
CSE (2017) Decentralized wastewater treatment-case studies. Centre for Science and Environment, India
Drechsel P, Scott CA, Raschid-Sally L, Redwood M, Bahri A (2010) Wastewater irrigation and health. Assessing and mitigating risk in low-income countries. IWMI-IDRC Earthscan. 432pp
Geetha Varma V, Swetti Jha L, Raju HK, Lalith Kishore R, Ranjith V (2022) A review on decentralized wastewater treatment systems in India. Chemosphere 300:134462., ISSN 0045-6535. https://doi.org/10.1016/j.chemosphere.2022.134462
Guidelines for Decentralised Wastewater Management, Prepared by IIT, Chennai for MoUD, GoI, December 2012
Gupta H (2015) A review on effectiveness of earthworms for treatment of wastewater. IJEDR 3(3) ISSN: 2321-9939:1–4
Iribarnegaray MA, Rodriguez-Alvarez MS, Moraña LB, Tejerina WA, Seghezzo L (2018) Management challenges for a more decentralized treatment and reuse of domestic wastewater in metropolitan areas. J Water Sanit Hyg Dev 8(1):113–122
Jeevitha P, Kiran BM, Manohara B, Sham Sundar KM (2007) Study on application and validation of term filtration in diary effluent treatment. Int J Innov Rers Sci Eng Technol. ISSN (Online): 2319-8753
Juwarkar AS, Oke B, Patnaik SM (1995) Domestic wastewater treatment through a constructed wetland in India. Water Sci Technol 32(3):291–294
Kamble SJ, Chakravarthy Y, Singh A, Chubilleau C, Starkl M, Bawa I (2017a) A soil biotechnology system for wastewater treatment: technical, hygiene, environmental LCA and economic aspects. Environ Sci Pollut Res. https://doi.org/10.1007/s11356-017-8819-6
Kamble S, Chakravarthy Y, Singh A, Chubilleau C, Starkl M, Bawa I (2017b) A soil biotechnology system for wastewater treatment: technical, hygiene, environmental LCA and economic aspects. Environ Sci Pollut Res Int 24. https://doi.org/10.1007/s11356-017-8819-6
Koottatep T, Surinkul N, Polprasert C, Kamal ASM, Koné D, Montangero A et al (2005) Treatment of septage in constructed wetlands in tropical climate: lessons learnt from seven years of operation. Water Sci Technol 51(9):119–126
Mohammadreza Kamali DP, Suhas ME, Costa IC, Aminabhavi TM (2019) Sustainability considerations in membrane-based technologies for industrial effluents treatment. Chem Eng J 1385–8947(19):30305–30305
National Institute of Urban Affairs (2019) Decentralized wastewater management: complementing centralized solutions for achieving 100% sanitation coverage–learning notes. National Institute of Urban Affairs, New Delhi

Njau KN, Minja RJ, Katima JH (2003) Pumice soil: a potential wetland substrate for treatment of domestic wastewater. Water Sci Technol 48:85–92

Novotny V, Imhoff KR, Olthof O, Krenkel PA (1989) Karl Imhoff's handbook of urban drainage and wastewater disposal. Wiley, New York, p 416

Opher T, Friedler E (2016) Comparative LCA of decentralized wastewater treatment alternatives for non-potable urban reuse. J Environ Manag 182:464–476. [CrossRef] [PubMed]

Pimentel D, Pimentel M (2008) Food, energy, and society. CRC Press, Boca Raton

Ramprasad C, Philip L (2016) Surfactants and personal care products removal in pilot-scale horizontal and vertical flow constructed wetlands while treating greywater. Chem Eng J 284:458–468

Raschid-Sally L, Jayakody P (2008) Drivers and characteristics of wastewater agriculture in developing countries: results from a global assessment, Colombo, Sri Lanka, IWMI research report 127. International Water Management Institute, Colombo. 35pp

Sharma AK, Tjandraatmadja G, Cook S, Gardner T (2013) Decentralised systems definition and drivers in the current context. Water Sci Technol 67(9):2091–2101

Shrivithal K, Pratik B, Vedant C, Abhilash R, Siddiqui R, Kalpesh kumar Sharma. (2021) Waste water treatment by soil biotechnology plant. Int Res J Mod Eng Technol Sci 03(05):2582–5208

Tietz A, Langergraber G, Watzinger A, Haberl R, Kirschner AK (2008) Bacterial carbon utilization in vertical subsurface flow constructed wetlands. Water Res 42(6):1622–1634

Torrens A, Molle P, Boutin C, Salgot M (2009) Impact of design and operation variables on the performance of vertical-flow constructed wetlands and intermittent sand filters treating pond effluent. Water Res 43(7):1851–1858

Torretta V, Schiavon M, Papa EA, Caruson P, Capodaglio AG (2016) Removal of odorous Sulphur compounds from industrial gases by biotrickling filters. Rev Ambient Água 11:499–507. [CrossRef]

Upadhyay AK, Bankoti NS, Rai UN (2016) Studies on the sustainability of simulated constructed wetland system for treatment of urban waste: design and operation. J Environ Manag 169:285–292

Zhang J, Huang X, Shao CF, Liu CX, Shi HC, Hu HY, Liu ZQ (2004) Influence of packing media on nitrogen removal in a subsurface infiltration system. J Environ Sci China 16:153–156

Zhang Y, Geißen SU, Gal C (2008) Carbamazepine and diclofenac: removal in wastewater treatment plants and occurrence in water bodies. Chemosphere 73(8):1151–1161

Zhang DQ, Jinadasa K, Gersberg RM, Liu Y, Ng WJ, Tan SK (2014) Application of constructed wetlands for wastewater treatment in developing countries: a review of recent developments. J Environ Manag 141:116–131

Chapter 7
Treatment of Emerging Contaminants Using Constructed Wetlands

Reshu Singh and Manoj Kumar Tiwari

Abstract The pervasive existence of emerging contaminants (ECs) in the aquatic environment holds the potential to significantly influence the environment as well as pose risk to human health. Conventional systems that are employed to treat wastewater are insufficient to effectively remove these micropollutants. This necessitates the exploration of alternative eco-friendly and cost-effective treatment systems. The main emphasis of the chapter is on the application of constructed wetlands (CWs) as a potential substitute for secondary wastewater treatment and can also be adopted as a decentralized sustainable treatment system for wastewater purification. However, the proper understanding of fate, transport and removal mechanisms are essential components that need to be understood before employing CWs for the eliminations of ECs. The effectiveness of CWs is additionally affected by the surface area of the substrate, the characteristics of the wastewater, and the physicochemical characteristics of the pollutants. It has been noted that substrates like activated carbon, zeolite, biochar, and palm mulch and macrophytes like *Eichhornia crassipes* and *Phragmites australis* exhibit considerable potential in effectively removing various ECs. The existing understanding regarding the removal efficiencies of ECs like PPCPs (pharmaceutical and personal care products) in CWs, the associated removal mechanisms, the toxicity pharmaceuticals may pose to CWs,

R. Singh
School of Water Resources, Indian Institute of Technology Kharagpur, Kharagpur, India

Department of Civil Engineering, Government Engineering College, Bhojpur, Bihar, India
e-mail: reshusingh5794@kgpian.iitkgp.ac.in

M. K. Tiwari (✉)
Department of Civil Engineering, Indian Institute of Technology Kanpur, Kanpur, Uttar Pradesh, India

Centre for Environmental Science and Engineering, Indian Institute of Technology Kanpur, Kanpur, Uttar Pradesh, India

Kotak School of Sustainability, Indian Institute of Technology Kanpur, Kanpur, Uttar Pradesh, India
e-mail: mktiwari@iitk.ac.in

S. S. Kumar, V. Kumar (eds.), *Nature-Based Solutions for Decentralised Liquid Waste Management*, Environmental Challenges and Solutions,
https://doi.org/10.1007/978-3-032-08362-3_7

and the impacts of crucial parameters such as configuration of flow regime, hydraulic loading, type of macrophytes and substrates etc. remains limited. This chapter encourages and advocates for further research on these topics to generate more robust and compelling evidence regarding the operation and effectiveness of bigger laboratory-scale, full-scale or pilot-scale CWs for the eradication of ECs from wastewater.

Keywords Constructed wetlands · Emerging contaminants · Wastewater treatment · Decentralized system

7.1 Introduction

With rapid increase in urbanization, the detection of novel organic compounds in land, soil, and water bodies has also paced. These novel organic compounds, known as emerging contaminants (ECs), are categorized as micropollutants and are found in trace concentrations in the environment. ECs are addressed by various nomenclatures such as micropollutants, contaminant of emerging concerns and emerging organic contaminant for their reference (Talib and Randhir 2017). These tend to pose serious health hazards, when available in environmentally relevant levels, owing to their toxic nature. The existence of ECs in nature were identified but the threats imposed by them were unknown. ECs that exist in nature could be either natural or anthropogenic which include pharmaceutical and personal care products (PPCPs), flame retardants, fertilizers, pesticides, lipid regulators, nanoparticles, endocrine-disrupting compounds, per- and polyfluoroalkyl substances (PFAS), microplastics and polycyclic aromatic hydrocarbons along with their transmuted products (Tran et al. 2018). Their ambient bioaccumulation worsens the aberrant hormonal regulation that results in reproductive dysfunction, lower fecundity, a rise in the occurrence of breast and testosterone malignancies, and enduring antibiotic resistance (Kumar et al. 2023). Due to their human-induced origin, ECs are primarily discharged from wastewater treatment plants (WWTPs), have demonstrated to exhibit inefficiency in their removal. The existing traditional approaches for wastewater treatment are not designed to eradicate these ECs completely. This has resulted in exploring different treatment approaches for the removal of ECs. Physical, biological, and chemical treatment technologies have been developed to deal with the removal of ECs. The commonly employed treatment methods for removal of ECs from wastewater include processes like photo-Fenton, reverse osmosis, photocatalysis, chemical oxidation, ion exchange, surface adsorption and membrane bioreactor (MBR), constructed wetlands (CWs), moving bed bioreactor (MBBR), and activated sludge process (ASP). However, the high treatment cost is a direct result of the high chemical and energy costs associated with current and effective procedures like advance oxidation processes (AOPs) (Ávila et al. 2010). To combat this, CWs can be used as an economically viable and eco-friendly solution for the eradication of ECs from polluted water.

In the past few decades, the utilization of CWs has garnered significant attention due to their notable capability to eliminate emerging contaminants such as personal care products, pesticides, pharmaceuticals etc. all while maintaining cost-effective fabrication, functioning and management (Overton et al. 2023). Often, CWs are widely employed for industrial and domestic treatment of wastewater to eliminate pollutants like phosphorus, nitrogen, heavy metals, and suspended solids (Vymazal 2013). More recently, CWs have demonstrated a remarkable efficacy in treating organic chemicals, particularly medicinal waste as well as microplastics, present in wastewater. The intricate removal processes in these wetlands involve various natural removal mechanisms such as physical (photochemical oxidation, volatilization, sorption, sedimentation), chemical (hydrolysis, reduction, oxidation, photolysis, cation exchange), biochemical mechanisms (microbiological degradation), and biological (metabolism and plant absorption), all contributing to the reduction of organic compound loads (Sánchez et al. 2022). Enhancing our comprehension of the main influences that impact the removal of pollutants of emerging concern within a CW setting would aid in selecting the most suitable configuration (e.g., horizontal, vertical and surface, and subsurface flow) and determining optimal operating parameters (e.g., conditions like aerobic/anaerobic, hydraulic retention time intermittent or discontinuous mode of operation etc.) (Saidulu et al. 2021). However, the probable effect of a specific configuration on the removal of individual ECs is reliant upon the inherent characteristics of the target EC (Li et al. 2014).

This chapter analyzes and evaluates the effectiveness of CWs for the removal of emerging pollutants, with a specific focus on the factors that influence their removal effectiveness. The role of different macrophytes, microorganisms, and substrates used to enhance the performance of CWs has also been discussed. The challenges and future perspectives related to the treatment and removal of ECs with the help of CWs is also presented in this chapter.

7.2 Genesis and Environmental Impact of ECs

ECs are introduced into the ecosystem through origins like municipal wastewater, agricultural runoff, industrial wastewater etc. The presence of emerging contaminants (ECs), including pharmaceuticals and personal care products (PPCPs), perfluoroalkyl and polyfluoroalkyl substances (PFASs), and endocrine-disrupting chemicals (EDCs) in aquatic environments poses a significant risk to water resources (Oluwole et al. 2020). Even at concentrations as low as ng/L, these contaminants can accumulate in human bodies via food chains with aquatic species, leading to mutagenic, reproductive, genotoxic, and carcinogenic consequences. This poses a significant risk to water resources as well as human health. Thus, effective measures are crucial to address this growing threat and safeguard our water sources along with human well-being. The removal of ECs thus becomes a complex topic to address when the target pollutant is unknown.

In particular, antibiotics are commonly employed in animal husbandry to prevent the proliferation of harmful microorganisms. However, their excessive usage contributes to environmental genetic selection, favoring the emergence of more dangerous and resistant microbes (Awasthi et al. 2022). This poses a significant concern for public health and ecosystem integrity, calling for responsible antibiotic use practices to mitigate potential risks. Exposure to antibiotics has been found to induce drug resistance in various pathogenic species, including bacteria (Katibi et al. 2021). Additionally, steroid hormones commonly used for birth control, contraception, and enhancing sexual performance can often disrupt the endocrine system or act as anti-androgenic ligands. At a dosage as low as 5 g/L, diclofenac exhibits adverse health effects on rainbow trout (Surana et al. 2022). Conversely, zebrafish exposed to a mixture of acetaminophen, carbamazepine, gemfibrozil, and venlafaxine at doses ranging from 0.5 g/L to 10 g/L experienced various effects, including tissue degradation, reduced embryo production, and elevated embryo mortalities (Galus et al. 2013).

For instance, PhACs are released in the environment usually by hospital wastewater. Personal care products can directly find their ways into the aquatic environment via wastewater originating at household levels (Olasupo and Suah 2021; Parida et al. 2021). Various point sources such as industries, hospitals etc. and non-point sources such as runoff from catchments and agricultural fields etc. also act as a source of ECs (Kasonga et al. 2021; Starling et al. 2019). Humans consume pharmaceutical drugs, artificial sweeteners which may be excreted out of the body if not completely metabolized. Most importantly, wastewater treatment plant effluents contribute significantly to the release of ECs in water environment as WWTPs are not designed to remove these persistent pollutants at low concentration (Mohapatra et al. 2016; Saidulu et al. 2021).

Pharmaceuticals have been the oldest known sources of ECs and the evidence of their occurrence dates back to 1970s–1990s. Subsequently the initial detection of PhACs in aquatic environments in the beginning years of 1980s, a variety of therapeutic compounds have been found in water (Majumder et al. 2021). These compounds can be categorized into several groups based on their medicinal properties. They have been classified as follows: (i) antibiotics (e.g., penicillins, macrolides, sulphonamides, tetracyclines), (ii) analgesics and anti-inflammatories (e.g., paracetamol, diclofenac), (iii) antidepressants (e.g., benzodiazepines), (iv) β-blockers (e.g., atenolol), (v) lipid lowering substances (e.g., fibrates), and (vi) other substances (e.g., antiseptics and narcotics) (Dhangar and Kumar 2020; Parida et al. 2021; Patel et al. 2019).

In China, the concentrations of ciprofloxacin in WWTP influents were stated to be in the range of 0.03–0.18 μg/L, which is lesser than the concentrations stated in Spain (0.16–13.6 μg/L) (Saidulu et al. 2021). However, a significantly higher concentration of ciprofloxacin (31,000 μg/L) was reported in India and 501.575 μg/L in South Africa (Faleye et al. 2019; Mutiyar and Mittal 2014). In Spain, ciprofloxacin was also found in water sourced from the deep confined aquifer, having a high level of 0.32 μg/L (Gracia-Lor et al. 2012).

Among the various analgesics found in the influent of WWTPs, ibuprofen was found to have a greater dose in North America (75.8 μg/L) in comparison to Europe (33.76 μg/L), and Asia (26.45 μg/L) and Australia (10.3 μg/L). Naproxen has been detected in countries including the USA, European countries, China, and India, in the range of 0.08–25 μg/L. Meanwhile, diclofenac, an anti-inflammatory drug, was found in municipal wastewaters in the array of 0.11–25.68 μg/L (Kasprzyk-Hordern et al. 2009; Mandaric et al. 2017; Metcalfe et al. 2003). In South Africa, the detected concentration of hormones, including 17 β-estradiol, ranged from 112 ng/L to 44.8 ng/L. Atenolol and propranolol, the often reported β-blockers, were found in WWTP influents across various European and Asian countries, with concentrations ranging from 0.001 μg/L to 300 μg/L (Estrada-Arriaga et al. 2016; Gani et al. 2020; Mohapatra et al. 2016).

Artificial sweeteners are widely consumed all around the globe and hence they are found in significant concentrations. In wastewater samples from Zurich, Switzerland, acesulfame was found in dose up to 43 μg/L (Buerge et al. 2009). Saccharin and sucralose have been found in WWTP influent in concentration of 389 μg/L and 46.1 μg/L in India and the USA respectively (Subedi et al. 2015; Subedi and Kannan 2014). Elsewhere, high concentrations of ECs originating from PCPs were detected in UK with maximum concentration of benzophenone-3 reported to be 3975 μg/L. Moreover, octocrylene was reported at the highest concentration in WWTP influent at 12 μg/L (Parida et al. 2021).

Pesticides are another type of ECs which have been consistently found in water environment due to runoff from agricultural fields. In the Yangtze River average atrazine concentration of 43–45 ng/L were detected. Triclosan was also detected in WWTP influents in China and the USA in average concentrations of 2.08 μg/L and 3.4 μg/L. Nonylphenol has also been found in WWTP influents in Spain, China, and UK in concentration range of 1.5–63 μg/L (Fair et al. 2009; Saidulu et al. 2021).

X-ray contrast media have been detected in the influents of wastewater in Asia and Europe with iohexol, iopamidol, and iopromide found in concentration of 125 μg/L, 45 μg/L, and 164 μg/L respectively. Various other ECs such as caffeine in range of 209 μg/L in Spain and 542 μg/L in UK have been reported. Elsewhere, plasticizers such as bisphenol-A and opioids like codeine have also been reported. It is evident that the presence of ECs is increasing in water environments hence their detection and removal is of significant interest (Parida et al. 2021; Saidulu et al. 2021).

7.3 Treatment of ECs in Wastewater

Despite these advancements in CWs, the precise primary mechanism accountable for the elimination of ECs in CWs remains unclear, mainly due to insufficient comprehensive mass balance data, particularly concerning the target trace organic pollutants (Chen et al. 2019). The ECs are being used on a wide scale which eventually contribute to the nonpoint pollution source. Thus, it becomes necessary to understand the fate of these pollutants in the environment since they pose a potential risk as they enter the ecosystem. The fate and transport of ECs remain a relatively unexplored area when industrial and urban wastewater are mixed within WWTPs. The potential influence of chemical and biological indicators on emerging contaminants yet needs to be discussed. Based on existing studies conducted, this chapter discusses some of the widely used treatment processes of existing emerging contaminants in the environment.

The various advanced treatment processes involved in the treatment and elimination of ECs from wastewater are activated sludge process (ASP), extended aeration (EA), sequential batch reactor (SBR), moving bed biofilm reactor (MBBR), fluidized aerobic bed reactor (FABR), up-flow anaerobic sludge blanket (UASB), waste stabilization pond (WSP), oxidation pond (OP) etc. (Dubey et al. 2023a). The conventional treatment processes like sedimentation, screening and filtration are not designed to treat the ECs from wastewater. The removal of ECs in WWTPs is a complex procedure, contingent on the biological and chemical characteristics of pollutants, such as solubility and hydrophilicity (Dubey et al. 2023b). When it comes to secondary treatment of wastewater, processes like ASP are eliminates the ECs through partition, biodegradation, biotransformation, and adsorption. In this process the removal efficiency of ECs is highly affected by parameters like adsorption capacity, sludge age, adsorption capacity, and the design of reactor (Lin et al. 2016). Since the molecular structure of different ECs vary so does their biodegradability and their removal efficiencies. It was reported that triclosan (TCS) showed a removal efficiency greater than 95% (McAvoy et al. 2002) while on the other hand diclofenac removal efficiency was less than 25% in secondary wastewater treatment plant (Wang et al. 2014). Numerous ECs such as PPCPs exhibit poor elimination during secondary treatment processes. Consequently, the implementation of tertiary treatment processes in STPs, such as AOPs, sand filtration and membrane separation, is often necessary to remove ECs before UV or chemical disinfection. Additionally, transformation of PPCPs has been observed during the chlorination process. Keeping in mind the ineffectiveness of conventional wastewater treatment systems, a need to adapt more advanced treatment options was called for. AOPs such as activated carbon adsorption, ozonation, photocatalysis, and membrane filtration process like reverse osmosis (RO) proved to be efficient in the removal of ECs from wastewater. In a pilot-scale experiment aimed at removing PPCPs such as ibuprofen and diclofenac using UV/H2O2 and ozone/H_2O_2, significant improvements in efficiency were observed. The removal efficiencies reached 92% and 97%, respectively, in comparison to the conventional treatment method, which only achieved a removal efficiency of 26% (Borikar et al. 2015).

Meanwhile, when talking about the fate of these ECs, two parameters which determine the fate and occurrence are (a) pH-dependent n-octanol-water distribution ratio (D_{ow}) that considers ion formation and hydrophobicity simultaneously; and (b) the portioning coefficient (K_{ow}) that is used to determine the hydrophobic portioning. Lower values of log K_{ow} (<1) indicate that the compound is water soluble and has high polarity which permits the contaminant to stay in the environment for longer periods (Barbosa et al. 2016; Parida et al. 2021). These results have indicated how elimination of ECs from wastewater can be enhanced by knowing the target contaminant and choosing the right treatment method.

7.4 Wetlands for the Treatment of ECs

The elimination mechanism of CWs for contaminants depends majorly on three critical mechanisms namely, adsorption of substrates, plant uptake, and decomposition by microbes (Fagan et al. 2016). CWs offer a heterogeneous environment that differs from many conventional wastewater treatment unit processes. Plant roots and microbiological layers both rely heavily on soil for their nourishment. In addition, the soil matrix plays a critical role in the hydraulic processes that take place. Within CWs, the physical substrate, plant roots, and diverse microbial populations create a variety of habitats that facilitate the formation of biofilms (Farooq et al. 2023). These biofilms play a crucial role in pollutant removal, as they provide a complex and interconnected network for the breakdown and transformation of contaminants, contributing to the overall effectiveness of CWs as a natural and sustainable wastewater treatment option (Song et al. 2015). In general, hydrophobic ECs compounds with high K_{ow} values are eliminated from the system through processes such as hydrolysis, adsorption, biosorption, and microbial degradation (Lv et al. 2016). Understanding the links between removal processes and the target emerging contaminant gives more insight into how CWs can help in the elimination process. As a result, it provides a larger and deeper knowledge basis for the design and management techniques that might optimize the elimination of contaminants in constructed wetlands.

7.4.1 Wetland Operations and Contaminant Removal Mechanisms

Constructed wetlands utilize various contaminant removal mechanisms, encompassing physical processes such as sedimentation and filtration, chemical processes like coagulation, adsorption, ion exchange, photolysis, and volatilization, as well as biological processes such as decomposition, uptake by plants, and fermentation methanogenesis (Vymazal 2019). The operation of CWs is also influenced by parameters like microbial community, soil chemistry, pH, hydraulic retention time as well as temperature (DuPoldt et al. 1996). Plants in CWs assist in sedimentation by capturing and allowing the sediment to settle under the influence of gravitational force, which is governed by the varying densities of the suspended particles. The plant roots grow within the compacted sediments, aiding in the decomposition of organic matter with the assistance of microorganisms. The limitation of this process is that it can only remove particles with a diameter greater than 1 mm, while heavy metals, dyes, and numerous other hazardous pollutants cannot be eliminated from the wastewater using this method (Ruppelt et al. 2020). Another physical process in CWs is filtration, which involves removing particulate materials from wastewater by passing it through the porous substrate of the system. The substrate bed typically consists of sand, gravel and soil, containing various voids. Suspended matter is filtered through the roots of plants and pores within the system bed. Precipitation/coagulation is another removal mechanism of CWs which helps in eliminating pollutants like phosphorus and certain pollutants of organic and inorganic nature. The effectiveness of precipitation largely relies on the wastewater's pH, the concentration and solubility product of metal ions, and the type of coagulant employed. The heavy metal removal is aided by adsorption on the surface of soil/sediment. However, this mechanism does not facilitate the removal of textile dyes. Adsorption in CWs helps in the removal of heavy metals and ECs based on the type of adsorbate used for the removal (Wang et al. 2020b). The most effective adsorbent used for pollutant removal from wastewater through adsorption is activated carbon derived from specific macrophytes. *Phragmites australis* activated carbon has been used for the removal of Rhodamine-B dye from textile wastewater (Sharma and Malaviya 2022).

One of the most commonly known removal mechanisms of CWs is plant uptake (Vymazal 2013). Macrophytes are known to be a critical component of CWs as they act as a storage unit for carbon and nutrients. Apart from being a storage unit they take part in phytodegradation, a mechanism where pollutants are directly up taken by plants. The process of pollutant degradation in the rhizosphere through substances secreted by plants is referred to as rhizodegradation. On the other hand, when pollutants enter the plant biomass and are later emitted through transpiration, the process is known as phytovolatilization. This mechanism is highly efficient in the removal of heavy metals and ECs like pharmaceuticals and textile dyes (Zhang et al. 2023a). Microbial degradation is another essential removal mechanism in CWs that aids in reducing the pollutant load of BOD, COD, as well as organic and inorganic pollutants from wastewater (Ercole et al. 2022). Under aerobic conditions, microorganisms use organic pollutants as a carbon source, producing CO_2, while under anaerobic conditions, they generate methane. Additionally, microorganisms create a biofilm on the bed material's surface, aiding in filtration. These microorganisms can also accumulate a small amount of heavy metals and convert them into nonmobile forms (Molle et al. 2005).

7.4.2 Elements Influencing the Efficiency of CWs in ECs Removal

7.4.2.1 Types of CWs

The flow regime among the different configurations of CWs available highly influence the removal mechanism of pollutants in wastewater (Verma and Suthar 2018). Based on the flow pattern, CWs can be classified as surface/subsurface flow, vertical flow and horizontal flow (Vymazal 2011). In subsurface flow constructed wetlands (SSFCW), water flows through substrate media beneath the surface bed. Horizontal subsurface flow constructed wetlands (HSSFCW) are constructed using materials like gravel, where water flows horizontally below the surface bed. On the other hand, vertical subsurface flow constructed wetlands (VSSFCWs) are designed for water to flow vertically in the subsurface, either in a downward or upward direction. These variations in design and flow pathways allow for diverse treatment capabilities, making CWs a valuable tool in wastewater remediation. By integrating various types of CWs together, researchers can optimize the treatment performance for specific pollutants by exploiting the unique characteristics from each individual system. Not only CWs can be combined with different flow regimes, they can also be combined with additional treatment methods like advanced oxidation processes (AOPs), photo-Fenton, ultra- and nanofiltration etc. in order to achieve better removal efficiencies of ECs. Table 7.1 shows combination of hybrid systems of CWs along with their removal efficiencies.

The configuration of CWs such as horizontal subsurface flow, vertical subsurface flow, and tidal flows affected the fate of ECs as well (Vymazal 2018). The removal efficacies of ciprofloxacin and oxytetracycline in vertical subsurface flow CW were higher when compared to the horizontal subsurface flow CW with a removal rate of 99% and 97% respectively (Liu et al. 2019). The elevated levels of based on the accumulated rate of antibiotics and abundances of antibiotic-degrading bacteria, it was indicated that substrate adsorption and biodegradation were the main pathways impacting the antibiotic elimination efficiency of CWs (Harrabi et al. 2019). Another study also concluded that VSSFCW tend to perform better removal of antibiotics in comparison other single CWs with an average elimination efficacy of 85–99% in a comparative experiment (Lv et al. 2022; Liu et al. 2019). The levels of antibiotics in the effluent of the three CWs exhibited the subsequent decreasing order: SF > HSF > VSF. The performance of CWs for the elimination of ECs can be enhanced by modifying parameters like HRT, temperature, innovating substrate media and by integrating other treatment process. Microbial degradation proved to be a key removal pathway for the elimination of antibiotics viz. sulfamethoxazole in integrated VFCWs. The microbial degradation was carried by microorganisms like *Pseudomonas*, *Bradyrhizobium*, *Sphingomonas*, and *Luteimonas* (Zheng et al. 2021).

An investigation showed removal of three different types of steroids, viz. oestrogens, testosterone, and progesterone at rates between 80% and 100% in the HSSFCW (Vymazal et al. 2015). When it comes to VSSFCW, the continuous flow of water in vertical direction allows the maintenance of aerobic conditions in the substrate. This helps the VSSFCW to facilitate more oxygen than HSSFCW. An investigation about partially saturated VSSFCW concluded the removal efficiency of trimethoprim and sulfamethoxazole as 99% and 64% respectively (Sgroi et al. 2018). Compared to HSSFCW systems, VSSFCW systems have demonstrated enhanced microbial decomposition capabilities. This improvement is attributed to the increased oxygenation facilitated by the vertical wastewater drainage through different soil layers. VSSFCW system demonstrates higher efficacy in removing antibiotics compared to HSSFCW systems. On average, VSSFCW achieves an

Table 7.1 The effectiveness of individual system CWs, hybrid CWs, and CWs integrated with other treatment units in addressing emerging contaminants

System configuration	Type of ECs	Type of WW	Macrophytes	Substrates	Target EC & their removal efficiency (%)	Remarks	Reference
TFCW with baffle and plants	PPCPs	Raw domestic sewage	*Canna indica L.*	Gravel	Biocides: 97.1 Antibiotics: 90.2 Steroid hormones: 99.8	CWs coupled with baffles and plants showed enhanced removal efficiency.	Cheng et al. (2021)
Aerated HSSF-CW	Pharmaceuticals	Partially treated sewage	*Phragmites australis*	Gravel	Acetaminophen: 90 Diclofenac: 97 Ketoprofen: >83 Bezafibrate: 88 Gemfibrozil: 100	Aeration enhanced the removal efficiency of ECs as well as conventional parameters of waste water.	Ávila et al. (2021)
CW-MFC	Pharmaceuticals (antibiotic)	Synthetic WW	*Ipomoea aquatic*	Gravel and granular activated carbon (GAC)	Sulfadiazine: 99	Anode accumulation rate of SDZ is higher when compared to cathode.	Song et al. (2018)
CW-MFC	Dye	Synthetic WW	*Canna indica*	Gravel and glass wool	Methylene blue dye: 93.15	Higher detention time leads to higher removal efficiency.	Yadav et al. (2012)
Electrolysis integrated TFCW	PPCPs	Synthetic WW	*Cyperus alternifolius*	Gravel and zeolite	Sulfamethazine: 57 Ibuprofen: 37 Diclofenac: 28 Carbamazepine: 68	Higher removal efficiency of target PPCPs was achieved after electrolysis and at higher HRT.	Liu et al. (2020)

VFCW	Pharmaceuticals	Recycled WW	*Phalaris arundinacea*	Gravel	Triclosan: 100 Sulfamethoxazole: 99.9	A planted mesocosm with higher HRT showed better removal efficiency for the target ECs.	Button et al. (2019)
VFCW	Antibiotics	Secondary WW	*Phragmites communis*	Mn-ore	Ciprofloxacin hydrochloride: 93 Sulfamethazine: 69	The type of substrate greatly affects the diversity in microbial community growth thus affecting the removal process.	Yuan et al. (2020)
Tropical CW	PFAS	Landfill leachate	*Typha angustifolia, Chrysopogon zizanioides* and *Cyperus papyrus*	Gravel bed	PFAA precursors: >80 PFOA: 59	Reed bed showed better removal capacity for PFAS.	Yin et al. (2017)

(continued)

Table 7.1 (continued)

System configuration	Type of ECs	Type of WW	Macrophytes	Substrates	Target EC & their removal efficiency (%)	Remarks	Reference
VFCW	Plastics and PCPs	Municipal WW	*Phragmites australis* and *Typha latifolia*	Gravel and sand	Bisphenol A: 60 Nonylphenol: 96.2	Among the two macrophytes used, *Typha latifolia* showed better removal efficiency. VFCW is better in removing ECs as compared to HFCW.	Papaevangelou et al. (2016)
VFCW	Pharmaceuticals, cytostatics and macrolides	Municipal WW	*Phragmites australis* and *Iris pseudacorus*	Sand enriched biochar	Diclofenac: 99.3 Ciprofloxacin: 94.2 Mecoprop: 94.8	Activated biochar proved to be a better substrate compared to nonactivated biochar.	Venditti et al. (2022)
IRCW	Pesticides	Recirculated WW	*Cyperus alternifolius*	Gravel and ceramics	Chlorpyrifos: 90 Diuron: 99	The amalgamation of plant and Fe-impregnated biochar can further boost the elimination of pesticides, making it a highly effective approach.	Tang et al. (2016)

IVCW	Antibiotic	Synthetic WW	*Phragmites australis*	Gravel	Sulfamethoxazole: 95	Ammonium oxidising bacteria (AOB), ANAMMOX, and functional bacteria are affected by prolonged antibiotic exposure.	Zheng et al. (2021)
Subsurface flow CW	Pharmaceutical	Industrial WW	*Acorus calamus*	Gravel and soil	Atrazine: 91.2	Pseudomonas and Arthrobacter were better atrazine degrading bacterial community.	Zhao et al. (2019)
HSSF-CW	Herbicide	Synthetic WW	*Phragmites australis*	Gravel	2,4-dichlor-ophenoxyacetic: 71–99	CW was efficient in the removal of organic matters as well.	Echeverri González et al. (2019)

impressive elimination efficiency of 85.9% for antibiotics, while HSSFCW system lags behind, removing only a few antibiotics with a 59% removal efficiency (He et al. 2021). The vertical subsurface flow design in VSSFCW contributes to the increased contact time between wastewater and the microbial communities, enhancing the biodegradation and removal of antibiotics. This notable difference in removal efficiency highlights the advantages of VSSFCW systems over HSSFCW systems for addressing antibiotic pollution in wastewater treatment. Another novel CW was developed to study the removal of surfactants from the effluent discharge of student accommodations on IIT Madras campus, Chennai, India. It was reported that green roof-top water recycling system (GROW) was able to remove surfactants like sodium do-decyl sulphate (SDS), propylene glycol (PG), and trimethyl amine (TMA) with an average efficiency of about 82% from the effluent (Ramprasad et al. 2017). Overall, based on the literature study conducted VFCWs proved to be more efficient and widely used for the elimination of ECs from wastewater. The vertical flow design promotes better aeration and oxygen availability to the microbial communities residing in the subsurface, fostering more efficient decomposition and degradation of pollutants in the wastewater.

7.4.2.2 Types of Macrophytes

Various studies have indicated how a vegetated CW shows greater performance efficacy as compared to an unvegetated one. A vegetated CW provides a more favorable habitat for the proliferation of microorganisms than unvegetated one (Brix 1994). Wetlands rely heavily on macrophytes, which perform a role in sediment stabilization and oxidation. The availability of vegetation means a better oxygen condition for expediting microbial community growth to facilitate emerging contaminant degradation. When dealing with wastewater treatment, the choice of macrophytes is critical. It has also been stated in studies that plant biomass growth is proportional to the concentration of organic pollutant available in wastewater (Manolaki et al. 2020). The macrophytes facilitate microbial community structure and diversity of the microbial population in CWs which perform a vital role in the efficient elimination of ECs (Chen et al. 2019). These microorganisms utilize ECs, particularly the ones rich in carbon source like antibiotics, which aids in the degradation of the parent compound, contributing to its elimination from the wastewater (Fernandes et al. 2015). The interplay of different microbial species within the biofilm creates a dynamic and adaptive system that enhances the overall pollutant removal capabilities of CWs, making them a valuable tool in addressing water and wastewater contamination. A study reported the removal of tetracycline and enrofloxacin can spike up to 94% due to the microbial accumulation and biofilm formation. CWs typically harbor a diverse range of bacterial species, with the most abundant ones including aerobic, anaerobic, denitrifying, nitrifying, and ammonia-oxidizing bacteria. These various types of bacteria play crucial roles in different stages of the wastewater treatment process, collectively contributing to the elimination of varied number of pollutants and the overall purification of the water. The combination of these bacterial populations creates a highly efficient and adaptable microbial community that enhances the effectiveness of CWs as a sustainable and ecologically friendly means of wastewater treatment.

In the treatment of ECs within CW systems, several plant species have been frequently employed. These commonly include *Phragmites australis*, *Typha angustifolia*, *Hydrilla verticillate*, *Typha latifolia*, *Ceratophyllum demersum*, *Canna indica*, and *Ipomoea aquatic* for the effective removal of emerging pollutants (Li et al. 2014), (Kumar et al. 2011). In an experiment conducted by (Kumar et al. 2011) it was concluded that floating macrophytes (*Eichhornia crassipes*) proved to be more effective in the removal of estrogens with an elimination efficiency of 69% as compared to submerged–emergent macrophytes (*Hydrilla verticillata* and *Oriza sativa*) which has an efficiency of 18%. In the CWs system with *Typha* sp., there was a notable improvement of 10–70% in the removal of carbamazepine (Log $K_{ow} = 1.5$, pKa = 14) and ibuprofen (Log $K_{ow} = 1.5$, pKa = 4.5) compared to the CWs system without *Typha* sp. (Sharma et al. 2016). These chemicals are likely to have been easily absorbed and transpired by plants due to their dissociated form or poor lipophilicity. *Phragmites australis* has been widely employed by researchers due to its effective removal of micropollutants. In other experimental studies *Eichhornia crassipes* was found to eliminate lead, caffeine, and ibuprofen with an efficiency of 93%, 97%, and 89% respectively in a VFCW suggesting this macrophyte to be a potential eliminator of certain ECs (de Oliveira et al. 2019), (Agarry et al. 2020).

Nevertheless, the efficiency of removal is not solely determined by a one factor, such as the variation in macrophytes, but is also heavily influenced by the type of CWs system and the Log K_{ow} or pKa values of the contaminants. A study was conducted to understand the behavior of 27 different ECs (pharmaceuticals like antibiotics, beta-blockers, analgesics etc.) and the removal effect of two macrophytes viz. *Phragmites australis* and *Iris pseudacorus* (Venditti et al. 2022). It was concluded that these macrophytes were able to remove diclofenac, carbamazepine, clarithromycin, and benzotriazole with an elimination efficiency of 99%, 92%, 96%, and 98% respectively. It was also addressed that when wetland plants are exposed to excessive ECs, it can hinder their growth and lead to the development of phytotoxic effects (Gahlawat and Gauba 2016). Studies have indicated that ECs could impede root cell growth while suppressing ammonium production by symbiotic N-fixing root nodules, crucial for macrophyte utilization.

7.4.2.3 Types of Substrates

Substrates are crucial components of CWs because they host the vast majority of physiochemical and biological reactions (Yang et al. 2018). The substrates, also referred as support matrix, are also responsible for expediting process the physiochemical and biological that are responsible for effective eradication of micro pollutants (Wang et al. 2020a). Critical to the efficacy of CWs, substrates offer a reaction zone for pollutant removal. These substrates also provide a favorable environment for the growth of microorganisms responsible for wastewater treatment

(Varsha et al. 2022). In general, the most commonly favored substrates employed in CW systems were gravel, sand, and local soil often extracted from nearby rivers or a combination of these materials. However, researchers have also successfully utilized multiple other substrate materials to effectively remove pollutants through CWs. These materials include clay-based substrates such as light expanded clay aggregates, shale and vermiculite, as well as siliceous-based substrates like perlite, zeolite, and diatomite (He et al. 2021). Additionally, activated carbon and activated alumina have also demonstrated their efficiency in pollutant removal in CWs (Dordio and Carvalho 2013).

The exclusion of antibiotics and other pollutants of emerging nature in CWs is influenced by various physiochemical characteristics of the substrates, which include particle size distribution, porosity, surface change and hydraulic conductivity. Substrates can be broadly categorized into two groups: natural substrates, derived from the environment, and man-made or artificial substrates, developed for specific purposes. A pilot-scale study was conducted to understand the role of activated and nonactivated biochar mixed with sand to act as a substrate in VFCW for the removal of ECs from municipal wastewater (Venditti et al. 2022). It was concluded that 15% activated biochar mixed with sand demonstrated to be effective in the exclusion of ECs like carbamazepine and diclofenac. Aminomethylphosphonic acid (AMPA) and cyclophosphamide were not as efficiently removed as other compounds as sorption was the driving removal mechanism of these compounds. Multiple mechanisms, including van der Waals interaction, hydrophobic partitioning, ion exchange, surface complexation and electrostatic interaction contribute to the sorption of contaminants onto the surface of a substrate.

The most commonly utilized material in CWs for pharmaceutical removal is gravel, and its implementation has not resulted in any detrimental impact on the wetlands' performance. Gravel has proven to be effective in sorption of pharmaceuticals that are resistant to biodegradation but possess relatively high hydrophobicity, such as carbamazepine (Verlicchi et al. 2013). Despite its efficiency, there is currently limited data available concerning a thorough grasp of gravel's sorption performance and the fundamental mechanisms involved in removing pharmaceuticals from wastewater using CWs. Based on the type of origin, physical and chemical properties of substrates, they can be further classified into organic and inorganic substrates. Organic substrates comprise of biochar and mulch while inorganic substrate include sand, gravel, and zeolite. Various inorganic substrates, including sand, shale, and gravel are employed in constructed wetland (CW) systems. Studies have indicated that CW systems utilizing sandy soil as a substrate are particularly efficient in eliminating polyether ionophore antibiotics. This is attributed to the sandy soil's higher hydraulic conductivity and infiltration rates, which promote enhanced interactions between the substrate and the antibiotics, leading to improved removal efficacy (Azfar Hussain et al. 2012). Organic substrates are mostly obtained in the form of biochar of vegetative media and palm mulch. Research findings indicate that palm mulch organic material serves as a superior carrier for the adsorption of steroids in comparison to both gravel and sand (Li et al. 2022).

In summary, while natural sand or gravel can remove ECs like antibiotics and steroids, their removal efficiency is currently low and requires enhancement. Nevertheless, these substrates offer good drainage capabilities. On the other hand, modified substrates such as zeolite, lapilli, activated carbon and palm mulch have demonstrated higher efficacy in eliminating antibiotics and steroids. However, it is worth noting that these modified substrates are more costly compared to natural sand or gravel. Therefore, a mixture of natural and modified substrates that works well together is the most obvious choice.

7.4.2.4 Types of Microorganisms

Various studies have been led across the globe to recognize the fate and transport of emerging pollutants due to their incomplete biodegradability in the treatment units. ARGs of tetracycline, sulfonamide, aminoglycoside, and multidrug was studied because of their positive detection rates and numbers were recorded to be at the peak towards the coasts and wastewater treatment plants. The application of catalytic ozonation facilitated the elimination of tetracycline and MLSB ARGs (macrolide-lincosamide-streptogramin B antibiotic resistance genes) in CWs (Hazra et al. 2022). The predominant microbial population responsible for the elimination of the aforementioned emerging contaminants in CWs were Actinobacteria, Bacteroidota, and Cyanobacteria. The removal of chloramphenicol, aminoglycoside, β-lactamase, sulfonamide, and multidrug ARGs was aided by denitrification, secondary sedimentation, and acid hydrolysis processes (Zhang et al. 2023b). Some of the volatile emerging pollutants can also be eliminated through the volatilization process, which involves changing them from soluble to gaseous state. Iron impregnated biochar which was made using *Cyperus alternifolius* for the removal of endosulfan, chlorpyrifos, diuron, and fenvalerate in integrated recirculating constructed wetlands (IRCW) (Tang et al. 2016). First-order kinetics was followed with half-lives of 1.5–11.6 h to see the fate of these pesticides. An investigation into mass balance revealed that IRCWs primarily undergo two main removal processes: sorption, accounting for 32.2–98.6% of the removal, and microbial degradation, responsible for 1.3–52.8% of the removal for the pesticides (Tang et al. 2016). In a separate study, aerobic bacteria of the Sphingomonos Ibu-2 species, isolated from activated sludge, were employed for the metabolic degradation of ibuprofen (Murdoch and Hay 2005). Ammonia-oxidizing bacteria (AOB) possess an enzyme called ammonia monooxygenase, which plays a vital role in the degradation of certain PhACs when exposed to various growth substrates. Additionally, autotrophic AOB demonstrate effectiveness in the degradation of numerous ECs, including bisphenol-A and ibuprofen (Czajka and Londry 2006). Cometabolic degradation occurs in both aerobic and anaerobic environments. Nevertheless, the majority of studies suggest that cometabolism of ECs is more favorable under aerobic conditions than under anaerobic conditions (Sarmah and Northcott 2008).

7.4.3 Contaminant Removal Mechanisms in CWs and Significance of Physicochemical Properties

Numerous methods such as plant uptake, photodegradation, breakdown by microbial action, hydrolysis etc. are involved in the elimination of micropollutants in CWs (Vymazal and Březinová 2015). The removal efficiencies are influenced by major physicochemical parameters of substrates, including porosity, hydraulic conductivity, particle size, surface charge, and pH. The molecular weight and structures of ECs are also responsible for the way ECs get eliminated. The breaking down of ECs can either be biotic and abiotic in nature. Biotic breakdown includes processes like biodegradation by microbes and plant uptake whereas abiotic processes include oxidation, photolysis, and hydrolysis which depend on factors like temperature, environmental conditions, physiochemical qualities etc. According to studies, microcontaminants with high Log K_{oc} ($\geq$3) and Log K_{ow} (>4) (indicating hydrophobic properties) can form robust bonds with soil particles, leading to a notably high elimination efficacy of the system concerning these micropollutants (Vystavna et al. 2017). The removal of these pollutants is significantly affected by processes like coagulation, flocculation, sedimentation, and adsorption. The micropollutants with low Log K_{ow} are eliminated by processes like biodegradation and photodegradation (Polianciuc et al. 2020) (Figs. 7.1, 7.2 and 7.3).

7.5 Potential for Field Application, Technological Challenges, and Directions for Innovation

The swift pace of growth of population and urbanization have resulted in a substantial increase in the global volume of wastewater generated from both industrial and residential sources. Poor wastewater management and inadequate control of wastewater sources have detrimental effects on both the environment and human well-being. The development of towns attracts investors and a growing population, which further strains the capacity of urban areas to fulfill the basic requirements of the migrating populace. Hence, promoting community awareness is crucial, alongside the implementation of cutting-edge, sustainable, cost-effective, and high-performance systems for wastewater management and treatment. While the development of viable and effective wastewater treatment technologies is a demanding and complex undertaking, it is imperative in light of the evident impacts of emerging contaminants (ECs) on the biota and ecosystems. Even though most ECs are only found at the trace level, the danger from these contaminants is severe. PhACs, pesticides, EDCs, and PCPs are only few of the many chemicals that have become pervasive in the aquatic environment and have severe harmful impacts on biota. This necessity calls for immediate attention and action for ECs remediation. Recent and effective techniques such as AOPs often require costly chemicals and energy for their implementation which consequently leads to a substantial increase in the

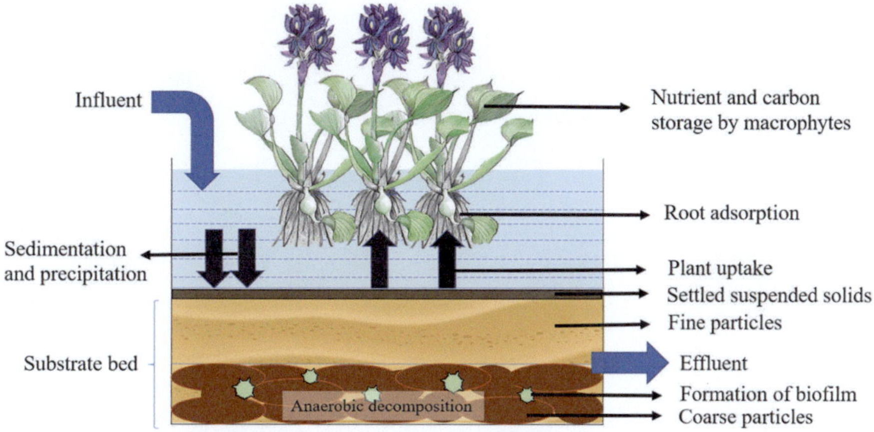

Fig. 7.1 Constructed wetland mechanisms involved in the removal of contaminants

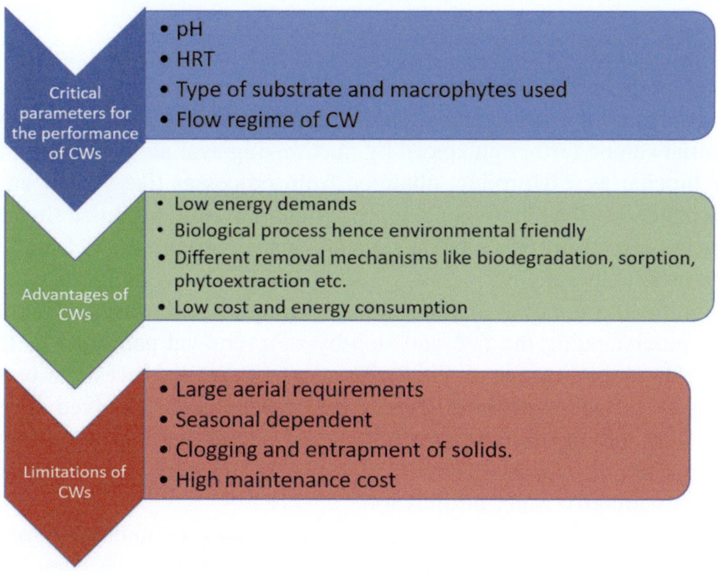

Fig. 7.2 Critical parameters, advantages and limitations of CWs

overall treatment cost. Hence, there has been a recent suggestion to integrate chemical techniques with biological processes to reduce the associated expenses. This approach aims to strike a balance between efficiency and affordability in wastewater treatment (Ahmed et al. 2017). Biochar integration with CWs has proved to be highly efficient absorbent for removal of pharmaceutical contaminants. Due to its low production cost, biochar emerges as a viable and outstanding adsorbent, presenting a promising alternative to GAC (granular activated carbon) and other conventional adsorbents for water treatment applications for the removal of ECs. For

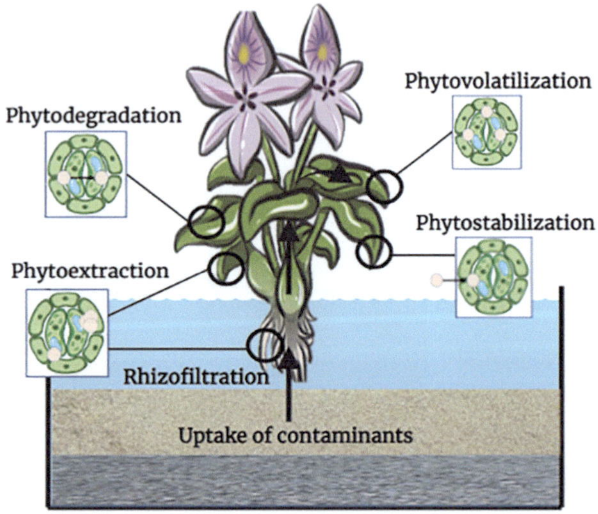

Fig. 7.3 Removal mechanisms of macrophytes in CWs

instance, in comparison to GAC and activated biochar, biochar exhibits approximately 20% higher removal efficiency for sulfamethoxazole, making it an attractive choice for enhancing water purification processes (Lin et al. 2017). The cost reduction potential can be further enhanced by maximizing availability through the utilization of biochar as a byproduct obtained from processes like syngas and biooil/biofuel production. This innovative approach can not only optimize the economic feasibility but also promote a more sustainable and environmentally friendly wastewater treatment solution.

One significant limitation of CWs is the scarcity of studies focused on comprehensively understanding the fate and step-by-step removal pathways of the target pollutants (Guedes-Alonso et al. 2020). This lack of detailed investigation hampers the ability to fully grasp the efficiency and mechanisms involved in pollutant removal within these natural treatment systems. Addressing this gap through more extensive research is crucial for optimizing CW performance and advancing their application in effective water purification.

The elimination of ECs in CW systems encompasses a combination of intricate physical, chemical, and biological processes. While there have been significant advancements in CWs, their widespread application is hindered by the substantial land area requirements. To mitigate the impact of land requirements and enhance ECs removal in CWs, extensive research has been conducted in recent years to explore diverse configuration designs and operational alternatives for these systems. However, most of the studies have only been conducted on specific ECs viz. pharmaceuticals and PCPPs estone, tetracycline, erythromycin etc. making it difficult for the application of CWs. Integration of artificial aeration with CWs, tidal flow CWs, enhancing the substrate materials with biochar, bioaugmentation, electroactive CWs and integration of algae-based biotechnologies with CWs are some of the innovative technologies that can be incorporated to enhance the removal efficiency of ECs. Indigenous microbial communities originating from estuarine environments possess the ability to address antibiotic-induced contamination and could be viable candidates for revitalizing impacted ecosystems using bioremediation approaches. CWs possess significant potential to function effectively as either a wastewater polishing treatment unit or a secondary wastewater treatment unit. Due to its technical simplicity and cost-effectiveness in wastewater management, CWs as a biological treatment is chosen even for a wide range of ECs. The removal efficiencies of these systems relied heavily on operational conditions like OLR, HRT, and SRT. The majority of studies conducted so far have been based on laboratory-scale experiments, but field studies are essential to yield more realistic results and data. To advance research on the removal of ECs by CWs, the following key recommendations for future studies have been identified:

- More CW field scale studies need to be performed for unexplored ECs like surfactants and pesticides for better applicability and to understand the specific effects of ECs.
- Research efforts should be directed towards investigating the mechanisms regarding electron transfer between different bacterial species, within the same species, and with plants during varied environmental conditions. These investigations will provide valuable insights into the intricate interactions within the wetland ecosystem and further improve our understanding of the treatment processes for ECs.
- Thoroughly examining the pollutant elimination methods of different macrophyte species and conducting extensive study on substrate variation and its associated challenges are crucial to unravel the underlying mechanisms effectively.
- The economical sustainability of these wetland systems should be explored by the policymakers as it could be one of the major concerns for the field application of CWs.
- The progress in genomic engineering is also essential to identify and improve the efficiency of microbial action in degrading emerging contaminants (ECs), ensuring the technical and financial viability of CW remediation method.
- The reliable analysis of emerging contaminants (ECs) in complex environmental faces challenges related to wastewater sample preparations, analytical techniques, and validation protocols.

The present chapter emphasized on how CWs can be utilized for the elimination of harmful ECs as a biological treatment unit for wastewater. The rising prevalence and health implications caused by ECs necessitate the establishment of a resilient system like CWs capable of effectively removing these persistent and hazardous compounds. While the adoption of hybrid systems at the field scale is gaining recognition for their effectiveness in removing ECs, conventional CW systems continue to be the prevailing approach in many developing countries. Conventional CWs can be modified by integrating chemical, biological, and other physical treatment processes like such as photo-Fenton, ozonation, filtration processes like RO, UF, and NF, etc. for enhanced performance. The hybrid systems have demonstrated notable improvements in the elimination of ECs, achieving elimination rates exceeding 95% and exhibiting enhanced suitability for treating high organic loads. Furthermore, these hybrid systems have shown to be viable techniques both at field and pilot scales for effectively removing ECs.

7.6 Conclusions

An adequate literature recommends that CWs can be efficient system for the removal of ECs such as insecticides, antibiotics, steroids, PCPPs from industrial as well as municipal wastewater. In CW treatment systems for emerging contaminants (ECs), gravel and zeolite are the two most commonly utilized substrates, while *Phragmites australis* stands out as the most widely employed plant species. The carbon content of the substrate can provide support and enhance the proliferation of microorganisms. The utilization of alternating feeding mode and intermittent aeration could potentially boost the removal of contaminants that are susceptible to aerobic biodegradation. Although this and other studies have investigated the impact of several parameter combinations on ECs removal, the practical consequences remain unclear. Using data acquired from prior studies, our primary component investigation indicated that the key factors contributing to an overall efficient removal of emerging contaminants (ECs) include the operation temperature, flow configuration, influent concentration of antibiotics and presence/species of plants, and influent concentration of antibiotics. Optimal flow configuration, appropriate plant species selection, and addressing low microbial activity at low temperatures are essential for consistently achieving efficient ECs removal. For targeted removal of specific ECs like antibiotics and pesticides, a crucial initial step involves investigating the principal elimination mechanisms of these pollutants which helps to identify the operational and design parameters required to enhance their elimination efficiency. Extensive literature indicates that biodegradation (both aerobic and anaerobic) is the prevailing removal process in CWs. Among these, aerobic biodegradation proves to be more efficient than anaerobic biodegradation, as exemplified by VSSFCW systems. Additionally, factors such as surface area, temperature, hydraulic retention time (HRT), and physicochemical parameters like effluent's pH and dissolved oxygen (DO) significantly impact the effectiveness of CWs in eradicating emerging contaminants (ECs) from wastewater.

References

Agarry SE, Oghenejoboh KM, Latinwo GK, Owabor CN (2020) Biotreatment of petroleum refinery wastewater in vertical surface-flow constructed wetland vegetated with Eichhornia crassipes: lab-scale experimental and kinetic modelling. Environ Technol 41(14):1793–1813. https://doi.org/10.1080/09593330.2018.1549106

Ahmed MB, Zhou JL, Hao Ngo H, Guo W, Thomaidis NS, Xu J (2017) Progress in the biological and chemical treatment technologies for emerging contaminant removal from wastewater: a critical review. J Hazard Mater 323:274–298. https://doi.org/10.1016/j.jhazmat.2016.04.045

Ávila C, Pedescoll A, Matamoros V, Bayona JM, García J (2010) Capacity of a horizontal subsurface flow constructed wetland system for the removal of emerging pollutants: an injection experiment. Chemosphere 81(9):1137–1142. https://doi.org/10.1016/J.CHEMOSPHERE.2010.08.006

Ávila C, García-Galán MJ, Uggetti E, Montemurro N, García-Vara M, Pérez S, García J, Postigo C (2021) Boosting pharmaceutical removal through aeration in constructed wetlands. J Hazard Mater 412. https://doi.org/10.1016/j.jhazmat.2021.125231

Awasthi SK, Kumar M, Sarsaiya S, Ahluwalia V, Chen H, Kaur G, Sirohi R, Sindhu R, Binod P, Pandey A, Rathour R, Kumar S, Singh L, Zhang Z, Taherzadeh MJ, Awasthi MK (2022) Multi-criteria research lines on livestock manure biorefinery development towards a circular economy: from the perspective of a life cycle assessment and business models strategies. J Clean Prod 341:130862. https://doi.org/10.1016/j.jclepro.2022.130862

Azfar Hussain S, Prasher SO, Patel RM (2012) Removal of ionophoric antibiotics in free water surface constructed wetlands. Ecol Eng 41:13–21. https://doi.org/10.1016/j.ecoleng.2011.12.006

Barbosa MO, Moreira NFF, Ribeiro AR, Pereira MFR, Silva AMT (2016) Occurrence and removal of organic micropollutants: an overview of the watch list of EU decision 2015/495. Water Res 94:257–279. https://doi.org/10.1016/j.watres.2016.02.047

Borikar D, Mohseni M, Jasim S (2015) Evaluations of conventional, ozone and UV/H2O2 for removal of emerging contaminants and THM-FPs. Water Qual Res J 50(2):140–151. https://doi.org/10.2166/WQRJC.2014.018

Brix H (1994) Functions of Macrophytes in constructed wetlands. Water Sci Technol 29(4):71–78. https://doi.org/10.2166/wst.1994.0160

Buerge IJ, Buser HR, Kahle M, Müller MD, Poiger T (2009) Ubiquitous occurrence of the artificial sweetener acesulfame in the aquatic environment: an ideal chemical marker of domestic wastewater in groundwater. Environ Sci Technol 43(12):4381–4385. https://doi.org/10.1021/es900126x

Button M, Cosway K, Sui J, Weber K (2019) Impacts and fate of triclosan and sulfamethoxazole in intensified re-circulating vertical flow constructed wetlands. Sci Total Environ 649:1017–1028. https://doi.org/10.1016/j.scitotenv.2018.08.395

Chen J, Deng WJ, Liu YS, Hu LX, He LY, Zhao JL, Wang TT, Ying GG (2019) Fate and removal of antibiotics and antibiotic resistance genes in hybrid constructed wetlands. Environ Pollut 249:894–903. https://doi.org/10.1016/j.envpol.2019.03.111

Cheng Y-X, Chen J, Wu D, Liu Y-S, Yang Y-Q, He L-X, Ye P, Zhao J-L, Liu S-S, Yang B, Ying G-G (2021) Highly enhanced biodegradation of pharmaceutical and personal care products in a novel tidal flow constructed wetland with baffle and plants. Water Res 193:116870. https://doi.org/10.1016/j.watres.2021.116870

Czajka CP, Londry KL (2006) Anaerobic biotransformation of estrogens. Sci Total Environ 367(2–3):932–941. https://doi.org/10.1016/j.scitotenv.2006.01.021

de Oliveira M, Atalla AA, Frihling BEF, Cavalheri PS, Migliolo L, Filho FJCM (2019) Ibuprofen and caffeine removal in vertical flow and free-floating macrophyte constructed wetlands with Heliconia rostrata and Eichornia crassipes. Chem Eng J 373:458–467. https://doi.org/10.1016/j.cej.2019.05.064

Dhangar K, Kumar M (2020) Tricks and tracks in removal of emerging contaminants from the wastewater through hybrid treatment systems: a review. Sci Total Environ 738:140320. https://doi.org/10.1016/j.scitotenv.2020.140320

Dordio AV, Carvalho AJP (2013) Organic xenobiotics removal in constructed wetlands, with emphasis on the importance of the support matrix. J Hazard Mater 252–253:272–292. https://doi.org/10.1016/j.jhazmat.2013.03.008

Dubey M, Vellanki BP, Kazmi AA (2023a) Emerging contaminants in conventional and advanced biological nutrient removal based wastewater treatment plants in India- insights into the removal processes. Sci Total Environ 894. https://doi.org/10.1016/j.scitotenv.2023.165094

Dubey M, Vellanki BP, Kazmi AA (2023b) Removal of emerging contaminants in conventional and advanced biological wastewater treatment plants in India-a comparison of treatment technologies. Environ Res 218:115012. https://doi.org/10.1016/j.envres.2022.115012

DuPoldt C, Edwards R, Garber L, Isaacs B, Lapp J (1996) A handbook of constructed wetlands: general considerations. Ecol Eng 1(1996):53. http://citeseerx.ist.psu.edu/viewdoc/download?doi=10.1.1.169.7471&rep=rep1&type=pdf%0Awww.unhabitat.org

Echeverri González AF, Zúñiga-Benítez H, Peñuela GA (2019) Removal of herbicide 2,4-D using constructed wetlands at pilot scale. Emerg Contam 5:303–307. https://doi.org/10.1016/j.emcon.2019.09.001

Ercole E, Adamo M, Lumini E, Fusconi A, Mucciarelli M (2022) Alpine constructed wetlands: a metagenomic analysis reveals microbial complementary structure. Sci Total Environ 822:153640. https://doi.org/10.1016/J.SCITOTENV.2022.153640

Estrada-Arriaga EB, Cortés-Muñoz JE, González-Herrera A, Calderón-Mólgora CG, de Lourdes Rivera-Huerta M, Ramírez-Camperos E, Montellano-Palacios L, Gelover-Santiago SL, Pérez-Castrejón S, Cardoso-Vigueros L, Martín-Domínguez A, García-Sánchez L (2016) Assessment of full-scale biological nutrient removal systems upgraded with physico-chemical processes for the removal of emerging pollutants present in wastewaters from Mexico. Sci Total Environ 571:1172–1182. https://doi.org/10.1016/j.scitotenv.2016.07.118

Fagan R, McCormack DE, Dionysiou DD, Pillai SC (2016) A review of solar and visible light active TiO2 photocatalysis for treating bacteria, cyanotoxins and contaminants of emerging concern. Mater Sci Semicond Process 42:2–14. https://doi.org/10.1016/J.MSSP.2015.07.052

Fair PA, Lee H-B, Adams J, Darling C, Pacepavicius G, Alaee M, Bossart GD, Henry N, Muir D (2009) Occurrence of triclosan in plasma of wild Atlantic bottlenose dolphins (Tursiops truncatus) and in their environment. Environ Pollut 157(8–9):2248–2254. https://doi.org/10.1016/j.envpol.2009.04.002

Faleye AC, Adegoke AA, Ramluckan K, Fick J, Bux F, Stenström TA (2019) Concentration and reduction of antibiotic residues in selected wastewater treatment plants and receiving waterbodies in Durban, South Africa. Sci Total Environ 678:10–20. https://doi.org/10.1016/j.scitotenv.2019.04.410

Farooq AJ, Chamberlain M, Poonja A, Mumford KG, Wallace S, Weber KP (2023) Peaks, pores, and dragon eggs: uncovering and quantifying the heterogeneity of treatment wetland biofilm matrices. Sci Total Environ 855. https://doi.org/10.1016/J.SCITOTENV.2022.158857

Fernandes JP, Almeida CMR, Pereira AC, Ribeiro IL, Reis I, Carvalho P, Basto MCP, Mucha AP (2015) Microbial community dynamics associated with veterinary antibiotics removal in constructed wetlands microcosms. Bioresour Technol 182:26–33. https://doi.org/10.1016/j.biortech.2015.01.096

Gahlawat S, Gauba P (2016) Phytoremediation of aspirin and tetracycline by *Brassica juncea*. Int J Phytoremediation 18(9):929–935. https://doi.org/10.1080/15226514.2015.1131230

Galus M, Jeyaranjaan J, Smith E, Li H, Metcalfe C, Wilson JY (2013) Chronic effects of exposure to a pharmaceutical mixture and municipal wastewater in zebrafish. Aquat Toxicol 132–133:212–222. https://doi.org/10.1016/j.aquatox.2012.12.016

Gani KM, Hlongwa N, Abunama T, Kumari S, Bux F (2020) Emerging contaminants in south African water environment- a critical review of their occurrence, sources and ecotoxicological risks. Chemosphere 269:128737. https://doi.org/10.1016/j.chemosphere.2020.128737

Gracia-Lor E, Sancho JV, Serrano R, Hernández F (2012) Occurrence and removal of pharmaceuticals in wastewater treatment plants at the Spanish Mediterranean area of Valencia. Chemosphere 87(5):453–462. https://doi.org/10.1016/j.chemosphere.2011.12.025

Guedes-Alonso R, Montesdeoca-Esponda S, Herrera-Melián JA, Rodríguez-Rodríguez R, Ojeda-González Z, Landívar-Andrade V, Sosa-Ferrera Z, Santana-Rodríguez JJ (2020) Pharmaceutical and personal care product residues in a macrophyte pond-constructed wetland treating wastewater from a university campus: presence, removal and ecological risk assessment. Sci Total Environ 703. https://doi.org/10.1016/j.scitotenv.2019.135596

Harrabi M, Alexandrino DAM, Aloulou F, Elleuch B, Liu B, Jia Z, Marisa C, Almeida R, Mucha AP, Carvalho MF (2019) Biodegradation of oxytetracycline and enrofloxacin by autochthonous microbial communities from estuarine sediments. Sci Total Environ 648:962–972. https://doi.org/10.1016/j.scitotenv.2018.08.193

Hazra M, Joshi H, Williams JB, Watts JEM (2022) Antibiotics and antibiotic resistant bacteria/genes in urban wastewater: a comparison of their fate in conventional treatment systems and constructed wetlands. Chemosphere 303. https://doi.org/10.1016/j.chemosphere.2022.135148
He Y, Zhang L, Jiang L, Wagner T, Sutton NB, Ji R, Langenhoff AAM (2021) Improving removal of antibiotics in constructed wetland treatment systems based on key design and operational parameters: a review. J Hazard Mater 407:304–3894. https://doi.org/10.1016/j.jhazmat.2020.124386
Kasonga TK, Coetzee MAA, Kamika I, Ngole-Jeme VM, Benteke Momba MN (2021) Endocrine-disruptive chemicals as contaminants of emerging concern in wastewater and surface water: a review. J Environ Manag 277:111485. https://doi.org/10.1016/j.jenvman.2020.111485
Kasprzyk-Hordern B, Dinsdale RM, Guwy AJ (2009) The removal of pharmaceuticals, personal care products, endocrine disruptors and illicit drugs during wastewater treatment and its impact on the quality of receiving waters. Water Res 43(2):363–380. https://doi.org/10.1016/j.watres.2008.10.047
Katibi KK, Yunos KF, Man HC, Aris AZ, Nor MZM, Azis RS, Umar AM (2021) Contemporary techniques for remediating endocrine-disrupting compounds in various water sources: advances in treatment methods and their limitations. Polymers 13(19):1–46. https://doi.org/10.3390/polym13193229
Kumar AK, Chiranjeevi P, Mohanakrishna G, Mohan SV (2011) Natural attenuation of endocrine-disrupting estrogens in an ecologically engineered treatment system (EETS) designed with floating, submerged and emergent macrophytes. Ecol Eng 37(10):1555–1562. https://doi.org/10.1016/j.ecoleng.2011.06.009
Kumar M, Sridharan S, Sawarkar AD, Shakeel A, Anerao P, Mannina G, Sharma P, Pandey A (2023) Current research trends on emerging contaminants pharmaceutical and personal care products (PPCPs): a comprehensive review. Sci Total Environ 859. https://doi.org/10.1016/j.scitotenv.2022.160031
Li Y, Zhu G, Ng WJ, Tan SK (2014) A review on removing pharmaceutical contaminants from wastewater by constructed wetlands: design, performance and mechanism. Sci Total Environ 468–469:908–932. https://doi.org/10.1016/j.scitotenv.2013.09.018
Li Y, Su J, Ali A, Hao Z, Li M, Yang W, Wang Z (2022) Simultaneous removal of nitrate and heavy metals in a biofilm reactor filled with modified biochar. Sci Total Environ 851(March). https://doi.org/10.1016/j.scitotenv.2022.158175
Lin T, Yu S, Chen W (2016) Occurrence, removal and risk assessment of pharmaceutical and personal care products (PPCPs) in an advanced drinking water treatment plant (ADWTP) around Taihu Lake in China. Chemosphere 152:1–9. https://doi.org/10.1016/j.chemosphere.2016.02.109
Lin L, Jiang W, Xu P (2017) Comparative study on pharmaceuticals adsorption in reclaimed water desalination concentrate using biochar: impact of salts and organic matter. Sci Total Environ 601–602:857–864. https://doi.org/10.1016/j.scitotenv.2017.05.203
Liu L, Li J, Fan H, Huang X, Wei L, Liu C (2019) Fate of antibiotics from swine wastewater in constructed wetlands with different flow configurations. Int Biodeterior Biodegrad 140:119–125. https://doi.org/10.1016/j.ibiod.2019.04.002
Liu X, Liang C, Liu X, Lu S, Xi B (2020) Intensified pharmaceutical and personal care products removal in an electrolysis-integrated tidal flow constructed wetland. Chem Eng J 394. https://doi.org/10.1016/j.cej.2020.124860
Lv X, Xiao S, Zhang G, Jiang P, Tang F (2016) Occurrence and removal of phenolic endocrine disrupting chemicals in the water treatment processes. Sci Rep 6. https://doi.org/10.1038/srep22860
Lv M, Zhang D, Niu X, Ma J, Lin Z, Fu M (2022) Insights into the fate of antibiotics in constructed wetland systems: removal performance and mechanisms. J Environ Manag 321:116028. https://doi.org/10.1016/j.jenvman.2022.116028
Majumder A, Gupta AK, Ghosal PS, Varma M (2021) A review on hospital wastewater treatment: a special emphasis on occurrence and removal of pharmaceutically active compounds,

resistant microorganisms, and SARS-CoV-2. J Environ Chem Eng 9(2):104812. https://doi.org/10.1016/j.jece.2020.104812

Mandaric L, Diamantini E, Stella E, Cano-Paoli K, Valle-Sistac J, Molins-Delgado D, Bellin A, Chiogna G, Majone B, Diaz-Cruz MS, Sabater S, Barcelo D, Petrovic M (2017) Contamination sources and distribution patterns of pharmaceuticals and personal care products in alpine rivers strongly affected by tourism. Sci Total Environ 590–591:484–494. https://doi.org/10.1016/j.scitotenv.2017.02.185

Manolaki P, Mouridsen MB, Nielsen E, Olesen A, Jensen SM, Lauridsen TL, Baattrup-Pedersen A, Sorrell BK, Riis T (2020) A comparison of nutrient uptake efficiency and growth rate between different macrophyte growth forms. J Environ Manag 274:111181. https://doi.org/10.1016/j.jenvman.2020.111181

McAvoy DC, Schatowitz B, Jacob M, Hauk A, Eckhoff WS (2002) Measurement of triclosan in wastewater treatment systems. Environ Toxicol Chem 21(7):1323–1329. https://doi.org/10.1002/ETC.5620210701

Metcalfe CD, Koenig BG, Bennie DT, Servos M, Ternes TA, Hirsch R (2003) Occurrence of neutral and acidic drugs in the effluents of Canadian sewage treatment plants. Environ Toxicol Chem 22(12):2872. https://doi.org/10.1897/02-469

Mohapatra S, Huang CH, Mukherji S, Padhye LP (2016) Occurrence and fate of pharmaceuticals in WWTPs in India and comparison with a similar study in the United States. Chemosphere 159:526–535. https://doi.org/10.1016/j.chemosphere.2016.06.047

Molle P, Liénard A, Boutin C, Merlin G, Iwema A (2005) How to treat raw sewage with constructed wetlands: an overview of the French systems. Water Sci Technol 51(9):11–21. https://doi.org/10.2166/wst.2005.0277

Murdoch RW, Hay AG (2005) Formation of catechols via removal of acid side chains from ibuprofen and related aromatic acids. Appl Environ Microbiol 71(10):6121–6125. https://doi.org/10.1128/AEM.71.10.6121-6125.2005/ASSET/4955614F-C1B9-4DAE-AA53-84759ECC2A10/ASSETS/GRAPHIC/ZAM0100559420003.JPEG

Mutiyar PK, Mittal AK (2014) Occurrences and fate of selected human antibiotics in influents and effluents of sewage treatment plant and effluent-receiving river Yamuna in Delhi (India). Environ Monit Assess 186(1):541–557. https://doi.org/10.1007/s10661-013-3398-6

Olasupo A, Suah FBM (2021) Recent advances in the removal of pharmaceuticals and endocrine-disrupting compounds in the aquatic system: a case of polymer inclusion membranes. J Hazard Mater 406(October 2020):124317. https://doi.org/10.1016/j.jhazmat.2020.124317

Oluwole AO, Omotola EO, Olatunji OS (2020) Pharmaceuticals and personal care products in water and wastewater: a review of treatment processes and use of photocatalyst immobilized on functionalized carbon in AOP degradation. BMC Chem 14(1):1–29. https://doi.org/10.1186/s13065-020-00714-1

Overton OC, Olson LH, Majumder SD, Shwiyyat H, Foltz ME, Nairn RW (2023) Wetland removal mechanisms for emerging contaminants. Land 12(2). https://doi.org/10.3390/LAND12020472

Papaevangelou VA, Gikas GD, Tsihrintzis VA, Antonopoulou M, Konstantinou IK (2016) Removal of endocrine disrupting chemicals in HSF and VF pilot-scale constructed wetlands. Chem Eng J 294:146–156. https://doi.org/10.1016/j.cej.2016.02.103

Parida VK, Saidulu D, Majumder A, Srivastava A, Gupta B, Gupta AK (2021) Emerging contaminants in wastewater: a critical review on occurrence, existing legislations, risk assessment, and sustainable treatment alternatives. J Environ Chem Eng 9(5). https://doi.org/10.1016/j.jece.2021.105966

Patel M, Kumar R, Kishor K, Mlsna T, Pittman CU, Mohan D (2019) Pharmaceuticals of emerging concern in aquatic systems: chemistry, occurrence, effects, and removal methods. Chem Rev 119(6):3510–3673. https://doi.org/10.1021/acs.chemrev.8b00299

Polianciuc SI, Gurzău AE, Kiss B, Ştefan MG, Loghin F (2020) Antibiotics in the environment: causes and consequences. Med Pharm Rep 93(3):231–240. https://doi.org/10.15386/mpr-1742

Ramprasad C, Smith CS, Memon FA, Philip L (2017) Removal of chemical and microbial contaminants from greywater using a novel constructed wetland: GROW. Ecol Eng 106:55–65. https://doi.org/10.1016/J.ECOLENG.2017.05.022

Ruppelt JP, Pinnekamp J, Tondera K (2020) Elimination of micropollutants in four test-scale constructed wetlands treating combined sewer overflow: influence of filtration layer height and feeding regime. Water Res 169. https://doi.org/10.1016/j.watres.2019.115214

Saidulu D, Gupta B, Gupta AK, Ghosal PS (2021) A review on occurrences, eco-toxic effects, and remediation of emerging contaminants from wastewater: special emphasis on biological treatment based hybrid systems. J Environ Chem Eng 9(4). https://doi.org/10.1016/J.JECE.2021.105282

Sánchez M, Ruiz I, Soto M (2022) The potential of constructed wetland systems and Photodegradation processes for the removal of emerging contaminants—A review. Environments—MDPI 9(9). https://doi.org/10.3390/ENVIRONMENTS9090116

Sarmah AK, Northcott GL (2008) Laboratory degradation studies of four endocrine disruptors in two environmental media. Environ Toxicol Chem 27(4):819–827. https://doi.org/10.1897/07-231.1

Sgroi M, Pelissari C, Roccaro P, Sezerino PH, García J, Vagliasindi FGA, Ávila C (2018) Removal of organic carbon, nitrogen, emerging contaminants and fluorescing organic matter in different constructed wetland configurations. Chem Eng J 332:619–627. https://doi.org/10.1016/j.cej.2017.09.122

Sharma R, Malaviya P (2022) Constructed wetlands for textile wastewater remediation: a review on concept, pollutant removal mechanisms, and integrated technologies for efficiency enhancement. Chemosphere 290:133358. https://doi.org/10.1016/j.chemosphere.2021.133358

Sharma VK, Johnson N, Cizmas L, McDonald TJ, Kim H (2016) A review of the influence of treatment strategies on antibiotic resistant bacteria and antibiotic resistance genes. Chemosphere 150:702–714. https://doi.org/10.1016/j.chemosphere.2015.12.084

Song X, Ding Y, Wang Y, Wang W, Wang G, Zhou B (2015) Comparative study of nitrogen removal and bio-film clogging for three filter media packing strategies in vertical flow constructed wetlands. Ecol Eng 74:1–7. https://doi.org/10.1016/J.ECOLENG.2014.08.008

Song HL, Li H, Zhang S, Yang YL, Zhang LM, Xu H, Yang XL (2018) Fate of sulfadiazine and its corresponding resistance genes in up-flow microbial fuel cell coupled constructed wetlands: effects of circuit operation mode and hydraulic retention time. Chem Eng J 350:920–929. https://doi.org/10.1016/j.cej.2018.06.035

Starling MCVM, Amorim CC, Leão MMD (2019) Occurrence, control and fate of contaminants of emerging concern in environmental compartments in Brazil. J Hazard Mater 372:17–36. https://doi.org/10.1016/j.jhazmat.2018.04.043

Subedi B, Kannan K (2014) Fate of artificial sweeteners in wastewater treatment plants in New York state, U.S.A. Environ Sci Technol 48(23):13668–13674. https://doi.org/10.1021/es504769c

Subedi B, Balakrishna K, Sinha RK, Yamashita N, Balasubramanian VG, Kannan K (2015) Mass loading and removal of pharmaceuticals and personal care products, including psychoactive and illicit drugs and artificial sweeteners, in five sewage treatment plants in India. J Environ Chem Eng 3(Part 4A):2882–2891. https://doi.org/10.1016/j.jece.2015.09.031

Surana D, Gupta J, Sharma S, Kumar S, Ghosh P (2022) A review on advances in removal of endocrine disrupting compounds from aquatic matrices: future perspectives on utilization of agri-waste based adsorbents. Sci Total Environ 826. https://doi.org/10.1016/j.scitotenv.2022.154129

Talib A, Randhir TO (2017) Managing emerging contaminants in watersheds: need for comprehensive, systems-based strategies. Sustain Water Qual Ecol 9–10:1–8. https://doi.org/10.1016/J.SWAQE.2016.05.002

Tang X, Yang Y, Tao R, Chen P, Dai Y, Jin C, Feng X (2016) Fate of mixed pesticides in an integrated recirculating constructed wetland (IRCW). Sci Total Environ 571:935–942. https://doi.org/10.1016/j.scitotenv.2016.07.079

Tran NH, Reinhard M, Gin KY-H (2018) Occurrence and fate of emerging contaminants in municipal wastewater treatment plants from different geographical regions-a review. Water Res 133:182–207. https://doi.org/10.1016/j.watres.2017.12.029
Varsha M, Senthil Kumar P, Rathi BS (2022) A review on recent trends in the removal of emerging contaminants from aquatic environment using low-cost adsorbents. Chemosphere 287:132270. https://doi.org/10.1016/j.chemosphere.2021.132270
Venditti S, Brunhoferova H, Hansen J (2022) Behaviour of 27 selected emerging contaminants in vertical flow constructed wetlands as post-treatment for municipal wastewater. Sci Total Environ 819. https://doi.org/10.1016/j.scitotenv.2022.153234
Verlicchi P, Galletti A, Petrovic M, Barceló D, Al Aukidy M, Zambello E (2013) Removal of selected pharmaceuticals from domestic wastewater in an activated sludge system followed by a horizontal subsurface flow bed—analysis of their respective contributions. Sci Total Environ 454–455:411–425. https://doi.org/10.1016/j.scitotenv.2013.03.044
Verma R, Suthar S (2018) Performance assessment of horizontal and vertical surface flow constructed wetland system in wastewater treatment using multivariate principal component analysis. Ecol Eng 116(July 2017):121–126. https://doi.org/10.1016/j.ecoleng.2018.02.022
Vymazal J (2011) Constructed wetlands for wastewater treatment: five decades of experience. Environ Sci Technol 45(1):61–69. https://doi.org/10.1021/es101403q
Vymazal J (2013) Emergent plants used in free water surface constructed wetlands: a review. Ecol Eng 61:582–592. https://doi.org/10.1016/J.ECOLENG.2013.06.023
Vymazal J (2018) Constructed wetlands for wastewater treatment. Encyclopedia Ecol 45(1):14–21. https://doi.org/10.1016/B978-0-12-409548-9.11238-2
Vymazal J (2019) Constructed wetlands for wastewater treatment. In: Encyclopedia of ecology. Elsevier, pp 14–21. https://doi.org/10.1016/B978-0-12-409548-9.11238-2
Vymazal J, Březinová T (2015) The use of constructed wetlands for removal of pesticides from agricultural runoff and drainage: a review. Environ Int 75:11–20. https://doi.org/10.1016/j.envint.2014.10.026
Vymazal J, Březinová T, Koželuh M (2015) Occurrence and removal of estrogens, progesterone and testosterone in three constructed wetlands treating municipal sewage in The Czech Republic. Sci Total Environ 536:625–631. https://doi.org/10.1016/j.scitotenv.2015.07.077
Vystavna Y, Frkova Z, Marchand L, Vergeles Y, Stolberg F (2017) Removal efficiency of pharmaceuticals in a full scale constructed wetland in East Ukraine. Ecol Eng 108:50–58. https://doi.org/10.1016/j.ecoleng.2017.08.009
Wang D, Sui Q, Lu SG, Zhao WT, Qiu ZF, Miao ZW, Yu G (2014) Occurrence and removal of six pharmaceuticals and personal care products in a wastewater treatment plant employing anaerobic/anoxic/aerobic and UV processes in Shanghai, China. Environ Sci Pollut Res 21(6):4276–4285. https://doi.org/10.1007/S11356-013-2363-9/METRICS
Wang Y, Cai Z, Sheng S, Pan F, Chen F, Fu J (2020a) Comprehensive evaluation of substrate materials for contaminants removal in constructed wetlands. Sci Total Environ 701:134736. https://doi.org/10.1016/j.scitotenv.2019.134736
Wang H, Xu J, Sheng L (2020b) Purification mechanism of sewage from constructed wetlands with zeolite substrates: a review. J Clean Prod 258:120760. https://doi.org/10.1016/j.jclepro.2020.120760
Yadav AK, Dash P, Mohanty A, Abbassi R, Mishra BK (2012) Performance assessment of innovative constructed wetland-microbial fuel cell for electricity production and dye removal. Ecol Eng 47:126–131. https://doi.org/10.1016/j.ecoleng.2012.06.029
Yang Y, Zhao Y, Liu R, Morgan D (2018) Global development of various emerged substrates utilized in constructed wetlands. Bioresour Technol 261:441–452. https://doi.org/10.1016/j.biortech.2018.03.085
Yin T, Chen H, Reinhard M, Yi X, He Y, Gin KYH (2017) Perfluoroalkyl and polyfluoroalkyl substances removal in a full-scale tropical constructed wetland system treating landfill leachate. Water Res 125:418–426. https://doi.org/10.1016/j.watres.2017.08.071

Yuan Y, Yang B, Wang H, Lai X, Li F, Salam MMA, Pan F, Zhao Y (2020) The simultaneous antibiotics and nitrogen removal in vertical flow constructed wetlands: effects of substrates and responses of microbial functions. Bioresour Technol 310. https://doi.org/10.1016/j.biortech.2020.123419

Zhang H, Wang XC, Zheng Y, Dzakpasu M (2023a) Removal of pharmaceutical active compounds in wastewater by constructed wetlands: performance and mechanisms. J Environ Manag 325:116478. https://doi.org/10.1016/j.jenvman.2022.116478

Zhang Y, Zhao Z, Xu H, Wang L, Liu R, Jia X (2023b) Fate of antibiotic resistance genes and bacteria in a coupled water-processing system with wastewater treatment plants and constructed wetlands in coastal eco-industrial parks. Ecotoxicol Environ Saf 252. https://doi.org/10.1016/j.ecoenv.2023.114606

Zhao X, Bai S, Li C, Yang J, Ma F (2019) Bioaugmentation of atrazine removal in constructed wetland: performance, microbial dynamics, and environmental impacts. Bioresour Technol 289. https://doi.org/10.1016/j.biortech.2019.121618

Zheng Y, Liu Y, Qu M, Hao M, Yang D, Yang Q, Wang XC, Dzakpasu M (2021) Fate of an antibiotic and its effects on nitrogen transformation functional bacteria in integrated vertical flow constructed wetlands. Chem Eng J 417:129272. https://doi.org/10.1016/j.cej.2021.129272

Chapter 8
Landfill Leachate Treatment Using Sustainable Technology: Current Practices and Perspectives

Sashi Yadav, Monika Dhawan, Rajesh Dhankhar, Abhilasha Yadav, Jyoti Rani, Ravina Rohilla, and Sunil Kumar Chhikara

Abstract Landfill leachate, a complex and highly contaminated liquid, poses significant environmental challenges. The need for sustainable and cost-effective treatment technologies has become paramount to mitigate the adverse impacts of leachate on water resources and ecosystems. By utilizing natural processes, minimizing energy consumption, promoting resource recovery, and supporting biodiversity, constructed wetlands offer a sustainable and environmentally friendly approach for landfill leachate treatment. Their cost-effectiveness, long-term performance, and compatibility with ecological systems make them an attractive option for managing the environmental impacts associated with landfill operations. Key aspects discussed include the selection of appropriate wetland vegetation, the role of microbial communities, and the importance of hydraulic control in optimizing treatment efficiency. Nutrient-rich plant biomass can be harvested and utilized for various purposes such as fertilizer production or renewable energy generation, contributing to a circular economy approach. Additionally, the aesthetic and community integration benefits of constructed wetlands are discussed, highlighting their potential for creating sustainable and socially inclusive spaces. The chapter concludes with perspective or future advancements, including hybrid systems, plant selection, and genetic modification, as well as the importance of optimizing treatment processes and evaluating the resilience of constructed wetlands in the face of climate change. It serves as a precious resource for the practitioner, researcher, and decision-maker interested in understanding the prospective and benefits of constructed wetland as a sustainable solution for the treatment of landfill leachate. It underscores the importance of

S. Yadav · M. Dhawan · R. Dhankhar · S. K. Chhikara (✉)
Maharshi Dayanand University, Rohtak, Haryana, India
e-mail: sunilchhikara.uiet@mdurohtak.ac.in

A. Yadav · J. Rani
Deenbandhu Chhotu Ram University of Science and Technology, Murthal, Haryana, India

R. Rohilla
Baba Mastnath University, Rohtak, Haryana, India

S. S. Kumar, V. Kumar (eds.), *Nature-Based Solutions for Decentralised Liquid Waste Management*, Environmental Challenges and Solutions,
https://doi.org/10.1007/978-3-032-08362-3_8

adopting environmentally sound approaches to address the challenges posed by landfill leachate and emphasizes the role of constructed wetlands in achieving sustainable waste management practices.

Keywords Constructed wetland · Landfill leachate · Sustainable · Vegetation · Pollution

8.1 Introduction

Municipal solid waste (MSW) generation has significantly increased due to increasing population and growing luxurious lifestyles. Urban area produce industrial and solid waste is ultimately disposed of in landfills (Fig. 8.1) (Akinbile et al. 2012). Municipal solid waste (MSW) is drastically expanding as the population and economy both increase (Zhao et al. 2021). By 2025, global municipal solid waste (MSW) generation is projected to exceed 2.2 billion metric tonnes annually, representing a major environmental and management challenge for sustainable waste treatment worldwide (Wijekoon et al. 2022). The most typical treatment is landfilling, which is a low cost, established processing procedure with great efficiency (Lin et al. 2020). Such accumulated waste causes a number of interconnected chemical reactions that result in landfill leachate (LL) that contains both macro- and micropollutants (Kumar and Bishnoi 2017; Waara and Wojciechowska 2019). LL develops with significant toxicity, nutritional imbalance, and complex composition over prolonged times of biochemical decomposition and rainwater percolation (Kurniawan and Lo 2009). Untreated landfill leachate discharge, therefore, poses a serious threat to the ecological balance of receiving water bodies and the ecosystem, and requires appropriate treatment before release. Landfill leachate contains high amount of organic matter with biodegradability and nonbiodegradability (recalcitrant) fraction, which include humic substance, inorganic salt, heavy metal, and microorganism (Zakaria

Fig. 8.1 Landfill site Bandhwari, Gurugram-Haryana (India)

and Aziz 2018). The landfill leachate also contains refractory materials all throughout. Commonly, landfill leachate is interpreted by conventional parameters including COD (chemical oxygen demand) (Baettker et al. 2020), fulvic acid (FA), humic acid (HA) (Iskander et al. 2019), total dissolved solid (TDS), total organic carbon (TOC), suspended solids, pH, ammoniacal nitrogen (NH_4^+-N) (Wdowczyk and Szymańska-Pulikowska 2021), biochemical oxygen demand (BOD) (Ancic et al. 2020), and heavy metal concentrations (Table 8.1). The presence of several contaminants in landfill leachate, such as XOCs (Xenobiotic Organic Compounds) and heavy metals, has detrimental consequences on the ecosystem and food chains, leading to the development of cancer, acute toxicity, and genotoxicity in living organisms.

In order to address the disposal limitations, many conventional treatment approaches are inefficient or need higher expenditures than the overall expenses of operating a landfill (Alver and Altaş 2017; Caroline Baettker et al. 2020). Biotreatment (e.g., activated sludge and constructed wetlands), physical-chemical processing (e.g., adsorption (Rohilla et al. 2022; Ravina and Chhikara 2023) and membrane) and chemical processing (e.g., Fenton, precipitation) were all effective LL treatment methods (Teng et al. 2021). In recent years, there has been a growing recognition of the need for sustainable and eco-friendly method for the landfill leachate treatment. Constructed wetlands have emerged as a promising solution that harnesses the natural processes of wetland ecosystems to effectively and sustainably treat landfill leachate. These engineered system mimic the functions of natural wetland, utilizing the power of vegetation, soil, and microbial communities for the removal of pollutants and improve water quality. LL had been treated using CWs, which are frequently described to as lower-energy-consumption, effective treatment technologies (Bakhshoodeh et al. 2020; Saeed et al. 2021). Free water surface flow constructed wetland (FWSCW) and subsurface flow constructed wetland are the two types of constructed wetland. French vertical flow (FVF) CW, horizontal flow (HF) CW, and hybrid type CW are the four categories used to classify subsurface flow (Parde et al. 2021). This objective of book chapter is to examine the use of

Table 8.1 Characteristics of landfill leachate

Location	pH	EC	TDS	BOD_5	COD	HA	FA	Ammonia Nitrogen	Chloride	References
Curitiba, Brazil	8.4	35,160		2500	7673	–	–	2885	2600	Baettker et al. (2020)
Virginia, USA	8.46	3065		860	10,750	765	740	2900	4940	Iskander et al. (2019)
Legnica, Poland	8.9	10,197	7026		2930	–	–	401.26	2132.3	Wdowczyk and Szymańska-Pulikowska (2021)
Mraclin, Croatia	8.09	–	–	288	1653	–	–	–	460	Ancic et al. (2020)

constructed wetlands as a sustainable method of treating landfill leachate. In the context of leachate treatment, the chapter seeks to provide an extensive understanding of the design principles, operational issues, performance evaluation, and sustainability aspects of constructed wetlands (Stanković 2017).

Constructed wetlands are categorized as one of the biological techniques that use phytoremediation (Kumar et al. 2017) to treat polluted liquids. They are described as designed systems that efficiently synergistically extract, convert, and degrade contaminants from wastewater through natural process using soil, vegetation, and microorganism. In CWs, phytofiltration, phytostabilization, phytoextraction, phytodesalination, phytovolatilization, and phytodegradation are the main mechanisms for pollutant removal (Yalçuk and Ugurlu 2020).

Constructed wetlands have been used for treatment of a wider range of wastewater, including mine drainage (Chen et al. 2021; Nguyen et al. 2022), stormwater runoff (Daukas et al. 2020; Sacco et al. 2021), domestic wastewater (Allen et al. 2022), industrial effluents (Sehar and Nasser 2019; Gholipour et al. 2020), agricultural drainage (Braschi et al. 2022; Chen et al. 2022) as well as landfill leachate (Saeed et al. 2021; Saeed et al. 2022a). Agricultural wastewater containing persistent organic pollutants and fertilizer enters in the food chain and poses a severe threat to human health, but wetland plant species that can bind these chemicals and safely deal with the problem (Ngweme et al. 2020; Parde et al. 2021).

8.2 Need for the Sustainable Solutions

The field of ecological engineering advanced during the twentieth century. In developed nations, the landfill leachate treatment has been done by conventional centralized facilities for almost a century. The massive installations of these conventional units, which have an average lifespan of 25–30 years and require further investment to create new ones to replace ageing ones, is energy-intensive. However, developing nations are often cannot afford to construct such massive centralized facilities or the technologies and having no idea how to operate them. Alternative treatment approaches are therefore required in both the developed and developing nations. Effective performance, cost-effectiveness, and of course the recently added sustainability component must all be incorporated into the techniques. The use of constructed wetlands and other natural remedies may be able to meet these requirements. The current state of the technology is evidence of its excellent performance. Since the majority of the resources used are renewable, they don't rely on expensive and energy-intensive processes, and no synthetic materials are needed. As an ecological alternative, they typically require natural ones, such as vegetation and substrate media (Solano et al. 2004).

Additionally, CWs recognize a number of challenges as a result of the world's increasingly stringent water safety regulations for wastewater management and for landfill leachate discharge in natural bodies. There is need of more innovation and research in the field of constructed wetland (Fig. 8.2).

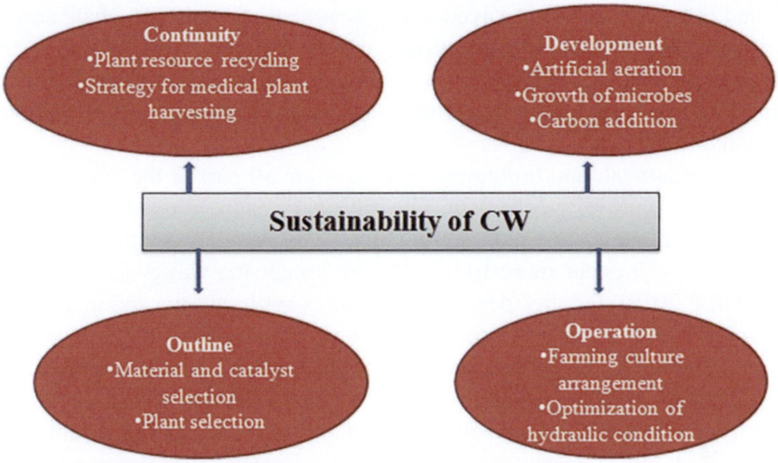

Fig. 8.2 A summary of the presumptions used in order to develop CWs sustainably

The choice of the plants and substrates media shows its significance for landfill leachate treatment. For the effective treatment and removal of contaminants various types of vegetation, substrate media and phytoplankton are required in order to boost the pollutant absorption, particularly for emerging contaminants; persistent organic compounds etc. and a strong emphasis should be made on the careful selection of macrophyte types, and the supply of oxygen and carbon compounds also enhanced. The operational and design characteristics are the primary component that defines the treatment's highest efficiency on ecological, hydrological, and operational circumstances. Therefore, these conditions enhancement required to improve the treatment efficiency, further research must be required in this field. Focus should also be placed on studying the fundamental process and mechanism underlying more effective pollution elimination (Zhang et al. 2021a).

The advancements in research and practical application in conventional CWs, many strategies developed for enhancing effectiveness of landfill leachate treatment. Artificial ventilation, phase feeding, extrinsic carbon incorporation, microbial community, vegetation, the combination of various media substrate, and hybrid CWs are typical examples of these advancements (Eslamian et al. 2019). Additionally, society always finds the idea of using vegetation as a filter for wastewater to be unique and intriguing, which can increase their social acceptability.

Natural systems offer decentralized treatment services, which can be considered a different approach to treating landfill leachate and sewage at or near the source. The idea of sustainable development is further satisfied by the fact that it produces the same outcome as by CWs technology such as more cost-effective, sustainably produced, and energy-efficiently. Environmental effect, resource conservation, climate change resilience, social acceptability and integration, economic viability, regulatory compliance, research, and innovation are a few factors underscoring the need for sustainability in constructed wetlands. To address the environmental, social, and economic issues involved in the treatment of landfill leachate and industrial wastewater, sustainable solutions in constructed wetlands are crucial. Constructed wetlands can offer long-term advantages while reducing their ecological impact and assisting in the transition to a more sustainable future by implementing sustainable practices.

Physical, chemical, and biological processes are all part of the constructed wetland's pollutant removal mechanism. The physical treatment results in the elimination of contaminants by sedimenting the wastewater's suspended particles. Sedimentation will occur more frequently the longer the wastewater remains suspended (DBT 2019). Gravity-dependent particles settle via the sedimentation process, and the particles are settle down at the tank bottom. The coliform bacteria are also eliminated during the sedimentation process, which also reduces organic debris. Due to the high adsorption capability of iron-rich cupola slag, adsorption is a crucial method for the removal of phosphorus, which is present in significant amounts (Meshram et al. 2023). Heavy metals are also eliminated by adsorption mechanism (Lin et al. 2023; Maftouh et al. 2023). Due to efficient adsorption, phosphorus removal is significant when zeolite media are utilized (Gao et al. 2020).

Constructed wetland media is beneficial for the accumulation of organic matter, sulphate, phosphorus, and arsenate as well as for the removal of infections (Dotro et al. 2015; Kumar and Singh 2019; Thathong et al. 2019; Ge et al. 2020; Gupta et al. 2020). Various types of media, including sugar bagasse (Alam et al. 2023; Rane et al. 2023), iron powder (Ma et al. 2023), marble chips (Dinakar et al. 2020; Prodanovic et al. 2023), Sylhet sand (Parde et al. 2021; Patil et al. 2022), soil (Jia et al. 2020), biochar made from rice husk (Jiang et al. 2022), rice straw (Wang et al. 2020), corn straw (Wang et al. 2023), fruit peels (Zhang et al. 2021b), coco-peat (Alam et al. 2023), cupola slag (Meshram et al. 2023), stones, recycled bricks (Saeed et al. 2022b), sand (Achak et al. 2019; Gupta et al. 2021), gravels (Cui et al. 2023), coal, fly ash (Gao et al. 2020), sawdust (Fahim et al. 2019), zero-valent iron (Wu et al. 2022; Kong et al. 2023), etc., can be used to improve the functioning of constructed wetland.

8.3 Types of Constructed Wetland

Free Water Surface Constructed Wetland A constructed wetland with a free water surface similar to the natural wetland where wastewater flows over the surface (Fig. 8.3) (Parde et al. 2021). According to Kadlec and Knight (1996), free water surface CWs are small (5–40 cm) basins covered with wetland macrophytes. There is no special necessity for the soil quality; the main role of soil is just to sustain the vegetation growth. The free water surface CW can use a variety of plants, including floating plants (*Eichornia*, *Lemna*), submerged plant (*Elodea*, *Potamogeon*), and emergent plants (*Typha*, *Phragmites*, *Scirpus*) (DBT 2019). The water column and the bottom layer of decomposing litter are treated (Bakhshoodeh et al. 2020). It has mean elimination efficiency for trace metal (45% for copper, 53% for iron, 52% for

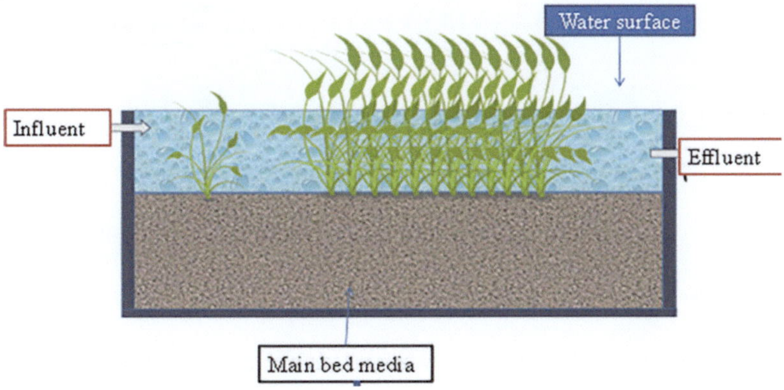

Fig. 8.3 Free water surface constructed wetland

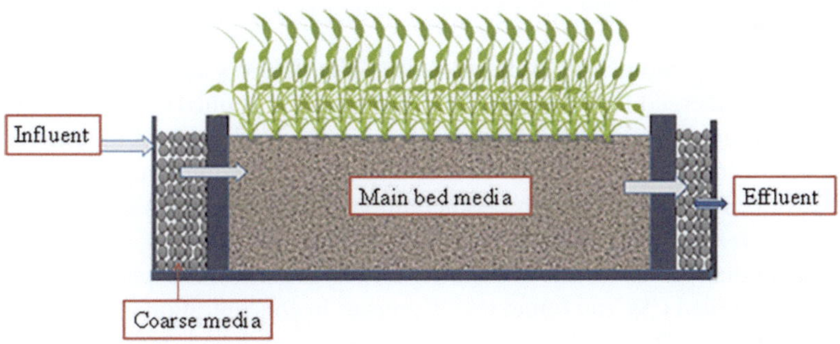

Fig. 8.4 Horizontal flow constructed wetland

zinc, and 52% lead), 50–60% for COD and BOD, 70–80% for total suspended solid, and nitrogen 50–65% (El-Sheikh et al. 2010; Parde et al. 2021). Waara and Wojciechowska (2019) studied that the removal of total phosphate and total nitrate was 98% and 89% from landfill leachate.

Horizontal Flow Constructed Wetland (HFCW) A horizontal flow CW is frequently referred to as a reed bed system (Fig. 8.4). The constructed wetland's bed has horizontal wastewater flow. The HFCW required land area is 5–10 m^2/PE (PE—population equivalent) (DBT 2019). Both aerobic and anaerobic conditions are introduced to the effluent. In horizontal flow CW, anaerobic conditions lead to the degradation of organic materials while aerobic conditions occur in the root zone (Vymazal 2005; DBT 2019). The HFCW is effective at removing pollutants such as COD, BOD, phosphate, ammoniacal nitrogen, and TSS (Jamwal et al. 2021; Angassa et al. 2022). Witthayaphirom et al. (2020a) treated landfill leachate for organic micropollutants and observed highest removal of 73% for hydrophilic compounds (such as DBP and DEP) followed by moderate hydrophobic compounds (2,6-DTBP and BHT) while more hydrophobic compound (DEHP) was less

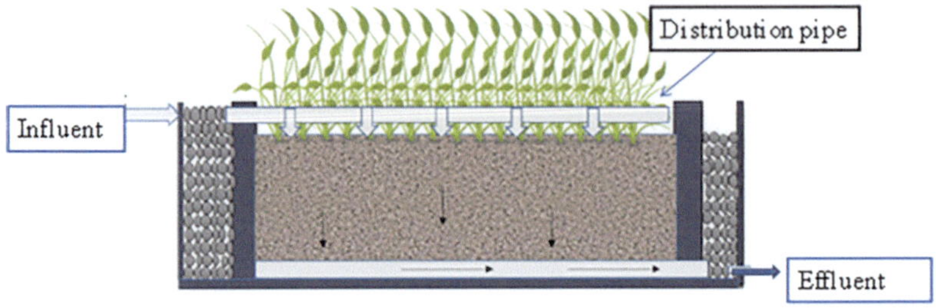

Fig. 8.5 Vertical flow constructed wetland

removed (58%). Different forms of wastewater, including industrial wastewater, agricultural wastewater, mine wastewater, etc., can be treated using HFCW. Although HFCW denitrifies more effectively than vertical flow constructed wetland, it requires a larger land area (Sudarsan et al. 2018).

Vertical Flow Constructed Wetland (VFCW) VFCWs is a wetland in which wastewater submerged from the top of the wetland and drained out from the bottom (Fig. 8.5) (Nguyen et al. 2020). Vertical flow constructed wetland requires 1–3 m^2 of land per PE (DBT 2019). VF CWs are supplied intermittently to enable for oxygen diffusion, when the filtration bed is empty and the water has percolated to the bottom. As a result, VF CWs have superior ammonia, BOD, COD, and other pollutants removal than HF CWs due to their higher oxygen transfer capacity. Wdowczyk et al. (2022) studied CW and found that decreases in concentration of AN (96–99%), TKN (75–88.5%), TP (84–88%), TN (62.5–70%), and heavy metals (Zn, Ni, Cu, and Cr) (41–56%).

French Vertical Flow Constructed Wetland The French-based CW is a two-stage VFCW set in a parallel way and works in series (Fig. 8.6). Although it is a sort of vertically constructed wetland, it is less expensive to build than vertical flow CW and having same removal efficiency. According to DBT (2019), the wetland needs 2.0–2.5 m^2 of land per PE. Due to changes in the hydrological loading and climate, wetland exhibits minor alterations (Paing et al. 2015). In order to remove biological oxygen demand (BOD), chemical oxygen demand (COD), total kjeldahl nitrogen (TKN), total phosphorous (TP), ammoniacal nitrogen (NH_3-N), total dissolved solid (TDS), and total volatile solid, Yadav et al. (2018) chose two different plant species (*Typha angustata* and *Canna indica*); at the first stage of the reactor, removal was 65%, 64%, 15%, 34%, 21%, and 54% and at second stage it is 88%, 90%, 50%, 58%, 52%, and 71%.

Hybrid Constructed Wetland According to Davis et al. (1995), a hybrid constructed wetland is a system with a different stages treatment with HFCW and VFCW (Fig. 8.7). In comparison to other types of constructed wetland, hybrid wetland has higher removal efficiency. The several forms of constructed wetland have variable removal effectiveness depending on anaerobic or aerobic circumstances

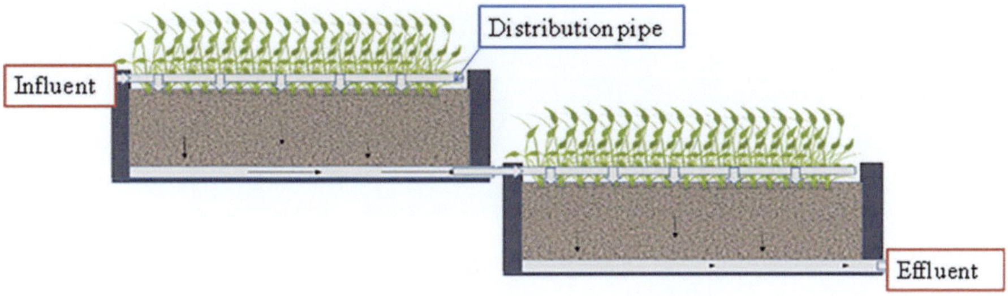

Fig. 8.6 French vertical flow constructed wetland

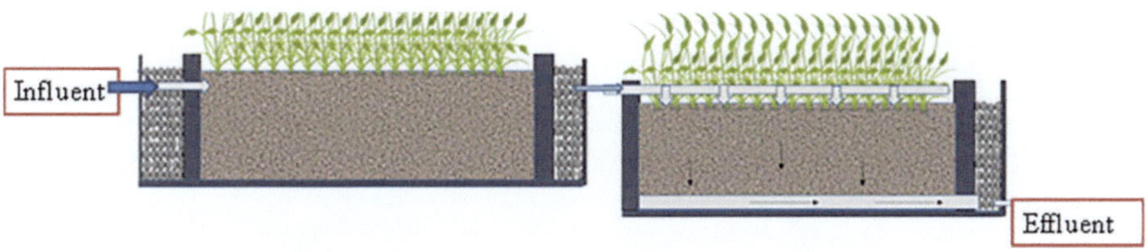

Fig. 8.7 Hybrid constructed wetland

(Yazdania and Golestanib 2019). The mean percentages of nitrogen (N), biochemical oxygen demand (BOD), phosphorus (P), and chemical oxygen demand (COD) removal during the absence of recirculation in the hybrid wetlands were 50–93%, 34–89%, 69–100%, and 55–76%, respectively (Saeed et al. 2020).

Baffled Subsurface Flow Constructed Wetland BSSFCW having vertical baffle beside the width of the wetland which manages the waste stream to flow ups and down via the bed of wetland (Fig. 8.8). With this approach, the wastewater has a longer path to travel and has more time to interact with the media. An essential characteristic that may be effectively attained in the baffled subsurface flow built wetland is the removal of nitrogen from wastewater (DBT 2019). Tee et al. (2012) planted baffled unit with rice husk media, which removed ammonia nitrogen (NH_3) at rates of 74%, 84%, and 99% during HRTs of 2, 3, and 5 days, respectively, compared to the conventional unit's rates of 55%, 70%, and 96%.

Aerated Constructed Wetland Aerated constructed wetland is a superior solution for fulfilling the need for oxygen (Fig. 8.9). The aerated constructed wetland required low energy as compared to traditional method. Due to the less accessibility of O_2, the present organic matters decompose at low rate. Aerators are consequently used in the wetland to increase its efficiency (Sánchez-Monedero et al. 2008). Both VFCW and HFCW could successfully remove organic matter and nitrogen when subjected to high frequency, medium intensity intermittent aeration (2 h of aeration, followed by 4 h of rest). The total nitrogen and COD removal efficiencies for HFCW were 73.22%, and 89.08% respectively, whereas the comparable figures for VFCW were 71.44% and 84.51% (Wang et al. 2021).

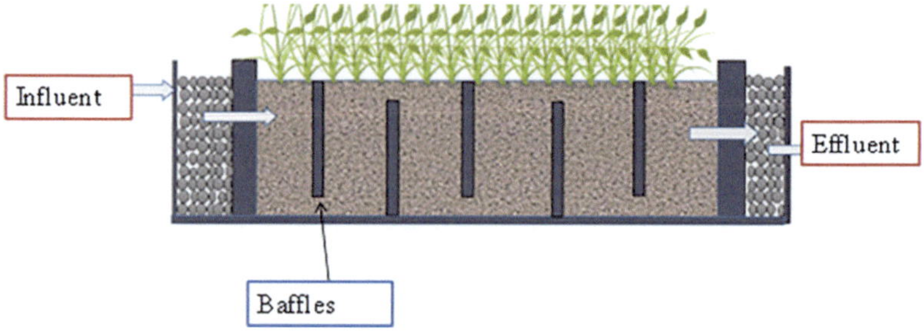

Fig. 8.8 Baffled subsurface flow constructed wetland

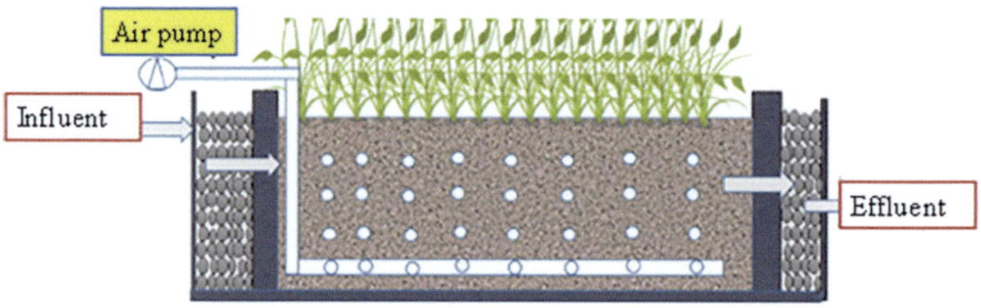

Fig. 8.9 Aerated constructed wetland

Fig. 8.10 Aerated constructed wetland

Multi-tropic Free Flow Engineered Wetland Aquatic plants that float grow in the wastewater (Fig. 8.10). In wetlands, floating plants like *Lemna* sp. and *Eichhornia crassipes* are used. Floating aquatic plants are effective in removing nutrients such as phosphates and nitrogen from wastewater; however, they exhibit relatively low efficiency in reducing biochemical oxygen demand (BOD) and chemical oxygen demand (COD). Regular harvesting is necessary for the free flow constructed wetland to function more effectively. Aquatic plants that float increase biomass, which can subsequently be utilized as a fertilizer, fuel, and feed for livestock etc. (Tilak et al. 2016).

8.4 Current Practices in Constructed Wetland Research

Constructed wetlands offer a sustainable and natural approach for the treatment of landfill leachate. They are engineered system that uses wetland vegetation, soil, and substrate media and related microbial communities to remove contaminants from leachate. Here some current practices used for landfill leachate treatment in constructed wetland using various substrate media and vegetation with their removal efficiency shown in Table 8.2.

Specific materials (like wetland substrate media), according to Saeed and Sun (2012), Khalifa et al. (2020), and Mlih et al. (2020), could supply the ingredients needed for microbial metabolism and chemical adsorption of wastewater contaminants. Degradation kinetics is carried out in the media, which also supports a microbial population and produces aerobic and anaerobic pores inside the bed matrix. Organic and construction materials are significant examples of such specific materials that enhanced chemical and microbiological pollution clearance kinetics in subsurface flow wetlands due to their unique physicochemical features (Saeed et al. 2018; Zhou et al. 2018). Such notable qualities of organic and building material may be used to increase the rate at which pollutants are removed by landfill leachate treatment wetland. Constructed wetland uses the reactive media have been developed for the treatment of LL. Sand, clay, and waste materials containing iron were used in this reactive medium to improve CW performance. Hazardous contaminants have been kept from migrating from a landfill site by using clay soil as a barrier. In order to target major or macropollutants in landfill leachate, the waste material of Fe in the form of slag or sludge, and zero-valent iron (ZVI), have been investigated (Chiemchaisri et al. 2015) or used as permeable reactive barriers.

Another element of wetlands that has garnered attention in earlier research is the plant. According to Saeed and Sun (2012) and Bakhshoodeh et al. (2020), plants increase the efficiency of constructed wetlands at removing pollutants by absorbing it; providing carbon, and acting as attachment sites for microbialgrowth surrounding the root area. Plant uptake, biological decomposition processes, and sedimentation and filtering of the suspended materials in the gravel bed are the major organic removal processes in CWs. Ammonia volatilization, plant absorption, and microbial transformation are the mechanisms in responsible for nutrient elimination (Saeed and Sun 2012). CWs' abilities to remove organic micropollutants have also been studied (Dan et al. 2017). In lab-scale CWs, Dan et al. (2017) found that the removal of 4-tert-butylphenol (4-t-BP) (18–100%), phenol (88–100%), and bisphenol A (BPA) (9–99%) from the synthetic LL. The removal of BPA and 4-t-BP were more effectively from *Phragmites australis*planted CWs compared to unplanted CWs, from older leachate with lower concentrations of acetate and propionate as readily biodegradable carbon source compared to younger leachate and in the dry season mode with longer retention time compared to the wet season mode with shorter retention time. The primary removal procedures for these phenolic compounds may involve adsorption at first step and subsequent biodegradation processes. The improvement of BPA- and 4-t-BP-degrading bacteria as well as the ability of microorganisms in CWs to utilize carbon sources is both benefited by the presence of plants.

Table 8.2 Various treatment media and vegetation used for landfill leachate treatment in different types of constructed wetlands

Treatment media	Vegetation	Constructed wetland type	Removal performance	References
Fly ash	*Canna indica*, *Phragmites australis*, and *Cyperus involucratus*	Three vertical flow CW	TN: 91.43% NH_3^+-N: 94.19% NO_3-N: 98.11%	Wang et al. (2023)
Gravel, sand, and an exchangeable layer: organic substrate + mineral substrate (zeolite, expanded clay)	*Phragmites australis*	Vertical flow	TP: 84–88% TKN: 75–88.5% AN: 96–99% TN: 62.5–70% Heavy metals (Zn, Ni, Cu, Cr): 41–56%	Wdowczyk et al. (2022)
Zeolite and Biochar	*Spartina alterniflora* and *Typha latifolia*	Hybrid	N: 79%	Yang et al. (2023)
Gravel, steel slag, concrete block, ceramic filters	*Phragmies* or *Vetiver*	Hybrid	NH_4–N and BOD: (55–59%) Organics: (66–90%) Nitrogen: (85–92%) Phosphorus: (97–100%)	Saeed et al. (2020)
Gravel (25 mm)	*Gyneriumsagittatum*, *Colocasia esculenta*, and *Heliconia psittacorum*	Horizontal subsurface flow	COD: 66% TKN: 67% NH_4^+-N: 72%	Madera-Parra et al. (2015)
Sand, solid compost and carriers	*A. plinii*, *Arundo donax*, and *Phragmites australis*	Hybrid	Nitrogen compounds, COD, chlorides and metals in the range 75–95%	Spiniello et al. (2023)
Brick	*Canna indica*, *Phragmites australis*, and *Cyperus involucratus*	Multistage subsurface flow	Hexachlorocyclohexane, dichlorodiphenyltrichloroetneby substrate adsorption (50.55–72.74%) and by microbial degradation (20.38–27.89%)	Yang et al. (2022)
ZELIAC and zeolite 2 mm (ZELIAC = Zeolite + activated carbon + limestone + rice husk ash + Portland cement)	(*Typha domingensis*)	Aerated constructed wetland	Color: 90.3% COD: 86.7% Ammonia: 99.2% Nickel: 86.0% Cadmium: 87.1%	Mojiri et al. (2016)

Coal, coco peat, concrete block and brick (5–10 mm), sand (0.075 mm)	*Phragmite* and *Vetiver*	Hybrid	Nitrogen: 50–93% Phosphorus: 69–100% Biochemical oxygen demand: 34–89% Chemical oxygen demand: 55–76%	Saeed et al. (2021)
Waste rubber tyre	*Phragmites australis*	Tidal flow CW	BOD: −35% and 76% COD: −22% and 76% AN: 10% and 90% TN: −0.2% and 71% P: 38% and 92%	Saeed et al. (2022a)
Sand, biochar, concrete particle, iron particle, and stone dust	*Phragmites australis*	Hybrid	Ammonium nitrogen: 91% and 98% Total nitrogen: 90% and 98% Total phosphorus: 97% and 99% BOD: 88% and 93% COD: 93% and 97%	Saeed et al. (2022b)
Sand/clay/iron powder of 60:30:10 (%w/w)	*Typha* sp.	Horizontal subsurface flow constructed wetland	DEHP: 58% DEP and DBP: 73%	Witthayaphirom et al. (2020b)
Fungal	*Heliconia psitacorum*, *Phragmites australis*	Horizontal subsurface flow	BPA and NP: 73.3% and 62.8% (*Heliconia*); 70.2% and 52.1% (*Phragmites*)	Toro-Vélez et al. (2016)
Gravel	*Phragmites australis*	Vertical flow	4-*t*-BP: 18–100% Phenol: 88–100% BPA: 9–99%	Dan et al. (2017)
Sand Zeolite Gravel	*Typha latifolia* and *Canna indica*	Vertical subsurface flow	COD: 81.0%; 84.0%, NH_4-N: 60.0%; 56.0%	Yalçuk and Ugurlu (2020)

In the vicinity of both active and dormant volcanoes, the mineral zeolite is a porous hydrated aluminosilicate (Montalvo et al. 2020). According to Makgabutlane et al. (2020), zeolite has a high CEC (cation exchange capacity) and is selective for ammonium. According to Halim et al. (2010), zeolite can absorb ammonia from LL at a rate of 17.45 mg/g. According to Wdowczyk et al. (2022), expanded clay modified VF CW (over 96%) and bark modified VF CW (over 97%) both had higher ammonia removal rates from landfill leachate than zeolite modified VF CW (over 99%). Additionally, linked nitrifying biofilms can bio-regenerate saturated zeolite in place, obviating the necessity for chemical addition or brine waste removal. Recent developments in CW development that aim to replace conventional media with substances that have higher adsorption capacities include the use of zeolites for ammonium removal (Mojiri and Ahmad 2017; Hou et al. 2018) and zeolites-slag for the adsorption and precipitation of heavy metal (He et al. 2017). Clay material has been used in CW as one of the natural adsorbents (Dordio and Carvalho 2013). It possesses an exchangeable cation-balanced negative charge. Clay support matrix, according to Wu et al. (2015), can enhance OMP removals of phthalates, phenoxy acids, and medicines. According to Boparai et al. (2011), and Shimizu et al. (2012), iron powder may also encourage oxidative reaction, adsorption, and precipitation. Ezzatahmadi et al. (2017) studied the usage of Fe_0 or ZVI also addresses the directional transfer of electrons from Fe to contaminants, which causes the latter to change into less harmful or nontoxic species.

Mojiri et al. (2016) conducted a study for co-treatment of LL and municipal wastewater by with aerated constructed wetland. *Typha domingensis* was transplanted two substrate layers of adsorbents media ZELIAC and zeolite were used. The removal efficiencies of color, ammonia, COD, nickel, and cadmium concentrations were 90.3%, 99.2%, 86.7%, 86.0%, and 87.1%, respectively, at the optimal contact duration (50.2 h) and leachate-to-wastewater mixing ratio (20.0%). Leachate treatment in hybrid CWs can be improved by using low-cost adsorbent materials such biochar and clinoptilolite, a naturally occurring zeolite mineral. A Florida (US) landfill was the site of two mesocosm hybrid CW installations. Zeolite was added to the VF-CW in the adsorbent-CW at a ratio of 10% (v/v) to improve nitrification, and biochar was added to the HF-CW at a ratio of 13% (v/v) to promote microbe's activity and vegetation development. A control-CW was made entirely of gravel. Both systems run for around two years at various hydraulic loading rates while being planted with cordgrass (*Spartina alterniflora*) and cattails (*Typha latifolia*). During the last six months of operation, a wood chip-containing HF-CW (Woodchip-CW) was introduced downstream of the adsorbent-CW to improve denitrification. Zeolite was added to the VF-CW, which resulted in cycles of ammonium adsorption and bioregeneration under intermittent loading circumstances, which significantly increased the nitrification rates by 16–93%. The addition of biochar promoted wetland plant growth and nitrogen removal due to improved plant growth and rhizosphere microbial activity. Additionally, it enhanced the removal of color (1–18%) and organic debris (21–26% to 29–44%). Woodchip-CW was added, which reduced nitrate buildup and allowed for up to 79% more nitrogen to be eliminated overall (Yang et al. 2023).

Wdowczyk et al. (2022) performed a study for treatment of leachate from municipal landfills using vertical flow constructed wetlands planted with *Phragmites australis* using different media such as sand gravel, and an exchangeable layer, depending on a variant: organic substrate or mineral media includes expanded clay and zeolite. The better reductions were obtained from zeolite media. In the study CW, removal achieved as AN 96–99%; TN 62.5–70%, TP 84–88%; TKN 75–88.5%; and Zn, Ni, Cu, Cr was 41–56%, respectively.

Madera-Parra et al. (2015) studied assess the accumulation of Cd (II), Cr (VI) Hg (II), and Pb (II) from synthetic landfill leachate with *Colocasia esculenta*, *Gyneriumsagittatum*, and *Heliconia psittacorum*planted and gravel media in horizontal subsurface flow constructed wetland. The removal efficiency for TKN, COD, and NH_4^+-N were 67%, 66%, and 72% and heavy metal removal varied from 92% to 98% in Cr (VI), Hg (II), Cd (II), and Pb (II). Saeed et al. (2021) studied hybrid constructed wetland planted *Phragmite* and *Vetiver* easily available in Bangladesh with substrate media coal, concrete block, coco peat, and brick (5–10 mm), sand (0.075 mm) and the mean removal efficiency for nitrogen; phosphorus; biochemical oxygen demand; and chemical oxygen demand was 50–93%, 69–100%, 34–89%, and 55–76%, respectively from landfill leachate.

Constructed wetland has also ability to remove pesticides such as organochlorine pesticides (OCPs) refractory organics (hexachlorocyclohexane (HCH) and dichlorodiphenyltrichloroethane (DDT)) from LL (Yang et al. 2022). The plant species used for OCPs removal was *Canna indica*, *Phragmites australis*, and *Cyperus involucratus*and the brick used as substrate. *Phragmites australis*, *Arundo donax*, and *A. plinii*have chlorides, nitrogen compounds, chemical oxygen demand (COD), and metals have removal in the range 75–95% (Spiniello et al. 2023).

According to Saeed et al. (2020), to co-treat landfill leachate and city wastewater, three parallel two-stage *Phragmites*- or *Vetiver*-based hybrid wetland mesocosms were constructed. The mesocosms were filled with steel slag, gravel, concrete blocks, and ceramic filters. The average daily inputs of organics, nitrogen, phosphorus, and other elements were 75 g N/m^2, 88 g BOD/m^2, 283 g COD/m^2, and 10 g P/m^2 for the early phases. The gravel-based VF wetland mesocosm absorbed higher BOD and NH_4-N (55–59%) during the landfill leachate treatment phase in comparison to co-treatment periods (12–52%). P removal (93%) in the slag-concrete-based VF wetland mesocosm was increased by media-based adsorption. In all VF wetland mesocosms, denitrification was constrained by carbon scarcity; TN removal was only about 32%. Higher organics (66–90%), nitrogen (85–92%), and phosphorus (97–100%) reductions were attained in second stage wetland mesocosms. The removal rate of nitrogen was over 90% in vertical flow constructed wetlands. Fly ash was used as the substrate, and the average removal efficiencies of TN, NH_3^+-N, and NO_3-N in T-VF were 91.43%, 94.19%, and 98.11%, respectively (Wang et al. 2022).

Heliconia psitacorum, *Phragmites australis*, and an unplanted horizontal subsurface flow constructed wetland (HSSF-CW) were all used as pilot projects for eight-week period. The removal efficiencies for BPA and NP in the *Heliconia* CW were 73.3% and 62.8%, respectively, but the *Phragmites* CW showed a similar removal for BPA 70.2% and less removal efficiency for NP 52.1%. The unplanted CW removed 62.2% of the BPA and 25.3 of the NP (Toro-Vélez et al. 2016). Utilizing a substrate-planted, vertical subsurface flow constructed wetland with *Typha latifolia* and *Canna indica*, landfill leachate was treated. The CWs were particularly operated in the *T. latifolia* reactor (R1), *C. indica* reactor (R2), and Control reactor (R3) at a flow rate of 5 l/day and a hydraulic retention time (HRT) of 22 days. According to Yalçuk and Ugurlu (2020), the average removal performance for R1, R2, and R3 was 60.0%, 56.0%, and 46.0% for NH_4-N, 81.0%, 84.0%, and 79.0% for COD, and 45.0%, 46.0%, and 32.0% for PO_4-P.

According to Saeed et al. (2022b) microbial fuel cell (MFC)-constructed wetlands, which were developed over the past 10 years as a result of the integration of electrochemical technologies, represents a significant improvement in constructed wetlands. Pollutants were eliminated from both synthetic and natural wastewaters by wetlands built by MFC. Employing unsaturated, partially saturated microbial fuel cells integrated tidal flow constructed wetlands, for the co-treatment of urban wastewater and landfill leachate is a novel technology. *Phragmites australis* was cultivated along with an abundance of organic or waste materials in the wetland habitats. Leachate and municipal wastewater were combined in a variety of volumetric ratios (i.e., 40%:60%; 20%:80%; 10%:90%) for co-treatment between 6 and 24 h. Mean phosphorus, total nitrogen, ammonium nitrogen, and ammonium nitrogen removal rates, respectively, varied from 35% to 76%, 10% to 90%, 22% to 76%, 0.2% to 71%, and 38% to 92%. The unsaturated, partially saturated microbial fuel cell integrated tidal flow wetlands developed a predominantly anoxic/anaerobic environment inside the media due to a higher influent pollutant load resulting from a high proportion of leachate in the mixed wastewater and longer retention time.

These environmental factors made it more difficult for electrochemically active, alternative removal pathways to remove organic and nutrient materials, especially in partially saturated wetlands. Leachate volumetric percentage decrease improved the removal of pollutants from unsaturated and partially saturated systems by adjusting the pollutant loads, carbon and other chemical composition, and aerobic gradient inside the media bed. The tidal flow constructed wetlands with integrated microbial fuel cells produced mean voltages between 4 and 152 mV and higher power densities between 50 and 527 mW/m^3. Production of power density was impacted by the physicochemical characteristics (voids, roughness, carbon composition) of the media. A study on most effective bioenergy producing system was a slightly saturated wetland made of waste rubber tyre material (Saeed et al. 2022a). The MFC-based hybrid wetland's bioenergy output declines as the input pollution load rises. The maximum power density production, or 294 mW/m^2, was attained by the single anode electrode-based VF wetland (Saeed et al. 2022c). The mean total nitrogen (TN), ammonium nitrogen (NH_4-N), total phosphorus (TP), chemical oxygen demand (COD) and biochemical oxygen demand (BOD), removal percentage within the hybrid system varied between 90% and 98%, 91% and 98%, 97% and 99%, 93% and 97%, and 88% and 93%, respectively, from landfill leachate (Saeed et al. 2022b).

Witthayaphirom et al. (2020a) built a horizontal subsurface flow constructed wetland (CW) containing *Typha* sp. for the three-year, non-plant-harvesting removal of organic micropollutants (OMPs) like phenolic compounds and phthalic acid esters from LL. As the CW medium advanced, it was found that the mass removal effectiveness of OMPs grew with time and averaged between 58% and 73%. The compounds with higher hydrophobicity, such as DEHP, exhibited the lowest removal rate of 58%. On the other hand, compounds with higher hydrophilicity, like DEP and DBP, showed the greatest removal rates of 73%. Moderately hydrophobic compounds, such as 2,6-DTBP and BHT, had removal rates that followed closely behind. During the summer, when the concentrations of influent were at their peak, there was a greater clearance of OMPs. The majority of hydrophilic DEP and DBP molecules were removed through biodegradation. Initially, DEHP was removed by adsorption, but eventually it underwent biodegradation, resulting in minimal accumulation in the medium. Specifically, after three years of operation, the accumulation was only 10% of the maximum adsorption capacity. According to microbial community study conducted by Witthayaphirom et al. (2020a, b), *Sphingomonas* sp. and *Rhodococcus* sp. are the primary microorganisms responsible for the biodegradation of OMPs.

One of the most significant environment pollutants on Earth today is microplastics (MP). Synthetic organic polymer plastics (SOPP) is widely used in MP in a variety of industries, including apparel, agricultural, and personal care products. The demand for plastic products has surged due to our increased reliance on plastic materials, which ultimately leads to environmental pollution. According to the research, MPs move between landfills for municipal solid waste to aquatic systems. Another emerging threat, MPs, is being effectively eliminated by natural wetlands. MPs are being effectively removed from aquatic streams by mangrove wetlands. CWs can be an effective treatment method for eliminating MPs from the environment before they enter natural water sources (Swarnakar et al. 2022).

The possibility for resource recovery offered by constructed wetlands is explored. As part of a circular economy strategy, nutrient-rich plant biomass that is taken from the wetland system can be used in a number of ways, including the production of organic fertilizer and renewable energy. There are advantages of constructed wetlands for aesthetics and community integration, their potential for developing environments those are environmentally sound and attractive to all people. The social aspect of constructed wetland as treatment initiatives is getting better and better. As they are more aesthetically pleasing and environmentally friendly than traditional treatment plants, CW facilities are more socially acceptable. As a way to boost their environmental ratings and integrate the CW installation into their corporate social responsibility strategy, many businesses (such as industries and municipal-private partnerships) choose for the CW technique for wastewater treatment produced on their site (Stefanakis 2020).

Approximately 166 constructed wetland systems that sustain public leisurely and educational activity have been discovered by Ghermandi and Fichtman (2015) throughout the world. According to Yang et al. (2008), sociological, cultural, and recreational factors must to be taken into account when determining a CW's economic value. The CWs are essential to the local communities in term of both education and recreation objectives because primary and middle school organizes field trip to the constructed wetlands in China for their students every year (Agaton and Guila 2023).

8.5 Perspectives

1. Plant selection and genetic modification: Research is being processed for identification of wetland vegetation species that are higher efficient in removing particular contaminants from landfill leachate. Additionally, genetic modification of plants to enhance their pollutant uptake capabilities is an emerging area of interest. These progresses could lead to the development of more effective and specialized plant species for landfill leachate treatment.
2. Hybrid approach: The integration of constructed wetlands with different treatment such as activated carbon adsorption, membrane filtration, microbial fuel cells or advanced oxidation can enhance the removal of specific pollutants, allowing for a more comprehensive treatment approach.
3. Long-term performance and maintenance: Assessment of the long-term performance and maintenance requirements of constructed wetlands is a key perspective. Understanding the longevity of treatment effectiveness, patterns of plant growth, and the need for periodic maintenance activities, such as proper management of vegetation and sediment removal, is crucial for sustainable operation and optimization of constructed wetlands.
4. Optimization of treatment: Current research focus on optimizing the parameters of design and operation of constructed wetlands. It also includes exploring various wetland configurations, compositions of media, and hydraulic regimes to accomplish high removal efficiencies for extensive range of contaminants. Mathematical modeling and simulation techniques can aid in optimizing treatment performance and reduces operational costs. Wastewater treatment facilities needed surface area or footprint may be reduced by considering design optimization essential.
5. Geo-physicochemical characteristics: The geo-physicochemical characteristics of CW under various operating settings will be long-term monitored, allowing for more study to be done to better understanding the main elements influencing CW optimization.
6. Climate change resilience: With changing climatic conditions, it is vital to assess the flexibility of constructed wetlands for leachate treatment. Knowing the effects of variations in temperature, patterns of precipitation, and extreme weather conditions on treatment performance can help design resilient systems that can continue to effectively treat landfill leachate under changing environment conditions.

8.6 Conclusion

Constructed wetlands provide a sustainable approach for landfill leachate treatment. CWs are engineered system that uses natural processes for the remove of contaminants from landfill leachate, offering numerous benefits in terms of cost-effectiveness, environmental compatibility, and resource recovery. CWs can be an effective aspect for the treatment of landfill leachate, having higher pollutant burden. Organic pollutants, nutrients (nitrogen and phosphorus), and harmful microbes are all effectively removed by CW. Heavy metals and very hazardous pollutants from landfill leachate can also be successfully treated by CWs. Many researchers have documented the elimination of toxic heavy metal as iron, lead, silver, arsenic, cadmium, mercury, copper, zinc, chromium, nickels, and manganese with CWs. Compared to traditional landfill leachate treatment methods, CWs offer a number of benefits. Additionally, CWs have the potential to treat landfill leachates that are challenging to treat, in addition to domestic and industrial wastewaters. Wetland systems can also reduce waterborne infections. CWs are systems for treating natural resources that are safe for the environment and improve public health. Additionally, CWs have the ability to eliminate developing pollutants that are challenging to eliminate using traditional treatment techniques. In order to remove MPs through CWs, more study is also necessary. All sorts of water pollutants can be removed with remarkable efficiency by CW, and wetland technology has advanced significantly in recent years. The availability of data on constructed wetland project's advantages does not convert into decision-support tools for legislators and project implementers. Therefore, future study should focus on how policies enable the use of built wetlands as a kind of natural water treatment.

References

Achak M, Boumya W, Ouazzani N, Mandi L (2019) Preliminary evaluation of constructed wetlands for nutrients removal from olive mill wastewater (OMW) after passing through a sand filter. Ecol Eng 136:141–151. https://doi.org/10.1016/j.ecoleng.2019.06.007

Agaton CB, Guila PMC (2023) Ecosystem Services Valuation of Constructed Wetland as a Nature-Based Solution to Wastewater Treatment. Earth 4(1):78–92

Akinbile CO, Yusoff MS, Ahmad Zuki AZ (2012) Landfill leachate treatment using sub-surface flow constructed wetland by Cyperus haspan. Waste Manag 32(7):1387–1393

Alam MK, Islam MA, Saeed T, Rahman SM, Majed N (2023) Life cycle analysis of lab-scale constructed wetlands for the treatment of industrial wastewater and landfill leachate from municipal solid waste: a comparative assessment. Water 15(5):909. https://doi.org/10.3390/w15050909

Allen DJ, Farrell M, Huang J, Reynolds C, Rupasinghe M, Mosley LM (2022) Long-term water quality response to increased hydraulic loadings in a field-scale free water surface constructed wetland treating domestic effluent. J Environ Manag 311:114858

Alver A, Altaş L (2017) Characterization and electrocoagulative treatment of landfill leachates: a statistical approach. Process Saf Environ Prot 111:102–111

Ancic M, Huđek A, Rihtaric I, Cazar M, Bacun-Druzina V, Kopjar N, Durgo K (2020) Physico chemical properties and toxicological effect of landfill groundwaters and leachates. Chemosphere 238:124574. https://doi.org/10.1016/j.chemosphere.2019.124574
Angassa K, Assefa B, Kefeni KK et al (2022) Brewery industrial wastewater treatment through mesocosm horizontal subsurface flow constructed wetland. Environ Syst Decis 42:265–275. https://doi.org/10.1007/s10669-022-09849-z
Baettker EC, Kozak C, Knapik HG, Aisse MM (2020) Applicability of conventional and non conventional parameters for municipal landfill leachate characterization. Chemosphere 251:126414. https://doi.org/10.1016/j.chemosphere.2020.126414
Bakhshoodeh R, Alavi N, Oldham C, Santos RM, Babaei AA, Vymazal J, Paydary P (2020) Constructed wetlands for landfill leachate treatment: a review. Ecol Eng 146:105725. https://doi.org/10.1016/j.ecoleng.2020.105725
Boparai HK, Joseph M, O'Carroll DM (2011) Kinetics and thermodynamics of cadmium ion removal by adsorption onto nano zero valent iron particles. J Hazard Mater 186:458–465
Braschi I, Blasioli S, Lavrnić S, Buscaroli E, Di Prodi K, Solimando D, Toscano A (2022) Removal and fate of pesticides in a farm constructed wetland for agricultural drainage water treatment under Mediterranean conditions (Italy). Environ Sci Pollut Res 29(5):7283–7299
Caroline Baettker E, Kozak C, Knapik HG, Aisse MM (2020) Applicability of conventional and non-conventional parameters for municipal landfill leachate characterization. Chemosphere 251:126414. https://doi.org/10.1016/j.chemosphere.2020.126414
Chen J, Li X, Jia W, Shen S, Deng S, Ji B, Chang J (2021) Promotion of bioremediation performance in constructed wetland microcosms for acid mine drainage treatment by using organic substrates and supplementing domestic wastewater and plant litter broth. J Hazard Mater 404:124125
Chen Y, Zhang J, Guo Z, Li M, Wu H (2022) Optimizing agricultural biomass application to enhance nitrogen removal in vertical flow constructed wetlands for treating low-carbon wastewater. Environ Res 209:112867
Chiemchaisri C, Chiemchaisri W, Witthayaphirom C (2015) Remediation of MSW landfill leachate by permeable reactive barrier with vegetation. Water Sci Technol 71(9):1389–1397
Cui E, Zhou Z, Gao F, Chen H, Li J (2023) Roles of substrates in removing antibiotics and antibiotic resistance genes in constructed wetlands: a review. Sci Total Environ 859:160257. https://doi.org/10.1016/j.scitotenv.2022.160257
Dan A, Daiki F, Satoshi S, Takashi M, Michihiko I (2017) Removal of phenol, bisphenol A, 4-tert-butylphenol from synthetic landfill leachate by vertical flow constructed wetlands. Sci Total Environ 578:566–576
Daukas P, Lowry D, Walker WW (2020) Design of wet detention basins and constructed wetlands for treatment of stormwater runoff from a regional shopping mall in Massachusetts. In: Constructed wetlands for wastewater treatment. CRC Press, pp 686–694
Davis L, Edwards R, Garber L, Isaacs B (1995) A handbook of constructed wetlands, general considerations, vol 1. US Environmental Protection Agency (EPA), Washington, DC
DBT C (2019) Manual on constructed wetland as an alternative technology for sewage management in India. Department of Bio Technology (DBT), New Delhi
Dinakar M, Tao W, Daley D (2020) Using hydrogen peroxide to supplement oxygen for nitrogen removal in constructed wetlands. J Environ Chem Eng 8(6):104517. https://doi.org/10.1016/j.jece.2020.104517
Dordio AV, Carvalho AJP (2013) Organic xenobiotics removal in constructed wetlands, with emphasis on the importance of the support matrix. J Hazard Mater 252–253:272–292
Dotro G, Fort RP, Barak J, Jones M, Vale P, Jefferson B (2015) Long-term performance of constructed wetlands with chemical dosing for phosphorus removal. In: The role of natural and constructed wetland in nutrient cycling and retention on the landscape. Springer, Cham, pp 273–292. https://doi.org/10.1007/978-3-319-08177-9_19

El-Sheikh MA, Saleh HI, El-Quosy DE, Mahmoud AA (2010) Improving water quality in polluted drains with free water surface constructed wetlands. Ecol Eng 36(10):1478–1484. https://doi.org/10.1016/j.ecoleng.2010.06.030

Eslamian S, Okhravi S, Eslamian F (2019) Constructed wetlands: hydraulic design. CRC Press

Ezzatahmadi N, Ayoko GA, Millar GJ, Speight R, Yan C, Li J, Li S, Zhu J, Xi Y (2017) Clay-supported nanoscale zero-valent iron composite materials for the remediation of contaminated aqueous solutions: a review. Chem Eng J 312:336–350

Fahim R, Lu X, Jilani G et al (2019) Comparison of floating-bed wetland and gravel filter amended with limestone and sawdust for sewage treatment. Environ Sci Pollut Res 26:20400–20410. https://doi.org/10.1007/s11356-019-05325-5

Gao J, Zhao J, Zhang J, Li Q, Gao J, Cai M, Zhang J (2020) Preparation of a new low-cost substrate prepared from drinking water treatment sludge (DWTS)/bentonite/zeolite/fly ash for rapid phosphorus removal in constructed wetlands. J Clean Prod 261:121110

Ge X, Cao X, Song X, Wang Y, Si Z, Zhao Y, Wang W, Tesfahunegn AA (2020) Bioenergy generation and simultaneous nitrate and phosphorus removal in a pyrite-based constructed wetland-microbial fuel cell. Bioresour Technol 296:122350. https://doi.org/10.1016/j.biortech.2019.122350

Ghermandi A, Fichtman E (2015) Cultural ecosystem services of multifunctional constructed treatment wetlands and waste stabilization ponds: time to enter the mainstream? Ecol Eng 84:615–623

Gholipour A, Zahabi H, Stefanakis AI (2020) A novel pilot and full-scale constructed wetland study for glass industry wastewater treatment. Chemosphere 247:125966

Gupta S, Srivastava P, Yadav AK (2020) Simultaneous removal of organic matters and nutrients from high-strength wastewater in constructed wetlands followed by entrapped algal systems. Environ Sci Pollut Res 27:1112–1117. https://doi.org/10.1007/s11356-019-06896-z

Gupta S, Nayak A, Roy C, Yadav AK (2021) An algal assisted constructed wetland-microbial fuel cell integrated with sand filter for efficient wastewater treatment and electricity production. Chemosphere 263:128132

Halim AA, Aziz HA, Johari MAM, Ariffin KS (2010) Comparison study of ammonia and COD adsorption on zeolite, activated carbon and composite materials in landfill leachate treatment. Desalination 262(1–3):31–35. https://doi.org/10.1016/j.desal.2010.05.036

He H, Duan Z, Wang Z, Yue B (2017) The removal efficiency of constructed wetlands filled with the zeolite-slag hybrid substrate for the rural landfill leachate treatment. Environ Sci Pollut Res 24:17547–17555

Hou L, Hu BX, He M, Xu X, Zhang W (2018) Effect of intermittent operation model on the function of soil infiltration system. Environ Sci Pollut Res 25:9615–9625

Iskander SM, Novak JT, He Z (2019) Reduction of reagent requirements and sludge generation in Fenton's oxidation of landfill leachate by synergistically incorporating forward osmosis and humic acid recovery. Water Res 151:310–317. https://doi.org/10.1016/j.watres.2018.11.089

Jamwal P, Raj AV, Raveendran L, Shirin S, Connelly S, Yeluripati J, Richards S, Rao L, Helliwell R, Tamburini M (2021) Evaluating the performance of horizontal sub-surface flow constructed wetlands: a case study from southern India. Ecol Eng 162:106170. https://doi.org/10.1016/j.ecoleng.2021.106170

Jia W, Sun X, Gao Y, Yang Y, Yang L (2020) Fe-modified biochar enhances microbial nitrogen removal capability of constructed wetland. Sci Total Environ 740:139534. https://doi.org/10.1016/j.scitotenv.2020.139534

Jiang S, Xu J, Wang H et al (2022) Study of the effect of pyrite and alkali-modified rice husk substrates on enhancing nitrogen and phosphorus removals in constructed wetlands. Environ Sci Pollut Res 29:54234–54249. https://doi.org/10.1007/s11356-022-19537-9

Kadlec R, Knight R (1996) Treatment wetlands. CRC, Baca Raton

Khalifa ME, El-Reash YGA, Ahmed MI, Rizk FW (2020) Effect of media variation on the removal efficiency of pollutants from domestic wastewater in constructed wetland systems. Ecol Eng 143:105668. https://doi.org/10.1016/j.ecoleng.2019.105668

Kong F, Wang J, Hou W, Cui Y, Yu L, Zhang Y, Wang S (2023) Influence of modified biochar supported sulfidation of nano-zero-valent-iron (S-nZVI/BC) on nitrate removal and greenhouse gas emission in constructed wetland. J Environ Sci 125:568–581. https://doi.org/10.1016/j.jes.2022.02.040

Kumar SS, Bishnoi NR (2017) Coagulation of landfill leachate by FeCl 3: process optimization using Box–Behnken design (RSM). Appl Water Sci 7:1943–1953

Kumar M, Singh R (2019) Assessment of pollutant removal processes and kinetic modelling in vertical flow constructed wetlands at elevated pollutant loading. Environ Sci Pollut Res 26:18421–18433. https://doi.org/10.1007/s11356-019-05019-y

Kumar SS, Kadier A, Malyan SK, Ahmad A, Bishnoi NR (2017) Phytoremediation and rhizoremediation: uptake, mobilization and sequestration of heavy metals by plants. In: Plant-microbe interactions in agro-ecological perspectives: volume 2: microbial interactions and agro-ecological impacts. Springer, Singapore, pp 367–394

Kurniawan TA, Lo W (2009) Removal of refractory compounds from stabilized landfill leachate using an integrated H2O2 oxidation and granular activated carbon (GAC) adsorption treatment. Water Res 43(16):4079–4091

Lin CJ, Chyan JM, Zhuang WX, Vega FA, Mendoza RMO, Senoro DB, Shiu RF, Liao CH, Huang DJ (2020) Application of an innovative front aeration and internal recirculation strategy to improve the removal of pollutants in subsurface flow constructed wetlands. J Environ Manag 256:109873

Lin G, Zeng B, Li J, Wang Z, Wang S, Hu T, Zhang L (2023) A systematic review of metal organic frameworks materials for heavy metal removal: synthesis, applications and mechanism. Chem Eng J 460:141710

Ma H, Kang Y, Li M, Dong J, Wang Y, Xiao J, Guo Z (2023) Enhancement of perfluorooctanoic acid and perfluorooctane sulphonic acid removal in constructed wetland using iron mineral: performance and mechanisms. J Hazard Mater 447:130819. https://doi.org/10.1016/j.jhazmat.2023.130819

Madera-Parra CA, Pena-Salamanca EJ, Pena MR, Rousseau DPL, Lens PNL (2015) Phytoremediation of landfill leachate with *Colocasia esculenta*, *Gynerumsagittatum*and *Heliconia psittacorum*in constructed wetlands. Int J Phytoremediation 17(1):16–24

Maftouh A, El Fatni O, El Hajjaji S, Jawish MW, Sillanpää M (2023) Comparative review of different adsorption techniques used in heavy metals removal in water. Biointerface Res Appl Chem 13:397

Makgabutlane B, Nthunya LN, Musyoka N, Dladla BS, Nxumalo EN, Mhlanga SD (2020) Microwave-assisted synthesis of coal fly ash-based zeolites for removal of ammonium from urine. RSC Adv 10(4):2416–2427

Meshram S, Raut SP, Ansari K, Madurwar M, Daniyal M, Khan MA, Hasan MA (2023) Waste slags as sustainable construction materials: a compressive review on physico mechanical properties. J Mater Res Technol 23:5821–5845

Mlih R, Bydalek F, Klumpp E, Yaghi N, Bol R, Wenk J (2020) Light-expanded clay aggregate (LECA) as a substrate in constructed wetlands—a review. Ecol Eng 148:105783. https://doi.org/10.1016/j.ecoleng.2020.105783

Mojiri A, Ahmad Z (2017) Ammonia, phosphate, phenol, and copper (II) removal from aqueous solution by subsurface and surface flow constructed wetland. Environ Monit Assess 189:337

Mojiri A, Ziyang L, Tajuddin RM, Farraji H, Alifar N (2016) Co-treatment of landfill leachate and municipal wastewater using the ZELIAC/zeolite constructed wetland system. J Environ Manag 166:124–130

Montalvo S, Huiliñir C, Borja R, Sánchez E, Herrmann CJBT (2020) Application of zeolites for biological treatment processes of solid wastes and wastewaters–a review. Bioresour Technol 301:122808

Nguyen XC, Tran TP, Hoang VH, Nguyen TP, Chang SW, Nguyen DD, Guo W, Kumar A, La DD, Bach Q (2020) Combined biochar vertical flow and free-water surface constructed wetland system for dormitory sewage treatment and reuse. Sci Total Environ 713:136404. https://doi.org/10.1016/j.scitotenv.2019.136404

Nguyen TT, Huang H, Nguyen TAH, Soda S (2022) Recycling clamshell as substrate in lab-scale constructed wetlands for heavy metal removal from simulated acid mine drainage. Process Saf Environ Prot 165:950–958

Ngweme GN, Al Salah DMM, Laffite A, Sivalingam P, Grandjean D, Konde JN, Poté J (2020) Occurrence of organic micropollutants and human health risk assessment based on consumption of Amaranthus viridis, Kinshasa in The Democratic Republic of the Congo. Sci Total Environ 754:142175. https://doi.org/10.1016/j.scitotenv.2020.142175

Paing J, Guilbert A, Gagnon V, Chazarenc F (2015) Effect of climate, wastewater composition, loading rates, system age and design on performances of French vertical flow constructed wetlands: a survey based on 169 full scale systems. Ecol Eng 80:46–52

Parde D, Patwa A, Shukla A, Vijay R, Killedar DJ, Kumar R (2021) A review of constructed wetland on type, treatment and technology of wastewater. Environ Technol Innov 21:101261

Patil R, Zahid M, Govindwar S, Khandare R, Vyavahare G, Gurav R et al (2022) Constructed wetland: a promising technology for the treatment of hazardous textile dyes and effluent. In: Development in wastewater treatment research and processes. Elsevier, pp 173–198

Prodanovic V, Zhang K, Zheng M, Hu S, Hong P, Yuan Z, Deletic A (2023) Nitrification potential of daily-watered biofiltration designs for high ammonium wastewater treatment. Sci Total Environ 863:160989. https://doi.org/10.1016/j.scitotenv.2022.160989

Rane NR, Kanojia A, Patil SM, Khandare R, Kodam KM, Jeon BH (2023) Constructed wetland system and its engineered designs for the treatment of textile industry effluent. In: Current developments in bioengineering and biotechnology. Elsevier, pp 601–626

Ravina YS, Chhikara SK (2023) Usability of Rice straw as biodsorbent for the removal of phosphate from aqueous solution: isotherms, kinetics and thermodynamics. Asian J Chem 2(35):461–467. https://doi.org/10.14233/ajchem.2023.26884

Rohilla R, Yadav S, Chhikara SK, Kumar P (2022) Usability of rice straw as adsorbent for the removal of nitrate from synthetic solution: isotherms, kinetics and thermodynamics. Zeichen J 8(1–12)

Sacco A, Cirelli GL, Ventura D, Barbagallo S, Licciardello F (2021) Hydraulic performance of horizontal constructed wetlands for stormwater treatment: a pilot-scale study in the Mediterranean. Ecol Eng 169:106290

Saeed T, Sun G (2012) A review on nitrogen and organics removal mechanisms in subsurface flow constructed wetlands: dependency on environmental parameters, operating conditions and supporting media. J Environ Manag 112:429–448. https://doi.org/10.1016/j.jenvman.2012.08.011

Saeed T, Muntaha S, Rashid M, Sun G, Hasnat A (2018) Industrial wastewater treatment in constructed wetlands packed with construction materials and agricultural by-products. J Clean Prod 189:442–453. https://doi.org/10.1016/j.jclepro.2018.04.115

Saeed T, Miah MJ, Majed N et al (2020) Pollutant removal from landfill leachate employing two-stage constructed wetland mesocosms: co-treatment with municipal sewage. Environ Sci Pollut Res 27:28316–28332. https://doi.org/10.1007/s11356-020-09208-y

Saeed T, Miah MJ, Majed N, Alam MK, Khan T (2021) Effect of effluent recirculation on nutrients and organics removal performance of hybrid constructed wetlands: landfill leachate treatment. J Clean Prod 282:125427. https://doi.org/10.1016/j.jclepro.2020.125427

Saeed T, Yadav AK, Miah MJ (2022a) Landfill leachate and municipal wastewater co-treatment in microbial fuel cell integrated unsaturated and partially saturated tidal flow constructed wetlands. J Water Process Eng 46:102633

Saeed T, Miah MJ, Yadav AK (2022b) Development of electrodes integrated hybrid constructed wetlands using organic, construction, and rejected materials as filter media: landfill leachate treatment. Chemosphere 303:135273

Saeed T, Majed N, Miah MJ, Yadav AK (2022c) A comparative landfill leachate treatment performance in normal and electrodes integrated hybrid constructed wetlands under unstable pollutant loadings. Sci Total Environ 838:155942

Sánchez-Monedero MA, Aguilar MI, Fenoll R, Roig A (2008) Effect of the aeration system on the levels of airborne microorganisms generated at wastewater treatment plants. Water Res 42(14):3739–3744

Sehar S, Nasser HAA (2019) Wastewater treatment of food industries through constructed wetland: a review. Int J Environ Sci Technol 16:6453–6472

Shimizu A, Tokumura M, Nakajima K, Kawase Y (2012) Phenol removal using zero valent iron powder in the presence of dissolved oxygen: roles of decomposition by the Fenton reaction and adsorption/precipitation. J Hazard Mater 201:60–67

Solano M, Soriano P, Ciria M (2004) Constructed wetlands as a sustainable solution for wastewater treatment in small villages. Biosyst Eng 87(1):109–118. https://doi.org/10.1016/j.biosystemseng.2003.10.005

Spiniello I, De Carluccio M, Castiglione S, Amineva E, Kostryukova N, Cicatelli A, Rizzo L, Guarino F (2023) Landfill leachate treatment by a combination of a multiple plant hybrid constructed wetland system with a solar photoFenton process in a raceway pond reactor. J Environ Manag 331:117211. https://doi.org/10.1016/j.jenvman.2022.117211

Stanković D (2017) Constructed wetlands for wastewater treatment. Gradevinar 69(08):639–652. https://doi.org/10.14256/JCE.2062.2017

Stefanakis AI (2020) Constructed wetlands: description and benefits of an eco-tech water treatment system. In: Waste management: concepts, methodologies, tools, and applications. IGI Global, pp 503–525

Sudarsan JS, Subramani S, Rajan RJ, Shah I, Nithiyanantham S (2018) Simulation of constructed wetland in treating wastewater using fuzzy logic technique. J Phys Conf Ser 1000(1):012137. IOP Publishing

Swarnakar AK, Bajpai S, Ahmad I (2022) Various types of constructed wetland for wastewater treatment-a review. IOP Conf Ser Earth Environ Sci 1032(1):012026. IOP Publishing

Tee H, Lim P, Seng C, Nawi MM (2012) Newly developed baffled subsurface-flow constructed wetland for the enhancement of nitrogen removal. Bioresour Technol 104:235–242. https://doi.org/10.1016/j.biortech.2011.11.032

Teng C, Zhou K, Peng C, Chen W (2021) Characterization and treatment of landfill leachate: a review. Water Res 203:117525

Thathong V, Tantamsapya N, Yossapol C, Liao C, Wirojanagud W, Padungthon S (2019) Role of Colocasia esculenta L. schott in arsenic removal by a pilot-scale constructed wetland filled with laterite soil. Heliyon 5(2):e01233. https://doi.org/10.1016/j.heliyon.2019.e01233

Tilak AS, Wani SP, Patil MD, Datta A (2016) Evaluating wastewater treatment efficiency of two field scale subsurface flow constructed wetlands. Curr Sci 110(09):1764–1772

Toro-Vélez AF, Madera-Parra CA, Peña-Varón MR, Lee WY, Bezares-Cruz JC, Walker WS, Cárdenas-Henao H, Quesada-Calderón S, García-Hernández H, Lens PNL (2016) BPA and NP removal from municipal wastewater by tropical horizontal subsurface constructed wetlands. Sci Total Environ 542:93–101

Vymazal J (2005) Horizontal sub-surface flow and hybrid constructed wetlands systems for wastewater treatment. Ecol Eng 25:478–490

Waara S, Wojciechowska E (2019) Treatment of landfill leachate in a constructed free water surface wetland system over a decade—Identification of disturbance in process behaviour and removal of eutrophying substances and organic material. J Environ Manag 249:109319. https://doi.org/10.1016/j.jenvman.2019.109319

Wang H, Xu J, Sheng L (2020) Preparation of straw biochar and application of constructed wetland in China: a review. J Clean Prod 273:123131

Wang Y, Wang WH, Zhang H et al (2021) Treatment of the actual landfill leachate in different constructed wetlands through intermittent and varied aeration mode. Environ Sci Pollut Res 28:64858–64870. https://doi.org/10.1007/s11356-021-15216-3

Wang H, Yang C, Wang B, He Z, Fu T (2022) Nitrogen removal performance and microbiological characteristics for the landfill leachate treatment in a three-stage vertical flow constructed wetlands system. Environ Technol Innov 28:102728. https://doi.org/10.1016/j.eti.2022.102728

Wang H, Zang S, Teng H et al (2023) Characteristic of KMnO4-modified corn straw biochar and its application in constructed wetland to treat city tail water. Environ Sci Pollut Res 30:49948–49962. https://doi.org/10.1007/s11356-023-25856-2

Wdowczyk A, Szymańska-Pulikowska A (2021) Analysis of the possibility of conducting a comprehensive assessment of landfill leachate contamination using physicochemical indicators and toxicity test. Ecotoxicol Environ Saf 221:112434. https://doi.org/10.1016/j.ecoenv.2021.112434

Wdowczyk A, Szymańska-Pulikowska A, Gałka B (2022) Removal of selected pollutants from landfill leachate in constructed wetlands with different filling. Bioresour Technol 353:127136. https://doi.org/10.1016/j.biortech.2022.127136

Wijekoon P, Koliyabandara PA, Cooray AT, Lam SS, Athapattu BCL, Vithanage M (2022) Progress and prospects in mitigation of landfill leachate pollution: risk, pollution potential, treatment and challenges. J Hard Mater 421:126627

Witthayaphirom C, Chiemchaisri C, Chiemchaisri W (2020a) Optimization of reactive media for removing organic micro-pollutants in constructed wetland treating municipal landfill leachate. Environ Sci Pollut Res 27:24627–24638. https://doi.org/10.1007/s11356-019-06010-3

Witthayaphirom C, Chiemchaisri C, Chiemchaisri W, Ogata Y, Ebie Y, Ishigaki T (2020b) Organic micro-pollutant removals from landfill leachate in horizontal subsurface flow constructed wetland operated in the tropical climate. J Water Process Eng 38:101581. https://doi.org/10.1016/j.jwpe.2020.101581

Wu Y, Si Y, Zhou D, Gao J (2015) Adsorption of diethyl phthalate ester to clay minerals. Chemosphere 119:690–696

Wu J, Zheng J, Ma K, Jiang C, Zhu L, Xu X (2022) Tertiary treatment of municipal wastewater by a novel flow constructed wetland integrated with biochar and zero-valent iron. J Water Process Eng 47:102777. https://doi.org/10.1016/j.jwpe.2022.102777

Yadav A, Chazarenc F, Mutnuri S (2018) Development of the "French system" vertical flow constructed wetland to treat raw domestic wastewater in India. Ecol Eng 113:88–93. https://doi.org/10.1016/j.ecoleng.2018.01.001

Yalçuk A, Ugurlu A (2020) Treatment of landfill leachate with laboratory scale vertical flow constructed wetlands: plant growth modeling. Int J Phytoremediation 22(2):157–166

Yang W, Chang J, Xu B, Peng C, Ge Y (2008) Ecosystem service value assessment for constructed wetlands: a case study in Hangzhou, China. Ecol Econ 68(1–2):116–125

Yang C, Wang B, Wang H, He Z, Pi Y, Zhou J, Liang T, Chen M, He T, Fu T (2022) Removal of organochlorine pesticides and metagenomic analysis by multi-stage constructed wetland treating landfill leachate. Chemosphere 301:134761. https://doi.org/10.1016/j.chemosphere.2022.134761

Yang X, Arias ME, Ergas SJ (2023) Hybrid constructed wetlands amended with zeolite/biochar for enhanced landfill leachate treatment. Ecol Eng 192:106990. https://doi.org/10.1016/j.ecoleng.2023.106990

Yazdania V, Golestanib HA (2019) Advanced treatment of dairy industrial wastewater using vertical flow constructed wetlands. Desalin Water Treat 162:149–155

Zakaria SNF, Aziz HA (2018) Characteristic of leachate at Alor Pongsu landfill site, Perak, Malaysia: a comparative study. IOP Conf Ser Earth Environ Sci 140(1):012013. IOP Publishing

Zhang B, Lin N, Chen X, Fan Q, Chen X, Ren T, Guo H (2021a) Nonlinear water quality response to numerical simulation of in situ phosphorus control approaches. Water 13(5):725

Zhang Y, Liu R, Tan L, Yang N, Wang H (2021b) Adsorption of Pb (II) by montmorillonite modified biochars and reduction Pb (II)-stress in plants of microcosms of constructed wetlands: mechanism and treatment performances. Desalin Water Treat 221:152–162

Zhao R, Liu J, Feng J, Li X, Li B (2021) Microbial community composition and metabolic functions in landfill leachate from different landfills of China. Sci Total Environ 767:144861

Zhou X, Liang C, Jia L, Feng L, Wang R, Wu H (2018) An innovative biochar-amended substrate vertical flow constructed wetland for low C/N wastewater treatment: impact of influent strengths. Bioresour Technol 247:844–850

Chapter 9
Microbial Technologies for Degradation of Microplastics in Wastewater

Manju Singh, Rachna Bhateria, Sunil Kumar, and Mona Sharma

Abstract Microplastics (MPs) have become a ubiquitous and concerning environmental issue in recent years. These tiny plastic particles, typically measuring less than 5 millimeters in size, have infiltrated various ecosystems worldwide including oceans, rivers, soils, and even the air we breathe. The pervasive use of plastics in numerous industries, coupled with improper waste management practices, has led to the widespread distribution of microplastics and their detrimental impacts on both the environment and human health. These particles persist in the environment for hundreds of years, as plastics are highly resistant to degradation. The effects of microplastics on ecosystems are multifaceted and concerning. Aquatic organisms, such as fish, marine mammals, and seabirds, mistake microplastics for food, leading to ingestion and subsequent health issues. The particles can also accumulate toxic chemicals from the surrounding environment, acting as carriers of pollutants and potentially transferring them up the food chain. In terrestrial environments, microplastics can disrupt soil ecosystems and affect plant growth and nutrient cycles. Addressing this issue requires innovative and sustainable approaches for plastic waste management. Recent studies have highlighted the potential of microbial degradation as a promising solution for the breakdown of microplastics (MPs). This chapter presents an overview of the present research on microbial degradation of microplastics, emphasizing recent data and findings. Researchers have identified various bacteria, fungi, and other microorganisms capable of enzymatically degrading different types of microplastics, including polyethylene (PE), polypropylene (PP), and polystyrene (PS). Recent studies have revealed the enzymatic pathways involved in the degradation process, providing insights into the mechanisms underlying microbial MP degradation. Recent data has demonstrated the effectiveness of

M. Singh · R. Bhateria (✉) · S. Kumar
Department of Environmental Sciences, Maharshi Dayanand University, Rohtak, Haryana, India

M. Sharma
Department of Environmental Studies, Central University of Mahendergarh, Jaat, Haryana, India

S. S. Kumar, V. Kumar (eds.), *Nature-Based Solutions for Decentralised Liquid Waste Management*, Environmental Challenges and Solutions,
https://doi.org/10.1007/978-3-032-08362-3_9

specific bacterial strains in degrading MPs, thereby offering a potential avenue for the development of practical approaches to plastic waste management. The use of microbial degradation not only contributes to the reduction of microplastic pollution but also holds promise for the development of sustainable waste treatment strategies.

Keywords Microbial degradation · Microplastics · Plastic waste management · Enzymatic activity · Bacterial strains · Sustainable solutions · Wastewater treatment

9.1 Introduction

In recent years, the discovery of emerging contaminants in various environmental matrices has raised significant concerns among scientists and policymakers globally (Thompson 2021). These contaminants are characterized by their wide distribution, bioaccumulation potential, toxicity, and persistence, posing a considerable challenge to environmental and human health (Hartmann et al. 2019). One particular contaminant that has garnered increased attention over the past decade is microplastic fragments.

Microplastics (MPs) are small plastic particles, typically measuring less than 5 millimeters in size, that have become a focal point of research and environmental concern (Rochman et al. 2019). They can originate from a range of sources, including the breakdown of larger plastic items, the shedding of microfibers from textiles, and the presence of microbeads in personal care products (Catarino et al. 2020). Their widespread occurrence in diverse ecosystems such as oceans, rivers, lakes, and even the atmosphere has raised alarm bells.

Studies have demonstrated the ubiquitous presence of microplastics in various environmental compartments and the potential risks they pose to wildlife and ecosystems (Horton et al. 2022). Aquatic organisms, including fish, seabirds, and marine mammals, are particularly vulnerable to the ingestion and entanglement of microplastics, leading to adverse effects on their health and survival (Rochman et al. 2019). Moreover, microplastics can enter the human food chain and raise concerns about potential impacts on human health (Schymanski et al. 2018). Research on microplastics has rapidly expanded, aiming to understand their sources, distribution, fate, and potential impacts. It has become evident that microplastic pollution is a complex and multifaceted issue that requires interdisciplinary approaches and effective mitigation strategies (Catarino et al. 2020). Ongoing studies continue to provide insights into the environmental occurrence and effects of microplastics, driving the need for comprehensive monitoring, regulations, and public awareness (Horton et al. 2022).

According to recent data, the environmental absorption of plastic waste stands at approximately 71%, while only 29% undergoes recycling or other forms of proper disposal, leading to an alarming rise in microplastic contamination (Jambeck et al. 2015; Lebreton et al. 2017). In terms of global plastic production, the figures indicate a significant increase. In 2016, approximately 335 million metric tons of plastic items was manufactured, while the production surged to approximately 348 million metric tons in 2020, reflecting a remarkable growth of nearly 4% within a single year (PlasticsEurope 2021; Statista 2021). Wastewater treatment plants (WWTPs) receive a diverse range of microplastics (MPs), including microfibers, microbeads, fragments, pellets, and foam, originating primarily from everyday products such as facial cleansers, cosmetics, sunscreen, shower gels, toothpaste, and more (Frias et al. 2022; Sadia et al. 2022). These MPs are also generated through the mechanical, biological, chemical, and photooxidative degradation of larger plastic items (Rochman et al. 2020). Global reports have highlighted the release of MPs into various environmental matrices, including water, soil, and deep-sea sediments, via WWTP effluent (Besseling et al. 2017; Rochman et al. 2019). Presently, the marine ecosystem receives an estimated eight million metric tons of plastic waste annually, with projections indicating a fourfold increase by 2050 (Jambeck et al. 2015). The accumulation and persistence of MPs from WWTPs have detrimental effects on aquatic organisms, primarily due to blockage of the digestive tract, leaching of chemicals from ingested MPs, and the introduction of toxic additives (Hurley et al. 2018). The small size of MPs makes aquatic species more susceptible to entanglement, leading to physical harm and asphyxiation (Gall et al. 2020).

Various technologies have been developed for the removal of microplastics (MPs) from aquatic environments, including physical, chemical, and biological methods. However, studies evaluating these techniques have not been able to establish their complete effectiveness and suitability. The main challenges in mitigating aquatic microplastic contamination are attributed to a lack of understanding and the selection of remediation approaches (Koelmans et al. 2021; Pitt et al. 2022).

In response to these challenges, green technologies aimed at reducing plastic consumption are being developed. These include pollution-free technologies, comprehensive environmentally friendly strategies, recycling of conventional plastics, and the ongoing development of biodegradable plastics (Gao et al. 2022a, b, c, d; Mohammad and Qusay 2023). This chapter provides a detailed study of microplastic pollution in aquatic environments and explores various techniques for their removal, with a particular emphasis on microbial biodegradation methods. The focus is on the environmental friendliness and sustainability of these MP removal technologies. However, it is important to note that biodegradation processes utilizing microorganisms in MP removal systems require significant advancements and further in-depth research in the future (Koelmans et al. 2021; Pitt et al. 2022).

9.2 Occurrence, Fate, and Transport of Microplastics

The occurrence, fate, and transport of microplastics (MPs) are influenced by various characteristics. MPs constitute a diverse class of materials, encompassing a wide range of polymer types, particle sizes spanning over six orders of magnitude, various forms (from spheres to fibers), and thousands of different chemical compositions, all of which are likely to be present in water (Cai et al. 2021; Koelmans et al. 2021). MPs can be classified into two main categories: primary and secondary. Primary MPs are intentionally manufactured in microsize, such as microspheres (500 nm or smaller) found in certain cosmetic products, sandblasting and shotblasting abrasives, and MPs used in medical applications and 3D printing. Secondary MPs are formed through the mechanical or photooxidative degradation of larger plastic polymers (Li et al. 2022).

According to approximately 1.5 million tons of primary MPs is discharged into water annually, providing an indication of the scale of water contamination from MPs. Major sources of MPs include personal care products, cleaning agents, synthetic textiles, airblast media, industrial raw materials, and inadequate waste management systems. Wastewater treatment plants (WWTPs) receive MPs from various domestic and industrial sources, including laundry, textile emissions, urban dust, tire wear, plastic pellet spills, road markings, and plastic-based products (Sadia et al. 2022). Subsequently, MPs can accumulate in aquatic systems through the application of sewage sludge as fertilizer on agricultural land, further impacting the digestive, circulatory, and reproductive systems of aquatic organisms, as well as human health (Hartmann et al. 2019; Li et al. 2022). The majority of microplastics (MPs) are considered resistant to biotic breakdown. However, some substances are susceptible to biodegradation by fungi and bacteria, and upon decomposition they can be ingested or absorbed by organisms at successive trophic levels, leading to gastrointestinal obstruction (Glaser 2019). Studies have shown that microplastics are present in organisms at various levels of the marine food chain (Setälä et al. 2018). The effects of microorganisms and microplastic interactions can vary depending on physiological and environmental conditions.

Ingesting microplastics can result in three types of potentially harmful consequences. Firstly, there are physiological consequences for the consuming species, including impaired development and alterations in feeding patterns, which become more pronounced with higher levels of MP ingestion (Horton et al. 2018). Secondly, there are risks associated with the release of hazardous substances. Additives used in plastic production, such as plasticizers, antioxidants, flame retardants, and pigments, can leach into body tissues, causing induced changes or bioaccumulation. The toxicity of these substances can vary depending on the composition and proportion of chemicals used in each plastic material (Botterell et al. 2019). Lastly, microplastics can unintentionally interact with absorbed pollutants. Due to their hydrophobic nature, increased surface area resulting from weathering, and extended exposure time, microplastics can adsorb pollutants at higher concentrations, serving as carriers for contaminants that can enter aquatic organisms (Li et al. 2021).

Microplastics can act as carriers for various pollutants, including polycyclic aromatic hydrocarbons (PAHs), polychlorinated biphenyls (PCBs), dichlorodiphenyltrichloroethane (DDT), organohalogenated insecticides, hexachlorocyclohexanes, and chlorinated benzenes. Persistent organic pollutants (POPs) like polybrominated diphenyl ethers (PBDEs) and PCBs have been found to possess endocrine-disrupting properties, mimicking natural hormones and interfering with reproductive and sexual functions (Gao et al. 2021; Syed et al. 2021). The specific polymer characteristics and the kinetics of the absorption of persistent organic pollutants into plastic materials play a significant role in the presence of contaminants on microplastics (Rosenfeldt et al. 2019). Humans can be exposed to microplastics through various routes, including the use of cosmetic products, dietary practices, inhalation of airborne particles, and the use of plastic products (Isaac and Kandasubramanian 2021).

9.3 Presence of Microplastic in Aquatic Environment

Microplastics are prevalent in various aquatic environments, including the water column, sediments, and different aquatic habitats. They have been detected in diverse ecosystems, including deep seas and alpine lakes, highlighting their status as global pollutants (Tagg et al. 2021). Plastic pollution has a widespread distribution in marine and oceanic systems due to the durability and buoyancy of microplastics (Horton et al. 2020). Consequently, these contaminants are considered a form of hazardous waste due to their persistence in the environment (Mehra et al. 2021a, b).

9.4 Microplastics in Freshwater Sources

The pollution of freshwater sources with microplastics is closely linked to the terrestrial environment, especially in areas where rivers pass through urban and agricultural regions. In urban areas, the discharge of sewage and wastewater from plastic production factories contributes to the contamination of freshwater sources. Various pathways, such as wastewater discharge, wastewater overflow, rainfall runoff, and littering, contribute to plastic pollution in rivers in metropolitan areas. Farmland is another major terrestrial source of microplastic pollution in freshwater. The use of fertilizers, plastic films, or greenhouses for crop protection and agricultural irrigation contribute to the presence of microplastics in agricultural soils. Through runoff from rain and irrigation, these microplastics can migrate into nearby rivers and surface waters. Additionally, studies have shown that sewage sludge containing synthetic fibers or embedded microplastics is commonly used as fertilizer or soil amendment (Ziajahromi et al. 2017).

Plastics are introduced into the terrestrial ecosystem in large quantities and can subsequently reach freshwater ecosystems through processes such as precipitation or land infiltration. In the European Union (EU), approximately 4 to 5 million tonnes of treated wastewater is used annually for agricultural purposes, leading to the introduction of microplastics into the soil (Lyare et al. 2020). Additionally, the use of mulch film in agriculture, particularly in facility agriculture, contributes to plastic contamination in the terrestrial environment. Unfortunately, the low recovery rate of agricultural film results in a significant amount of it remaining in the environment after use (Ma et al. 2020). The direct utilization of wastewater containing microplastics for irrigation purposes also leads to the accumulation of microplastics in the soil, which can eventually migrate into freshwater bodies. The type of microplastics, such as microbeads and microfibers, can interact with soil differently, influencing their movement in soil matrices (Lyare et al. 2020).

Another source of microplastics in freshwater environments is the release of microplastics into the air. Studies have shown that air dust in metropolitan regions contains fibers composed of synthetic materials, which can subsequently be transported and deposited in aquatic systems through processes like transit and sedimentation (Dris et al. 2016). Sewage treatment facilities, while capable of removing a portion of microplastics, should not be disregarded as a source of contamination. Despite their removal efficiency, a significant proportion of microplastics ends up in sludge, which is commonly used in the production of crop fertilizers or directly applied to crops, thus becoming a significant source of microplastics in the environment (Lyare et al. 2020).

9.5 Microplastics in Marine Sources

Microplastics in marine sources have been extensively studied, revealing their presence in various marine environments worldwide and throughout the water column. Understanding the distribution of microplastics vertically and horizontally in oceans, between geographical regions, and within different compartments is crucial for assessing their potential impacts on marine biota and ecosystems (Lusher et al. 2021; van Sebille et al. 2020). While the exact origins of microplastics are challenging to pinpoint due to their fragmented and heterogeneous nature, several land-based sources have been identified as major contributors. These include urban and storm runoff, industrial activities, improper waste disposal, and beachgoers (Koelmans et al. 2017; Lebreton et al. 2017). Storm sewers, wind, and currents play a significant role in transporting microplastics into the marine environment, as does the degradation of macroplastic waste from marine recycling ports and landfills (Browne et al. 2011; Cózar et al. 2014). Additionally, the application of sewer sludge, which contains a higher concentration of microplastics compared to effluent, can lead to the introduction of microplastics into aquatic ecosystems (Besseling et al. 2014).

Microplastics have been found in a wide range of marine organisms, including shorebirds, fish, mollusks, mammals, and crabs, as well as in beaches, seabed sediments, and surface waters (Gall and Thompson 2015; Phuong et al. 2016; Wright et al. 2013). Studies have demonstrated that microplastics can enter the ocean through the waste of zooplankton, highlighting the indirect ingestion of microplastics through the consumption of fecal pellets. Estimates suggest that a significant proportion of microplastics sink to the seafloor, with around 92% of the 5.25 trillion particles globally being microplastics (Kooi et al. 2017). Contrary to expectations, microplastics are found to be 100 times less abundant on the sea surface than estimated (Kane et al. 2020). The Arctic Sea has been identified as a global sink for microplastics, with some being discovered frozen within ice formations (Obbard et al. 2014). Microplastics have a wide geographic reach, and studies have primarily focused on surface waters in the Northern Hemisphere. Evidence from various regions, including South China, coastal and open oceans, and contained and semi-closed seas, confirms the presence of microplastics at the ocean's surface (Cózar et al. 2017; Su et al. 2021).

9.6 Hazards Caused to Living Ecosystems

Microplastics (MPs) pose several hazards to aquatic life and can also impact human health when contaminated food, including aquatic animals, is consumed or when individuals are exposed to areas with aquatic MP contamination (Issac and Kandasubramanian 2021a, b). While the full extent of the impact is still not fully understood, studies have revealed significant life cycle alterations among microbes due to MP contamination. Some changes may have neutral effects on other living systems, but it is concerning that MP contamination can promote the growth of pathogenic microbes, leading to indirect adversities in aquatic environments (Arias-Andres et al. 2019a, b).

One of the primary hazards of ingesting microplastics is the physical blockage, internal abrasions, and internal and external wounds they can cause in organisms. This can result in the affected organisms needing to expend energy on digestion, suffering from malnutrition and debilitation, or even perishing (Badola et al. 2021a, b; Roy et al. 2022a, b). Furthermore, microplastics can release hazardous additives, such as plasticizers, stabilizers, colors, fillers, and flame retardants, which can be toxic to organisms. These compounds can not only poison the organisms but also exhibit carcinogenic or endocrine-disrupting effects, impacting the species' ability to reproduce (Chen et al. 2022a, b, c, d). Microplastic's hydrophobic nature and high surface-to-volume ratio make them capable of absorbing various contaminants, including heavy metals, chlorinated and aromatic chemicals, and potentially persistent organic pollutants. This process can lead to an increased accumulation of toxic substances in the aquatic environment (Westphalen and Abdelrasoul 2018a, b; Chen et al. 2022a, b, c, d).

9.7 Status of Microplastic Usages and Disposal Systems in India

India has witnessed significant growth in its plastics sector, with Western India being the largest user of plastics in the country. The annual plastic waste generation in India is estimated to be around 26 million metric tons, and the average plastic consumption per person is approximately 11 kg per year. This places India among the countries with significant plastic waste production (Central Pollution Control Board 2020). Regarding microplastic contamination in marine sediments, studies have documented the presence of microplastics in various coastal regions of India, including Gujarat, Kerala, Goa, Pondicherry, Tamil Nadu, Maharashtra, Andaman and Nicobar Islands, Karnataka, and Lakshadweep (Perumal and Muthuramalingam 2021a, b). A pioneering study by Reddy et al. (2006) investigated microplastic contamination in intertidal deposits at a ship-breaking yard in Gujarat. The study identified nylon, polyester, polyurethane, and polystyrene (PS) as the main polymeric materials in the sediments, commonly used in shipbuilding.

Recent investigations on microplastic contamination in India's marine environment have found that the majority of microplastics are irregularly shaped fragments. The abundance of microplastics in a specific area appears to be influenced by both anthropogenic and natural activities. For instance, a study in Puducherry revealed a significant association between microplastics found on six beaches and the volume of tourism activity (Dowarah and Devipriya 2019). As of the latest available information, India's plastics sector continues to expand rapidly, making it one of the largest users of plastics globally. Western India, which includes states like Maharashtra, Madhya Pradesh, Gujarat, Daman and Diu, Chhattisgarh, and Dadra and Nagar Haveli, accounts for a significant portion (47%) of the country's overall plastic usage. India generates approximately 26 million metric tons of plastic waste annually and has an average per capita plastic consumption of about 11 kg per person each year, making it a major contributor to plastic waste production.

Regarding microplastic usage and disposal systems in India, the situation is still an area of concern and is not fully explored. Studies have identified microplastics in marine sediments of various regions in India, including Gujarat, Kerala, Goa, Pondicherry, Tamil Nadu, Maharashtra, Andaman and Nicobar Islands, Karnataka, and Lakshadweep (Perumal and Muthuramalingam 2021a, b).

Recent studies on microplastic contamination in India's marine environment have found that a significant majority of microplastics are irregularly shaped fragments. The abundance of microplastics in specific areas appears to be influenced by both anthropogenic (human activities) and natural factors. For example, a study in Puducherry found a correlation between the presence of microplastics on beaches and the volume of tourism activity (Dowarah and Devipriya 2019). However, it is important to note that the status of microplastic usage and disposal systems in India may change over time as the country continues to develop and implement measures to tackle plastic waste pollution. Ongoing research and monitoring are crucial to fully understand the extent of microplastic contamination in different regions of India and to implement effective measures to mitigate its impact on the marine ecosystem and human health.

Moreover, Karthik et al. (2018) examined 25 beaches along Tamil Nadu's south coast and discovered that coastlines close to rivers had higher microplastic (MP) abundances, indicating that the majority of these fragments were carried by rivers from land-based sources. The study highlights the significant role of rivers in transporting microplastics to coastal areas. While studies on microplastics in Indian freshwater systems have primarily focused on surface water and sediment components of lakes and rivers, there have been limited investigations on groundwater samples. Recent research explored the presence of microplastics in Indian lakes and identified the fragmentation of macroplastic debris as a major source of these synthetic polymers (Vaid et al. 2021a, b). Leakage of primary microplastics from personal care products or industries, riverine transport, and runoff activities during rainfall and dry deposition are other contributing factors to microplastic pollution in lakes.

Research conducted by Napper et al. (2021) revealed that the Ganges, Brahmaputra, and Meghna rivers are estimated to discharge between 1 and 3 billion microplastics into the Bay of Bengal every day. While not all microplastics found in rivers may reach the oceans, some of them accumulate in river sediments, which serve as an important sink for microplastics.

9.8 Microplastic Detection Methods

(a) *Sampling*. The first step in measuring microplastics is to choose an appropriate sampling technique. Field sampling, water sampling, porous medium sampling, and biota sampling are examples of general sampling techniques. As microplastics are buoyant, airborne dust is often included in field samples. Typically, manta trawls or nets are employed to sample surface water. Sampling the sediment and the benthic population is especially crucial for detecting microplastics with a density greater than water, which are more likely to sink into a deeper sea environment (Arthur et al. 2020). Porous media sampling, such as using sediment cores or grab samples, is appropriate for obtaining microplastic samples from the lake or ocean floor as well as from the top layer of a beach or lakeshore (Löder et al. 2017; Mintenig et al. 2020). Biota sampling involves the collection of organisms to track the presence of microplastics in the aquatic environment. Fish, seagulls, shellfish, and plankton are commonly used for biota sampling (Paul-Pont et al. 2020).

(b) *Identification*. Once the samples are collected, identification of microplastics is essential. Several techniques are commonly used for this purpose, including visual identification, microscopy, spectroscopy, and molecular techniques. Visual identification involves the manual sorting and identification of particles based on their size, shape, color, and texture (Frias et al. 2014). Microscopy techniques, such as optical microscopy and electron microscopy, provide higher magnification and resolution for detailed analysis of microplastics (Claessens et al. 2013; Song et al. 2021). Spectroscopy techniques, such as Fourier transform infrared (FTIR) spectroscopy and Raman spectroscopy, are used to analyze the chemical composition of microplastics (Gündoğdu et al. 2020; Yang et al. 2019). Molecular techniques, including polymerase chain reaction (PCR) and DNA barcoding, can be applied to identify microplastics based on their genetic information (Harrison et al. 2011; Nel et al. 2020).

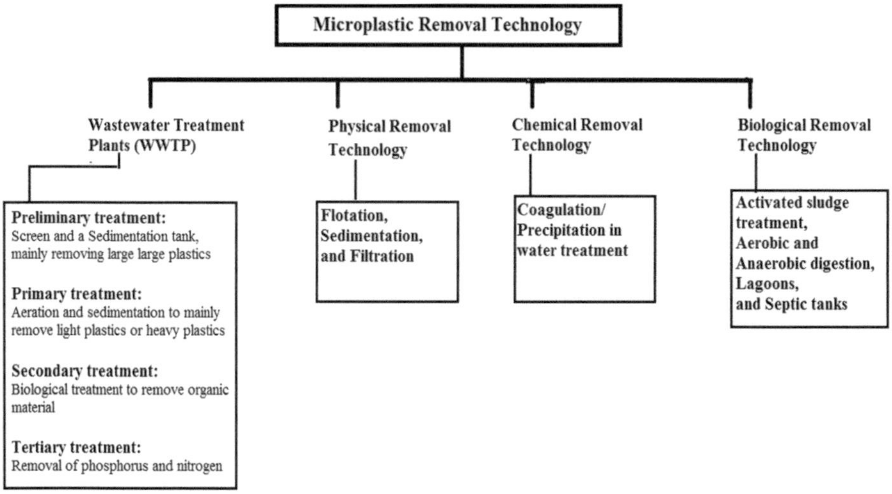

Fig. 9.1 Microplastic removal techniques. (Source: Mitrano et al. 2020)

9.9 Techniques of Microplastic Removal from Water

Methods for the removal of microplastics from water can be categorized into physical, chemical, and biological techniques as shown in Fig. 9.1. These methods aim to prevent microplastics from entering different ecosystems and are primarily developed for wastewater treatment.

Physical Removal methods involve the separation of microplastics from water through various physical processes. Some common physical techniques include filtration, sedimentation, flotation, and sieving. Filtration involves the use of filters with specific pore sizes to trap and remove microplastics from the water. Sedimentation relies on gravity to allow microplastics to settle at the bottom of a container or tank for removal (Li et al. 2018). Flotation techniques utilize the buoyancy of microplastics to separate them from water by air or gas bubbles (Krause et al. 2020). Sieving involves the use of sieves or meshes with specific openings to physically separate microplastics based on their size (Dehaut et al. 2016).

Chemical Methods for microplastic removal involve the use of chemical agents or processes to facilitate the separation or degradation of microplastics. Some examples include coagulation/flocculation, oxidation, and advanced oxidation processes. Coagulation/flocculation techniques involve the addition of chemicals that cause microplastics to aggregate and form larger particles for easier removal (Schwarz et al. 2021). Oxidation methods use chemical oxidants to degrade microplastics into smaller fragments or mineralize them into harmless by-products (Ragab et al. 2018). Advanced oxidation processes, such as photocatalysis or ozonation, utilize powerful oxidants generated by UV light or ozone to break down microplastics (Mitrano et al. 2020).

Biological Techniques focus on the use of microorganisms or biological agents for microplastic degradation. Biodegradation processes utilize microorganisms that can enzymatically degrade microplastics into smaller fragments or break down their chemical bonds. Bioaugmentation, biofilm development, and bio-inspired materials are some examples of biological techniques being explored for microplastic removal (Zhao et al. 2021).

Overall, a combination of these physical, chemical, and biological techniques may be necessary to effectively remove microplastics from water and mitigate their impact on ecosystems.

9.10 Role of Wastewater Treatment Plants (WWTPs) in Preventing Microplastic Contamination

Wastewater treatment plants (WWTPs) play a crucial role in preventing microplastic contamination by removing microplastics from wastewater before it is discharged into the environment. While WWTPs are already effective in removing a significant percentage of microplastics, further measures are necessary to reduce their release. Microplastics present in wastewater primarily originate from sources such as synthetic garments, cleansers, and cosmetics. Despite the existing treatment processes, the final effluent still contains a considerable number of microplastics, leading to their release into rivers and oceans. This highlights the need for improved procedures to effectively separate microplastics from both the sludge and the effluent to minimize their environmental impact (Calero et al. 2021a, b).

To address this issue, WWTPs should implement measures to limit recontamination of the processed sludge and effluent. This can involve adopting advanced treatment technologies or modifying existing treatment processes to enhance microplastic removal efficiency (Arpia et al. 2021a, b). The standard wastewater treatment process in WWTPs consists of four phases: preliminary treatment, primary treatment, secondary treatment, and tertiary treatment. The initial stage, known as "pretreatment," includes both preliminary and primary treatment steps. During this stage, bulky objects such as rags and sticks are removed to prevent damage to pumps and disruptions in subsequent purification steps (Chen et al. 2022a, b, c, d).

It is essential for WWTPs to continually improve their treatment processes and explore innovative technologies to enhance microplastic removal efficiency. This may involve the implementation of advanced filtration systems, optimized sedimentation processes, and the incorporation of additional treatment steps specifically designed for microplastic removal. By effectively managing and treating wastewater, WWTPs can significantly reduce the release of microplastics into the environment, contributing to the overall protection of ecosystems and human health.

During the first treatment stage of a wastewater treatment plant (WWTP), the presence of large flocs of fat, oil, and grease in the wastewater can aid in capturing microplastics (MPs). This leads to the removal of suspended or floating MPs along with other insoluble contaminants, resulting in an efficiency of 35–59% for MP elimination at this step. In the primary clarifiers, surface skimming and gravity separation techniques are employed to partially remove the remaining MPs. The heavier MPs can settle or become trapped in sludge flocs, while the lighter floating MPs can be captured during surface skimming by the grease layer. As a result, after the primary treatment, typically 50–98% of MPs can be eliminated. The removal efficiency of MPs during the pretreatment procedure is influenced by factors such as the size, chemical composition, shape, and other properties specific to MPs.

Pretreatment techniques involving photocatalytic and biological disintegration have shown potential for reducing the release of microplastics into the environment and improving treatment efficiency. Activated sludge has also been historically used for MP separation, but the disposal of sludge poses sustainability challenges. Membrane filtration has demonstrated promising results in removing MPs, particularly when combined with other treatment methods. However, membrane fouling remains a challenge. To address these issues, additional techniques such as electrocoagulation, magnetic extraction, and adsorption are being explored for MP removal. However, the commercial application of these approaches is presently limited by factors such as higher energy consumption, potential for secondary microplastic contamination, and longer process duration (Dey et al. 2021a, b, c, d, e).

In the secondary treatment stage of a WWTP, activated sludge and a clarifier tank are used together to further remove any remaining suspended particles and dissolved organic contaminants from the water. The microorganisms in the activated sludge not only help in capturing MPs but also enhance their removal efficiency. The development of biofilms on the surface of MPs alters their relative densities, facilitating their sinking and subsequent separation from the suspended particles. Although a small number of bacteria, including *Bacillus*, *Rhodococcus*, and *Nocardia asteroides*, are known to be capable of breaking down MPs, their biodegradation rates are low, and their contact times with activated sludge are short, which has little impact on the breakdown of MPs during secondary treatment (Harrison et al. 2020). In comparison to the pretreatment effluent, the overall removal effectiveness of MPs is reported to be in the range of 86–99.8%, and this is often accompanied by a reduction in the average size of the MPs (Wang et al. 2022).

The third step of the wastewater treatment procedure, tertiary treatment, also known as advanced treatment, is widely used to produce high-quality drinking water. Various techniques such as membrane bioreactors (MBRs), sand filtration, disc filters, dissolved air flotation (DAF), and gravity filtration are employed to reduce the proportion of suspended solid impurities, organic contaminants, heavy metals, and pathogens in water (Alam et al. 2020). These advanced treatment methods have shown high effectiveness in removing MPs, with reported removal efficiencies ranging from 98% to 99.9%. It is worth noting that the removal of fibrous MPs can be particularly challenging using these methods (Alam et al. 2020).

In the final stage of wastewater treatment, after the traditional activated sludge (CAS) approach, several cutting-edge technologies are employed as final-stage wastewater treatment technologies. Membrane bioreactors (MBRs) are utilized to treat primary wastewater, combining membrane separation with activated sludge (CAS) to greatly enhance the quality of the final effluent. Secondary effluent is often treated with rapid (gravity) sand filters (RSF) and dissolved air flotation (DAF), both of which contribute to further purification (Dong 2018).

9.11 Physical MP Removal Approaches

Physical MP removal approaches, including filtration, sedimentation, absorption, and flotation, are widely used for microplastic removal (Fryczkowska and Przywara 2021a, b). Among these approaches, filtration techniques such as microfiltration (MF), ultrafiltration (UF), reverse osmosis (RO), dynamic membranes (DM), and membrane bioreactors (MBRs) have shown promise in producing high-quality water from primary or secondary effluent (Vahvaselkä and Winquist 2021a, b).

Filtration using asymmetric membranes with micrometer- or nanometer-sized pores is effective in removing contaminants such as bacteria, protozoa, viruses, and suspended particles. However, despite the good efficacy of UF and RO in removing MPs in wastewater treatment plants (WWTPs), millions of MPs can still be present in the effluent (Vahvaselkä and Winquist 2021a, b). To enhance MP removal efficiency, membrane technologies have been combined with other processes. The development of membrane bioreactors (MBRs) has shown promise, as it combines a biological reactor with a membrane system. MBRs utilize biologically activated sludge with a high biomass concentration and a long solid retention time to effectively remove refractory organic contaminants. The combined suspension is separated using a semi-crossflow filtration system, where the membrane catches the suspended solids and concentrates them in the retentate. The combination of bioreaction mechanisms and porous membranes in MBRs has demonstrated higher effectiveness in treating MPs in contaminated wastewater (Chen et al. 2022a, b, c, d).

However, the biodegradation of chemically inert MPs during the MBR process is limited, similar to the biological process in secondary treatment. Challenges such as membrane fouling and abrasion can impact the purification effectiveness of MBRs. Nonetheless, MBRs hold significant potential for MP removal in tertiary treatment.

Sedimentation is another physical method employed for the removal of microplastics (MPs) from wastewater. The wide range of polymer types in wastewater, characterized by varying densities, shapes, and sizes, can influence their physical separation from water. Sedimentation has been shown to remove an average of 72% of MP particles in sewage influent, although the smallest particles (<27 μm in diameter) are more likely to pass through without effective removal (Tursi et al. 2022a, b). Studies have demonstrated the effectiveness of sedimentation in reducing the overall size and quantity of MP fragments in wastewater. For example, Liu et al. (2018) reported a reduction of 58.8% in size and 40.7% in quantity of MP pieces in the water of a Chinese WWTP after initial screening and sedimentation. Additionally, the presence of MP adhering to suspended solids in the water facilitated the physical removal of finer particles and fibers, contributing to higher removal efficiency (Tursi et al. 2022a, b).

However, conventional sedimentation techniques have limitations, as they convert impurities from liquid to concentrated solid waste and may not effectively capture the smallest MP particles. Improper disposal or recycling of sludge can lead to the reintroduction of MPs into the water system, posing significant environmental concerns. To enhance the density separation of MPs, the use of salt solutions in laboratory-scale studies has shown promise for isolating MP samples in challenging matrices like sewage or natural water. However, scaling up this technique is challenging due to the requirement of large quantities of expensive salt solutions. Alternatively, increased settling through coagulation is a prospective approach that can be widely applied for MP removal (Tursi et al. 2022a, b).

Adsorption is an effective method for the removal of nanoplastics (NPs) and small microplastics (MPs), particularly those smaller than 10 μm. The adsorption mechanism involves electrostatic interactions, hydrogen bond interactions, and π-π interactions. Researchers have utilized granular or pulverized adsorbents for MP removal (Gao et al. 2022a, b, c, d). For example, synthesized Zn-Al layered double hydroxide (LDH) using a coprecipitation technique and found that the LDH's Zn-Al layers exhibited a high adsorption capacity of 164.49 mg/g for 55 nm polystyrene (PS) MPs in deionized water, with a 96% removal rate for all MPs.

Adsorbents with enhanced porosity can be immobilized on a carrier to effectively absorb and remove MPs from water. The interconnected pore structure of the carriers facilitates rapid solvent transit and ensures sufficient interaction between MPs and adsorbents. Developed foam materials based on zirconium metal-organic frameworks, which achieved removal rates of 88.2 ± 1.7%, 85.7 ± 4.8%, and 90.1 ± 2.1% for polymethyl methacrylate (PMMA), PS, and polyvinylidene fluoride (PVDF) MPs in water, respectively. In recent developments, magnetic separation has emerged as a technique for MP removal. Large-surfaced magnetic nanoparticles are employed as adsorbents during the removal process, enabling them to bind with MPs. The magnetized MPs can then be swiftly and easily separated from water using magnetic force.

The effectiveness of magnetic removal of microplastics (MPs) can be influenced by various factors such as the type and size of MPs, water pH, quantity of magnetic particles, and other variables. A recent study has shown that nano-Fe_3O_4 can be used for magnet adhesion and magnetization to remove MPs. The average removal rates for the most common forms of MPs, including PE, PP, PS, and PET, ranged from 62.83% to 86.87%. However, magnetic materials often require superparamagnetization, which leads to their desorption and regeneration. To ensure effective removal, a significant amount of magnetic adsorbents needs to be added, surpassing the quantity of MPs in the water. In laboratory settings, centrifugation is commonly employed to achieve this, but implementing it in large-scale drinking water treatment plants (DWTP) and wastewater treatment plants (WWTP) is challenging. For example, multi-walled carbon nanotubes (MCNTs) demonstrated the removal of approximately 80% of MPs after four cycles of use through heat treatment at 600 °C. To validate the efficacy of this approach, its large-scale implementation is necessary. However, it is crucial to address the potential environmental hazards associated with magnetic adsorbent residue and take measures to minimize their impact.

Flotation methods are employed in various forms for wastewater treatment, including dissolved air flotation (DAF), froth flotation, induced air flotation (IAF), electrolytic flotation, and vacuum flotation. These methods rely on the formation of bubbles to facilitate the removal of pollutants. Flotation has proven to be an effective technique for removing small plastic granules from soil or sediment in dense liquids.

Dissolved air flotation is particularly effective in removing algae and low-density particles from wastewater. However, this technology comes with high operation and maintenance costs compared to the sedimentation process. A moveable density flotation method was presented which demonstrated an average effectiveness of 95.8% in extracting microplastics from sediments with particle sizes ranging from 100 μm to 10 mm (Fahimeh Hooriabad Saboor 2022). In wastewater treatment facilities, flotation and agglomeration techniques are frequently employed to create larger aggregates of particles that are easier to separate. These procedures involve the use of coagulants such as Fe- and Al-based salts, which initiate uptake-complexation mechanisms and facilitate the formation of solid connections between waste particles (Alrbaihat 2022a, b, c).

9.12 Chemical MP Removal Approaches

Mostly, the chemical processes of MP removal processes make use of coagulation-flocculation-sedimentation, oxidation, and photocatalytic techniques as discussed below.

Coagulation-Flocculation-Sedimentation Several DWTPs use the coagulation-flocculation-sedimentation (CFS) method to remove particles and colloidal compounds, as well as MPs and NPs. Particles are separated from the aqueous phase during the removal by coagulation/flocculation processes, primarily by the use of metallic salts (iron and aluminum) and polymeric flocculant aids like polyacrylamide (PAM) (Pan et al. 2023a, b). Rajala et al. recently investigated the usage of commonly used inorganic and organic coagulants for the removal of MPs in tertiary wastewater treatment (2020). MP removal was studied by adding known quantities of PS spheres with a diameter of less than 10 μm to secondary WWTP effluent. Coagulation/flocculation was achieved using ferric chloride, polyaluminum chloride, and cationic polyamine, followed by the settling of the mixture. MP concentrations were measured using flow cytometry (Vahvaselkä and Winquist 2021a, b). Alum CFS treatment was found to more readily remove PSMPs that were shorter than 90 μm, as observed during the validation of the methodology. Alum flocculation has been proposed as the primary method for removing MPs from drinking water, as it can remove MPs from solutions containing surfactants. Additionally, when ferric and aluminum sulfates were used as coagulants, approximately 80% of MPs were eliminated. Under ideal coagulation conditions (ferric sulfate at 20 mg/L and pH 7 or aluminum sulfate at 40 mg/L and pH 7), polyvinyl chloride (PVC) MPs smaller than 50 μm were coagulated by ferric and aluminum sulfate, and MPs no smaller than 15 μm were completely removed. The number of studies on this technique for MPs is presently scarce, particularly for wastewater treatment systems, according to Bui et al. (2020). Future research should focus on identifying the ideal coagulant/flocculant aids and conditions for their use in MP removal.

Electrocoagulation (EC) is a process that utilizes metal electrodes to generate coagulant electrically, rather than chemically. It has been demonstrated to efficiently remove colors, heavy metals, oils, and pharmaceuticals from wastewater. Perren et al. (2018) utilized EC to treat artificial wastewater containing PE microbeads at various concentrations. The study investigated the effects of initial pH, NaCl content, and current density on the removal efficiency. It was found that EC effectively removed MPs by over 90% across pH ranges from 3 to 10. The highest removal efficiency of 99.2% was achieved at a pH of 7.5 (Dey et al. 2021a, b, c, d, e). Herborg et al. (2018) developed a novel method for MP removal from water using organosilanes, which are compounds with one organic group and three reactive groups. In the first stage of the fixation process, organosilanes bind to the surface of MP particles and aggregate them due to the interaction between the organic group and the MP surface. In the second step, the three reactive groups combine to form a solid hybrid silica gel through a water-induced sol-gel process, chemically incorporating and fixing the MPs. The larger agglomerates formed float on the water's surface, enabling easier removal from other materials (Vahvaselkä and Winquist 2021a, b). Organosilanes show great promise as a chemical class for MP removal methods, as the surface chemistry of the polymer can be adjusted through the organic group. Increasing the concentration of organosilanes is another approach to enhance efficiency.

9.13 Oxidation Methods for the Removal of Microplastics from Water

The method of chemical oxidation, based on either mineralization or the breakdown of MPs into tiny molecules, has gained attention for its potential in removing MPs from water. Among the various oxidation methods, advanced oxidation method

AOPs have emerged as effective strategies for the mineralization of organic contaminants by oxidizing them using radicals such as SO_4•- and HO•-. AOPs have also been recently explored for reducing MPs in water. For instance, developed an electro-Fenton-like method using TiO_2/graphite (TiO_2/C) cathode to degrade polyvinyl chloride (PVC), resulting in a remarkable mean elimination ratio of 75%.

During the oxidation process, several short-chain organic molecules, including alkenes, alcohols, monocarboxylic acids, dicarboxylic acids, and esters, are formed as by-products in the posttreatment solution. Additionally, high MP concentrations can lead to the accumulation of MPs on the working electrode, hindering mass transfer and blocking active sites on the TiO2/C catalyst, thereby reducing the removal efficiency.

Earlier studies have highlighted that the choice of anode materials, electrolyte types, and their concentrations significantly influence the practical removal efficiency of MPs. For example, demonstrated that the use of boron-doped diamond anodes improved the removal efficiency of MPs compared to other anode materials. Similarly, the selection of appropriate electrolyte types and concentrations can enhance the oxidation process, leading to efficient MP removal (Gottschalk et al. 2018).

Despite the promising results of oxidation methods for MP removal, certain limitations should be considered. These techniques often require high energy consumption and demand sophisticated reaction mechanisms. Moreover, the presence of complex matrices in water, such as dissolved organic matter, can interfere with the oxidation process and affect the removal efficiency of MPs.

9.14 Photocatalysis for the Degradation of Microplastics in Water

Photocatalysis, an environmentally friendly method that harnesses solar energy, has gained significant attention for its efficacy in degrading various pollutants, including antibiotics, pesticides, and dyes in water. Recently, it has emerged as a promising technique for the removal of microplastics (MPs) from water. Additionally, the influence of catalysts on the photocatalytic process and the potential of photocatalysis as a cost-effective and sustainable method for complete mineralization of MPs are discussed. Photocatalysis, utilizing the energy from sunlight, has garnered considerable interest as an eco-friendly approach for water purification. In the context of microplastics, photocatalysis has demonstrated its effectiveness in oxidizing and degrading MPs present in water. For instance, TiO2 nanoparticle coatings exposed to UV light irradiation for 12 h resulted in a remarkable degradation of 98.4% of PS MP solutions (Zhang et al. 2020). The influence of different catalysts on the photocatalytic process and the varying degradation efficiencies of different types of MPs have been explored to enhance the efficacy of this method.

Research findings have shown that the photocatalytic degradation efficiency varies for different types of MPs. For example, polymethyl methacrylate (PMMA) exhibited faster mineralization compared to polystyrene (PS) particles of similar sizes, potentially due to the breakage of C = O and C-O bonds in PMMA being more facile than C-C, C=C, and C-H bonds in polystyrene (Chen et al. 2020). These variations highlight the importance of considering the specific characteristics of MPs when employing photocatalytic degradation strategies. The choice of catalysts plays a crucial role in the photocatalytic process for MP degradation. Comparative studies on PS particles coated with TiO_2 nanoparticles, with and without the presence of Triton X-100 (TXT), revealed significant differences in the degradative effects. The findings indicated that the photocatalytic process can be greatly influenced by the choice of catalysts.

Photocatalysis has gained recognition as a cost-effective and environmentally friendly method for the complete mineralization of MPs and the reduction of MP pollution. For instance, zinc oxide nanorods exposed to visible light for 2 weeks achieved a remarkable photocatalytic destruction efficiency of 65% for polypropylene (PP) MPs adsorbed on the surface. This highlights the potential of photocatalysis as a sustainable solution for addressing the challenges associated with MP contamination in water. Photocatalysis has emerged as a promising method for the degradation of microplastics in water, driven by its effectiveness in harnessing solar energy to oxidize and degrade MPs. The choice of catalysts and the specific characteristics of MPs play pivotal roles in determining the efficiency of the photocatalytic process. Moreover, the cost-effectiveness and environmental sustainability of photocatalysis make it a valuable approach for the complete mineralization and reduction of MP pollution.

9.15 Mechanism of Microbial Degradation of Microplastics

Bioremediation, particularly microbial degradation, has emerged as a promising strategy for the removal of microplastics due to its potential to utilize various bacterial and fungal species or their enzymes for enzymatic hydrolysis of plastics. This article explores the mechanisms involved in the microbial degradation of microplastics, focusing on the adhesion of microorganisms to the polymer surface, enzymatic hydrolysis, depolymerization, and the mineralization process. The role of exoenzymes in breaking down complex polymers into simpler molecules and the subsequent utilization of these molecules as carbon and energy sources by microorganisms are discussed. Additionally, the aerobic and anaerobic degradation pathways and the generation of final by-products, including CO_2, H_2O, and CH_4, are elucidated (Fig. 9.2).

The biodegradation of microplastics primarily involves two key mechanisms: microbial adhesion to the polymer surface and subsequent enzymatic hydrolysis. Microorganisms colonize the exposed surface of the microplastic, and the attached enzymes catalyze the hydrolytic cleavage, leading to polymer degradation through hydrolysis. The resultant breakdown products include simpler molecules such as oligomers, dimers, and monomers, which serve as carbon and energy sources for microorganisms (Miloloža et al. 2022a, b, c, d).

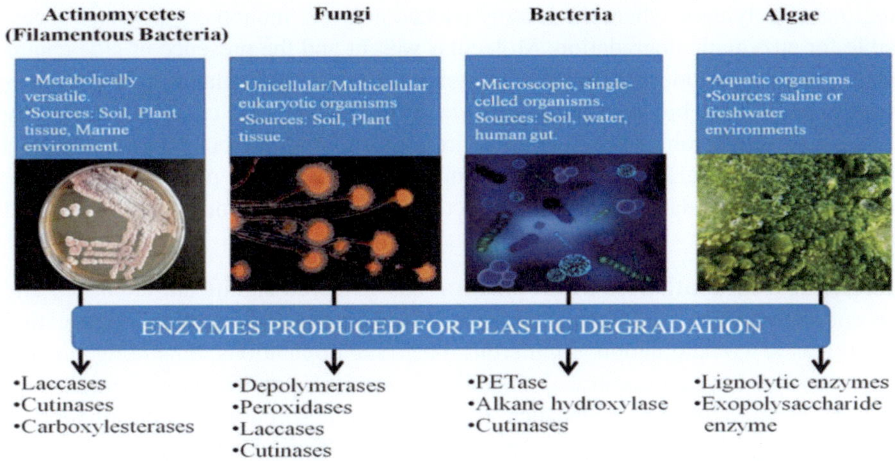

Fig. 9.2 Microbial enzymes causing plastic degradation. (Kaushal et al. 2021)

During microbial degradation, exoenzymes produced by microorganisms play a crucial role in breaking down complex polymers. This process, known as depolymerization, leads to the formation of shorter chains of molecules. These monomers, dimers, and oligomers can be further utilized by microorganisms for their carbon and energy needs. Ultimately, microbial mineralization occurs, converting the degraded products into CO_2, H_2O, and CH_4 as final by-products (Juliana Oliveira 2020; Venkatesan et al. 2022a, b). Microorganisms can mineralize the generated monomers, dimers, and oligomers through either aerobic or anaerobic pathways. Aerobic bacteria employ oxygen (O_2) as an electron acceptor to convert complex microplastic components into simpler molecules. This process results in the release of CO_2, water, and biomass. In contrast, anaerobic degradation occurs in the absence of oxygen, where microorganisms utilize alternative electron acceptors to break down microplastics, leading to the production of methane (CH_4) as a by-product (Dey et al. 2022a, b, c).

Anaerobic biodegradation of microplastics is a significant process in which bacteria break down complex polymers in the absence of oxygen, resulting in the production of by-products such as CH_2, CO_2, H_2O, and biomass. Recent studies have highlighted the role of anaerobic microbes in utilizing alternative electron acceptors for biodegradation, including nitrate, sulfate, iron, manganese, and carbon dioxide (Iram et al. 2019). These alternative electron acceptors enable microorganisms to carry out biodegradation under anaerobic conditions. Surface characterization studies have indicated that microorganisms have a higher affinity for colonizing hydrophilic surfaces, which poses a challenge when considering the highly hydrophobic nature of microplastics. Recent research has shown that modifying the surface properties of microplastics to enhance hydrophilicity can promote microbial colonization and biodegradation. Additionally, the degree of crystallinity of the polymer influences the rate of biodegradation, with more crystalline materials being more resistant to microbial breakdown. Microorganisms primarily target the amorphous regions of polymers, where the loosely packed polymer monomers are more accessible for enzymatic degradation. Molecular weight and the presence of side chains also impact the biodegradability of polymers. Higher molecular weight polymers tend to have lower biodegradability, while polymers with side chains are generally less prone to microbial degradation. Furthermore, environmental factors such as moisture content, pH variations, and temperature fluctuations play crucial roles in microbial biodegradation processes, affecting the efficiency of microplastic removal (Iram et al. 2019).

Bacterial Degradation of Microplastics has garnered significant attention, and recent research has shed light on the role of specific bacterial strains in mitigating microplastic contamination. Numerous bacterial populations inhabiting aquatic environments have demonstrated the ability to settle and multiply on microplastics, thereby accelerating processes such as adsorption, desorption, and elimination of microplastics (Peng et al. 2021; Peng et al. 2022). These bacteria possess unique metabolic capabilities that facilitate the breakdown of microplastics and contribute to their overall removal from the environment. Efficient degradation of organic matter by bacteria and their capacity to aggregate and settle are crucial for effective waste removal from water systems. Microorganisms present in inactivated sludge can form solid aggregates, known as flocs, which are large enough to settle at the bottom of settling basins. Enhanced flocculation properties of sludge promote improved settling and wastewater treatment efficiency (Yang et al. 2020a, b, c; Hu et al. 2021).

The consumption of polymer materials by microorganisms is essential for their carbon acquisition, as they can only thrive in environments with an available nutrient supply. The biodegradation of polymers by bacteria leads to morphological and chemical changes in the microplastics, resulting in a decrease in dry weight, average molecular weight, and molecular dispersion (Liu et al. 2021; Su et al. 2022a, b). Recent studies have provided valuable insights into the diverse bacterial strains capable of degrading microplastics and the underlying mechanisms involved in their degradation processes.

9.16 MP Specific Bacterial Biodegradation Method

Recent studies have provided valuable insights into the specific bacterial biodegradation methods for microplastics, highlighting the role of different bacteria in degrading specific types of microplastics.

Chen et al. (2020) conducted an analysis of polystyrene (PS) microplastics and found that hyperthermophilic bacteria in high-temperature composting (hTC) played a crucial role in the effective biodegradation of PS microplastics. Through strong bio-oxidation capabilities, bacteria belonging to the *Thermus*, *Bacillus*, and *Geobacillus* genera were identified as the predominant bacteria responsible for the efficient biodegradation during hTC. This research demonstrates the potential of hyperthermophilic bacteria in removing microplastics from the environment (Chen 2020). In addition to hyperthermophilic bacteria, other bacteria have shown effectiveness in breaking down specific types of microplastics. Shahadat Hossain (2022) highlighted the role of heterotrophic bacteria in fish and shellfish culture systems, where these bacteria consume organic substances and efficiently reduce toxic nitrogen loading and decompose microplastics such as polyethylene (PE), polystyrene (PS), and polypropylene (PP). This indicates the potential of utilizing specific bacteria for the remediation of microplastics in aquatic environments (Hossain 2022).

Furthermore, the impact of microplastics on wastewater treatment efficiency was investigated by Liu (2019). The study focused on polyester (PES), polyethylene (PE), and polyvinyl chloride (PVC) microplastics and their effects on ammonium-oxidizing bacteria (AOB), nitrite-oxidizing bacteria (NOB), denitrifiers, and polyphosphate-accumulating organisms (PAOs). The findings suggested that microplastics in the studied size range and concentration did not significantly affect the activities of AOB, NOB, denitrifiers, and PAOs. Therefore, the influence of microplastics on wastewater treatment efficiency may not require excessive concern (Liu 2019). The biodegradation of microplastics by bacterial strains can be influenced by various physicochemical and environmental conditions, as highlighted in recent studies. A study by Auta et al. (2017) investigated the ability of *Bacillus cereus* and *Bacillus gottheilii*, isolated from mangrove sediments, to remove microplastics made of different polymers. Morphological and structural changes of the microplastics were examined using electron microscopy and FTIR analysis, and the rate of biodegradation was determined based on weight loss. The results showed that *B. cereus* exhibited the fastest mass decrease (0.0019/day) and the lowest half-life of degradation (363.16 days) when degrading polystyrene microplastics, while *B. gottheilii* on polyethylene achieved a mass decrease rate of 0.0016/day and a half-life of 431.25 days (Auta et al. 2017; Lichtfouse 2020).

The particle size of bioplastics can also affect their biodegradation rate. Larger particle sizes provide less surface area for bacterial biodegradation, resulting in longer durations for aerobic or anaerobic biodegradation. García-Depraect et al. (2022) investigated the biodegradation of different polyhydroxyalkanoates (PHAs) and polycaprolactone (PCL) with varying particle sizes. Anaerobic biodegradation of PHB and PHBV reached a plateau after 35, 42, and 65 days for particle sizes of 100–250 μm, 250–500 μm, and 500–1000 μm, respectively. Under aerobic conditions, the biodegradation of PHAs took about 52 days to plateau for the smallest particle size and up to 3 months for the largest particle size. Similarly, the aerobic biodegradation of PCL (100–250 μm) reached a plateau after 65 days, while particle sizes of 250–500 μm and 500–1000 μm required 79 and 86 days, respectively. These findings demonstrate that the physicochemical properties of microplastics, such as polymer type and particle size, can significantly influence bacterial biodegradation rates. Understanding these factors is essential for developing effective strategies to mitigate microplastic pollution.

Bacterial Biodegradation of Microplastics in Wastewater Treatment Plants (WWTPs) has gained attention as a potential method for mitigating microplastic pollution. Recent studies have focused on identifying bacteria that produce enzymes capable of breaking down microplastics and evaluating their effectiveness in WWTPs.

Enzymes like PETase and MHETase, produced by certain bacteria, have shown promise in degrading microplastics. Research efforts are directed toward identifying additional microorganisms that can produce these enzymes to enhance the decomposition and reduction of microplastic pollution (Dey et al. 2021a, b, c, d, e). *Bacillus cereus* and *Bacillus gottheilii* are bacteria that have been reported to possess the capability to degrade microplastics and nanoparticles (MPs/NPs). Under anaerobic conditions, the degradation of microplastics leads to the production of carbon dioxide, water, methane, and hydrogen sulfide. In aerobic conditions, the further breakdown of monomers occurs, resulting in the generation of carbon dioxide and water. Incorporating these microbial strains into WWTP systems through modern bioengineering approaches can aid in the remediation of MPs/NPs as shown in Table 9.1.

It is important to note that while bacterial biodegradation is environmentally friendly, it is often less efficient and time-consuming. Microorganisms can be sensitive to various factors, requiring further research to optimize their application on a larger scale in WWTPs (Sudharshan Reddy 2022). These findings highlight the potential of bacterial biodegradation as a strategy for microplastic remediation in WWTPs. Further research and development are necessary to enhance the efficacy and practical application of this approach in tackling the issue of microplastic pollution. In marine environments, the presence of complex hydrocarbon-degrading bacteria has been observed in association with plastics. These bacteria possess genes encoding enzymes such as monooxygenases, peroxidases, and dehydrogenases, which have the potential to initiate the degradation of plastics. Amplicon sequencing of the 16S rRNA gene has revealed the presence of *Flavobacteriaceae* and *Rhodobacteraceae* on various types of plastic materials including PET bottles, PE sheets, PVC, PE and PP dolly ropes, and PE, PP, and PS plastics (Vaksmaa et al. 2021a, b, c).

Recent studies have shed light on the bacterial biodegradation of microplastics in marine environments. For example, a study by Su et al. (2022a, b) investigated the microbial degradation of polyethylene microplastics in seawater. The researchers identified bacterial strains belonging to the genera *Pseudomonas*, *Bacillus*, and *Vibrio* that were capable of degrading polyethylene. The degradation process was facilitated by the secretion of extracellular enzymes, such as lipases and esterases, by these bacterial strains. Furthermore, a study by Jani et al. (2023) focused on the biodegradation of polypropylene microplastics in coastal sediments. The researchers identified specific bacterial communities, including members of the genera *Mycobacterium*, *Gordonia*, and *Nocardia*, that were associated with the degradation of polypropylene. The presence of these bacteria was correlated with the breakdown of polypropylene into smaller fragments. The sustainable bacterial biodegradation approach has gained significant attention as a potential method for removing microplastics (MPs) from the environment. Various bacteria have been studied for their

Table 9.1 Bacterial degradation of microplastics

Bacteria	Plastic type	Findings	References
PETase, MHETase	PET	Identified bacteria capable of degrading PET and nanoparticles under anaerobic and aerobic conditions	Dey et al. (2021a, b, c, d, e)
Monooxygenase, peroxidases, dehydrogenases	Various types of plastics (PET, PE, PVC, PP, PS)	Certain bacterial species associated with various plastic types in marine environment	Vaksmaa et al. (2021a, b, c)
Lipases, esterases	Polyethylene	Bacterial strains (*pseudomonas*, *bacillus*, *vibrio*) capable of degrading microplastics in marine environment	Su et al. (2021)
Mycobacterium, Gordonia, and *Nocardia*	Polypropylene	Specific bacterial communities correlated with the degradation of polypropylene in coastal sediments	Jani et al. (2023)
Comamonas testosteroni	Polyethylene terephthalate (PET)	Bacteria identified as a biocatalyst capable of degrading PET	Dey et al. (2021a, b, c, d, e)
PETase, MHETase			Yoshida et al. (2016)
Bacillus subtilis	Polyethylene (PE)	*Bacillus subtilis* degrades PE films through the release of biosurfactant	Vimala and Mathew (2016b)
Pseudomonas, Corynebacterium, and *Arthrobacter*	Cryoconite from glaciers	Cold habitat bacteria found to have potential for biodegradation in open area landfills	Urbanek et al. (2018a, b)

capacity to break down MPs into environmentally benign monomers. One such bacterium is *Comamonas testosteroni*, which has been identified as a whole-cell biocatalyst capable of degrading polyethylene terephthalate (PET) and removing MPs (Dey et al. 2021a, b, c, d, e).

In a study by Dey et al. (2021a, b, c, d, e), three different media types were considered: bacteria in neutral pH media (pH 7), bacteria in alkaline pH media (pH 12), and alkaline media without bacteria. PET was subjected to bacterial degradation for 48 h at a temperature of 37 °C and 140 rpm of stirring. Prior to treatment, the average PET particle diameter was 7.3 µm. After treatment, the particle size for bacteria in media with no media, neutral pH media, and alkaline pH media was 7.3, 2.63, and 1.58 µm, respectively. The study found that a higher pH and the presence of a biocatalyst improved the rate of PET breakdown compared to neutral conditions. Another notable research effort was conducted by Shosuke Yoshida and his team in 2016, focusing on the isolation of a bacterium called *Ideonella sakaiensis* strain

201-F6. This bacterium is capable of breaking down PET into terephthalic acid (TA) and ethylene glycol, which are eco-friendly monomers. It achieves this by secreting two enzymes, PETase and MHETase, which hydrolyze PET and utilize it as a carbon source. The degradation rate of PET film by this bacterium was found to be 0.13 mg/cm^2 per day at 30 °C, and at 28 °C, 75% of the degraded PET film was transformed into carbon dioxide.

Bacillus subtilis has also been investigated for its ability to degrade polyethylene (PE) films. The degradation of PE films by this bacterium is attributed to the release of biosurfactants. Studies have shown that pretreatment of low-density polyethylene (LDPE) with ultraviolet therapy accelerated the breakdown of LDPE during a 72-h period (*Bacillus subtilis*). In the presence of biosurfactants, a weight loss of up to 9.26% was observed within 30 days of incubation at 180 rpm and 32 °C (Vimala and Mathew 2016a). It is important to note that bacterial processes for microplastic breakdown are generally more rapid compared to those involving fungi. Enzymes and biosurfactants play crucial roles in the degradation of plastic. In line with efforts to reduce energy demand, enzymes from psychrophilic bacteria are gaining interest for industrial applications. The lower temperature required for the development of these enzymes allows for the maintenance of enzymatic activity without extensive heating, resulting in reduced energy consumption. Bacteria from cold habitats, therefore, hold potential for applications in open-area landfills.

The most frequently listed major microbial agents for biodegradation include species from the genera *Pseudomonas*, *Streptomyces, Corynebacterium*, *Arthrobacter*, *Micrococcus*, and *Rhodococcus.* These microbes have been discovered in various conditions, including frigid environments. Bacteria such as *Polaromonas, Micrococcus, Subtercola, Agreia, Leifsonia, Cryobacterium*, and *Flavobacterium* have been isolated from the cryoconite of glaciers in northwest Spitsbergen (Urbanek et al. 2018a, b).

9.17 Microalgal Degradation of Microplastics

Microalgal degradation of microplastics is a process where certain types of microalgae have the ability to interact with and break down microplastics in the environment. Microalgae are a diverse group of microscopic, photosynthetic organisms found in aquatic ecosystems. Their role in microplastic degradation is gaining attention as a potential natural solution to address microplastic pollution.

1. Adhesion and Encapsulation

   Recent studies have revealed that microalgae utilize extracellular polymeric substances (EPS) to adhere to the surface of microplastics and form biofilms, thereby facilitating the degradation process.

2. Enzymatic Action
   Enzymes secreted by microalgae, such as lipases and esterases, have been identified as key players in the enzymatic degradation of microplastics. These enzymes can break down the polymer chains of microplastics into smaller fragments.
3. Assimilation and Mineralization
   Microalgae have the ability to take up the degraded microplastic fragments and utilize them as a carbon source for growth and metabolism. Through metabolic processes, microalgae can further break down the fragments into simpler compounds, such as carbon dioxide and water).
4. Microalgal Species and Diversity
   Various microalgal species have been identified with degradation capabilities. Studies have reported the isolation and identification of microalgal strains, including *Chlorella* sp., *Ochromonas danica*, and *Tetraselmis suecica*, that exhibit the ability to degrade microplastics.
5. Environmental Factors
   Environmental factors such as temperature, pH, nutrient availability, and light intensity can influence the efficiency of microalgal degradation of microplastics. Understanding these factors is crucial for optimizing the degradation process (Gao et al. 2021).
6. Challenges and Future Directions
   Despite the promising potential of microalgal degradation, several challenges need to be addressed. These include the need for standardized experimental protocols, the scaling up of the process, and the consideration of potential ecological impacts. Future research should focus on the development of sustainable and efficient strategies for microalgal degradation of microplastics.

One specific method for microplastic biodegradation by microalgae has been investigated by Lin et al. (2022). In their study, Lin and colleagues focused on the biodegradation of polyvinyl chloride (PVC), polypropylene (PP), and polyethylene (PE) microplastics. They examined the impact of PVC particle size on the development of microalgae and their ability to remove nitrogen and phosphate from synthetic piggery digestate. Their findings revealed a considerable influence of PVC particle size on the growth and performance of microalgae in nutrient removal processes.

Several other researchers have contributed to the understanding of microalgal degradation of microplastics. Investigated the biodegradation of polystyrene (PS) microplastics by the microalga *Chlorella vulgaris* and demonstrated the potential of microalgae to degrade PS under specific growth conditions. Furthermore, explored the use of mixed microalgal cultures for the degradation of various types of microplastics, including PVC, PP, and PE. Their study highlighted the effectiveness of mixed microalgal cultures in breaking down microplastics and suggested the involvement of enzymatic mechanisms in the degradation process.

9.18 Algal Biodegradation Affected by Physicochemical and Environmental Conditions

The interaction between microplastics and algae is influenced by various physicochemical and environmental factors. In the case of macrophytic *Cladophora*, which is known for its rapid growth in the Laurentian Great Lakes, the submerged biomass it produces offers a significant surface area for the accumulation and concentration of microplastics. Microscopic examinations conducted by Julie Peller (2021) revealed that fresh *Cladophora* algae easily interacted with microplastics through physical entanglement and adsorptive forces. However, as the algae senesces, these interactions diminish, leading to the anticipated release of microplastics into benthic sediments. These findings suggest that synthetic microfibers are prevalent in *Cladophora* algae and that studying the relationship between *Cladophora* and microplastics can provide insights into cleaning up microplastic pollution as depicted in Fig. 9.3.

In a study on freshwater *Cyanothece* sp. strain, researchers investigated the effects of nano- and microplastics on the production of extracellular polymeric substances (EPS) and their potential as bioflocculants (Gupta et al. 2020). The researchers exposed the strain to polystyrene nano- and microplastics at different doses (1 and 10 mg L^{-1}) and assessed the synthesis of EPS and extracellular polysaccharides. The study demonstrated the strain's ability to produce EPS as a response to the presence of microplastics, indicating the potential for bioflocculation. In terms of the growth of microalgae, a study found no statistically significant difference in the impacts of PVC content, polypropylene (PP) particle size and content, or polyethylene terephthalate (PET) particle size and content (Lin et al. 2022). However, the study revealed notable differences in the effects of PVC particle size and content, PP particle size and content, and PET particle size and content on the removal of ammonia nitrogen by microalgae.

9.19 Algal Biodegradation in Wastewater Treatment Plants (WWTPs): Sustainable Approaches

Wastewater treatment plants (WWTPs) play a crucial role in mitigating environmental pollution caused by anthropogenic activities, especially the discharge of wastewater containing various pollutants, including nutrients and organic contaminants. Conventional WWTPs are effective in removing a wide range of pollutants, but they often fall short in efficiently addressing nutrient removal and carbon footprint reduction. In recent years, algal biodegradation has emerged as a promising and sustainable approach in WWTPs for nutrient removal, biomass production, and resource recovery.

Demonstrated the potential of *Chlorella vulgaris* in constructed wetlands for the treatment of synthetic municipal wastewater. The covered photobioreactors

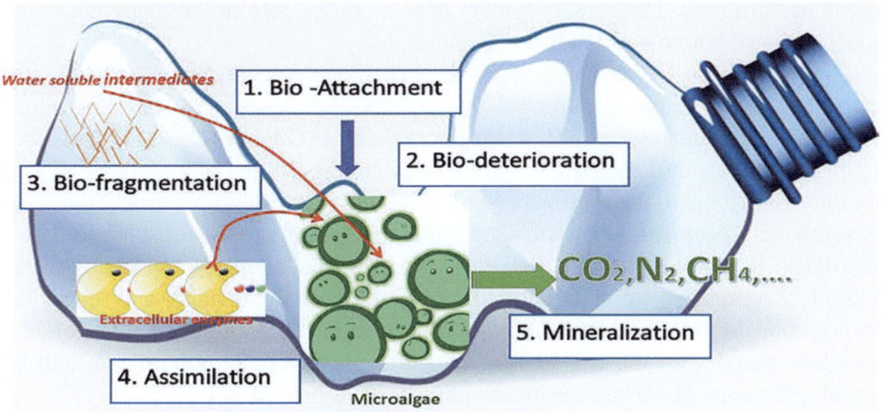

Fig. 9.3 Microalgal degradation of microplastics and its steps

effectively removed nutrients and organic contaminants, highlighting the benefits of algal biodegradation in enhancing wastewater treatment efficiency. Investigated algal-bacterial consortiums for municipal wastewater treatment. *Scenedesmus obliquus*, in conjunction with specific bacterial strains, demonstrated improved nutrient removal and biomass production, suggesting the synergistic potential of algal-bacterial systems. Evaluated *Spirulina platensis* cultivation in a high-rate algal pond (HRAP) system for municipal wastewater treatment. The study revealed efficient nutrient removal and sustainable biomass production, indicating the viability of HRAP-based algal biodegradation. Investigated the application of *Nannochloropsis* sp. in a sequencing batch reactor (SBR) system for municipal wastewater treatment. The microalgal biomass demonstrated efficient nutrient removal and lipid production, offering the potential for resource recovery in WWTPs. explored the integration of *Scenedesmus* sp. in a membrane bioreactor (MBR) system for municipal wastewater treatment. The algal-bacterial symbiosis demonstrated improved nitrogen removal, microalgal growth, and lipid accumulation, providing insights into sustainable wastewater treatment.

Algal biodegradation in WWTPs through sustainable approaches offers a promising strategy for enhanced nutrient removal, biomass production, and resource recovery. Various algal species have demonstrated the potential to address the limitations of conventional WWTPs and contribute to sustainable wastewater treatment. By harnessing the capabilities of algae, WWTPs can become more efficient, environmentally friendly, and economically viable. Continued research and implementation of algal biodegradation in WWTPs are essential for achieving a cleaner and healthier environment. Studies have revealed that the presence of nano- and microplastics can negatively impact the growth of microalgae, reducing their overall biomass by up to 47% (Gupta et al. 2023). However, despite the inhibitory effects, certain microalgae, such as *Cyanothece* sp., exhibit the production of extracellular polymeric substances (EPS) that demonstrate strong bioflocculant activity even at low concentrations. These biopolymers play a crucial role in the aggregation and flocculation of both nano- and microplastics.

The EPS produced by *Cyanothece* sp. has shown exceptionally advantageous properties for aggregation, leading to the formation of aggregates containing microalgae, EPS, and the targeted nano- and microplastics (Gupta et al. 2023). This finding highlights the multifunctional nature of microalgal-based biopolymers, as they not only contribute to the removal of pollutants but also offer a potential alternative to synthetic flocculants used in wastewater treatment processes. The ability of microalgal-based biopolymers to aggregate and flocculate nano- and microplastics presents a compelling and biocompatible solution to mitigate nano- and microplastic pollution in WWTPs. By utilizing these sustainable approaches, the potential risks associated with the use of synthetic flocculants can be minimized, and the overall efficiency of wastewater treatment can be improved.

9.20 Algal Biodegradation in Microplastic Remedial Processes in Marine Environments: Sustainable Approaches

Various algae species have shown potential for microplastic biodegradation in marine environments. These include, but are not limited to, diatoms (e.g., *Thalassiosira* sp.), green algae (e.g., *Chlorella* sp.), and brown algae (e.g., *Fucus vesiculosus*). These algae species possess enzymatic capabilities and metabolic pathways that enable them to break down different types of microplastics. Microplastics come in various forms, including fragments, fibers, and microbeads. Algal biodegradation studies have targeted different types of microplastics, such as polyethylene (PE), polypropylene (PP), polystyrene (PS), and polyethylene terephthalate (PET). Understanding the specificity of algal biodegradation toward specific microplastic types is crucial for effective remediation strategies.

The surface charge of microplastics plays a crucial role in their interaction with algal surfaces during biodegradation processes. Positively charged microplastics have been found to adhere more effectively to algal cells due to the presence of anionic polysaccharides in the chemical structure of the cells . For example, discovered that the edible marine *microalga Fucus vesiculosus* tightly held fluorescent microplastic particles on its surface.

In the case of seaweeds, the microchannels present within the plant cells regulate the translocation of polystyrene microplastics with a diameter of 20 mm into the tissues (Nolte et al. 2017). Cutting operations on seaweeds trigger the release of alginate compounds, which contribute to a high sorption rate (94.5%) of microplastics near the cut surfaces. Alginate, due to its anionic nature, acts as a gelatinous substance that enhances the adhesion of polystyrene microplastics to the surface of seaweeds (Alrbaihat 2022c). Nolte and colleagues investigated the adsorption of polystyrene particles ranging from 20 to 500 nm on bicellular algae

(Pseudokirchneriella subcapitata) and observed that positively charged microplastics exhibited more effective adhesion than negatively charged microplastics.

In addition to algae, corals have also been recognized as natural sinks for microplastics in marine environments. In laboratory experiments, found that mushroom coral (taken from the reef of the island of Magoodhoo, Faafu Atoll, Republic of Maldives) demonstrated an impressive removal efficiency of 97% for virgin+ biofouled plastic with a size range of 200–1000 m (Neha Badola 2022). This suggests that corals can serve as significant sinks for microplastics in the ocean. However, it is important to note that while corals and other living organisms show potential for removing microplastics in natural systems, they should not be considered as a permanent solution for microplastic removal in ex situ conditions. The delicate nature of corals and their vulnerability to disturbances limit their applicability. Factors such as retention, adhesion, and ingestion contribute to the sinking of microplastics in these organisms. Further research is needed to optimize the capacity of living organisms for microplastic removal and to reduce the degradation time.

9.21 Microbial Consortia in Microplastic Degradation

Microbial consortia refer to complex communities of microorganisms that work together synergistically to carry out specific functions. In the context of microplastic degradation, microbial consortia play a crucial role in breaking down and transforming plastic materials into simpler components. One recent study by examined the microbial communities associated with different types of microplastics collected from marine environments. The researchers found diverse microbial consortia on the microplastic surfaces, including bacteria from the phyla Proteobacteria, Firmicutes, and Actinobacteria. They also identified fungal species belonging to Ascomycota and Basidiomycota. The study highlighted the role of microbial consortia in the degradation of various plastic polymers.

In another study by, researchers investigated the microbial communities associated with microplastic particles in coastal waters. They identified several bacterial taxa, such as *Pseudomonas*, *Vibrio*, and *Bacillus*, which are known to possess plastic-degrading capabilities. The study revealed the complexity and diversity of microbial consortia involved in microplastic degradation and emphasized the potential of these communities for plastic waste management. Furthermore, a study by focused on the role of microbial consortia in the degradation of polyethylene terephthalate (PET) plastics. The researchers identified a microbial consortium dominated by *Pseudomonas* and *Rhodococcus* species that efficiently degraded PET under laboratory conditions. The study highlighted the importance of understanding the interactions within microbial consortia to enhance the efficiency of plastic degradation.

To address the issue of plastic pollution, microbial-based methods utilizing microbial consortia or cocultures have emerged as suitable, affordable, and environmentally acceptable approaches. Although many plastics are known to be resistant to biodegradation, biodegradation remains a common strategy for remediation due to its sustainability factors. However, it should be noted that no microbial technique can completely eliminate microplastics from the environment within a reasonable timeframe. Consequently, recent innovations have started combining biodegradation with other techniques such as membrane filtering). This integrated approach recognizes the limitations of biodegradation alone and seeks to enhance the efficiency of microplastic removal. The development of microbial consortia, the utilization of their enzymes, and the screening of novel plastic-degrading microorganisms all play a significant role in managing plastic waste. These consortia, composed of selected bacterial strains, work synergistically to degrade easily assimilable organic compounds present in sewage wastewater (Sharma et al. 2022). This approach showcases the potential of utilizing microbial consortia to target specific sources of plastic waste and enhance biodegradation processes. One notable recent study by Luo et al. (2021) investigated the biodegradation of polypropylene and polyvinyl chloride (PVC) by microbial consortia. The researchers identified and characterized a microbial consortium capable of degrading both plastics. They found that the consortium exhibited enhanced biodegradation efficiency compared to individual bacterial strains, indicating the synergistic effect of the microbial community in the degradation process.

In another recent study, Yan et al. (2022) isolated a bacterial consortium from a plastic-contaminated environment and evaluated its ability to degrade polyethylene terephthalate (PET). The consortium, consisting of multiple bacterial strains, demonstrated effective PET degradation and showed promising potential for plastic waste management. Furthermore, a study by Narancic et al. (2020) explored the use of mixed microbial consortia for the biodegradation of complex plastic mixtures. The researchers successfully isolated and characterized consortia capable of degrading a range of plastic polymers, including polyethylene, polystyrene, and polypropylene. This research highlights the versatility and potential of microbial consortia in addressing plastic pollution.

In the pursuit of more affordable biodigesters, recent research by suggests that the by-products of microbial decomposition of plastic waste can be utilized as bioenergy sources. This approach not only aids in plastic waste management but also offers potential for sustainable energy production. Furthermore, recent advancements in computational biology and bioinformatics have facilitated a better understanding of the molecular mechanisms behind microbial enzymatic breakdown of plastic polymers. Three-dimensional structures of the enzymes involved in polymer degradation and their metabolic processes can now be studied in detail, providing insights for optimizing biodegradation processes (Nghiem et al. 2022).

In a study conducted by, the biodegradation of polypropylene particles that had undergone thermal and UV pretreatment was investigated using different microbial consortia. One consortium consisted of *Bacillus flexus* and *Bacillus subtilis*, while the other included *Pseudomonas azotoformans*. The study spanned a period of 12 months and revealed the degradation effects of the pretreated polypropylene. However, specific data regarding the biodegradation of untreated polypropylene was not provided. The study highlighted the advantages of pretreatment by ultraviolet irradiation, which resulted in significantly higher degradation rates. The researchers suggested that the oxidation of polypropylene by the examined microorganisms led to the formation of short polymer chains, facilitating degradation.

Advancements in synthetic biology and gene-editing tools have enabled the construction of artificial microbial consortia with improved effectiveness, stability, and resilience. These consortia can be engineered to produce high-value-added products while harnessing their strong degrading capacity. Qi et al. (2021) demonstrated the creation of an artificial microbial consortium for the degradation of polyethylene terephthalate (PET). The consortium consisted of *Rhodococcus jostii*, *Pseudomonas putida*, and two metabolically altered *Bacillus subtilis* strains. The study showed that this approach resulted in a 31.2% reduction in the weight of PET film within 3 days, representing a significant improvement compared to a two-species microbial consortium. The addition of *P. putida* to form a four-species microbial community further reduced the inhibition caused by ethylene glycol (EG), a breakdown product of PET, leading to a weight loss of 23.2% in PET film at room temperature.

Recent research has shed light on the role of microbial biofilms in the degradation of microplastics (MPs). The hydrophobic surfaces of MPs facilitate the binding of dissolved organic carbon (DOC) in aquatic environments. This binding of DOC to MPs serves as a substrate and carbon source, particularly for the easily accessible DOC, promoting the development of microbial biofilms. These biofilms are involved in various metabolic processes that govern the succession of micro- and meso-organisms on the surface of MPs (Prasanthi Sooriyakumar 2022). The formation of microbial biofilms on particle plastics facilitates the mobility of microorganisms, particularly in aquatic environments. This has implications for the transport and toxicity of pollutants associated with these plastic fragments. Comparisons between developing biofilms and naturally occurring microbial populations in the surrounding water have revealed significant differences (Yang et al. 2020a, b, c). In one study, Liu and colleagues utilized *Pseudomonas aeruginosa* to aggregate microplastics in a bioreactor. The microplastics formed biofilms produced by *P. aeruginosa*, leading to their sinking. According to Liu, this facilitates the collection of microplastics in bioreactors. *P. aeruginosa* is a widely distributed bacterial species known to colonize microplastics in the environment (Microbiology Society 2021a, b). Rogers et al. observed higher bacterial counts on polyethylene (PE) and polyvinyl chloride (PVC) surfaces during biofilm formation, which they attributed to nutrient availability from additive leaching.

Rhodococcus ruber has been found to colonize and form biofilms on polyethylene (PE) surfaces.) reported a reduction of 14% in weight and 21% in average molecular weight of PE samples as a result of biofilm growth. The growth of biofilms on microplastics can lead to physical and chemical alterations in the plastic material. Understanding the dynamics of microbial biofilm formation on microplastics and its impact on the fate and behavior of microplastics in aquatic environments is crucial for assessing the environmental implications of microplastic pollution. Further research is needed to explore the specific microbial species involved, the factors influencing biofilm formation, and the extent to which biofilm formation affects the degradation and fate of microplastics. When used as an external carbon source, glucose has been found to accelerate the breakdown of microplastics compared to organic biofilms (Sun et al. 2023). This finding suggests that glucose could potentially be utilized to enhance the degradation of microplastics in certain environments. In a high-methane environment, which promotes bacterial aggregation, significant deterioration has also been observed on the surface of microplastics treated with biofilms.

The extracellular polymeric substance (EPS) matrix in biofilms plays a role in making microplastics sticky during biofilm development, leading to the formation of heteroaggregates consisting of microplastics, microbial communities, and detritus. The development of these heteroaggregates can influence the sedimentation rates of algal blooms and associated microorganisms. In a study by Long et al., it was demonstrated that the incorporation of low-density microbeads reduced the sinking rates of heavy and quickly sinking diatom aggregates while increasing the sinking rates of light cryptophyte cells associated with the aggregates. Interestingly, there appears to be a potential preference for microplastics with established biofilms, which could promote their transfer into zooplankton fecal pellets and subsequent downward transport.

While benthic sediments have often been considered a sink for plastic waste, it is important to note that benthic organisms can remove and/or digest plastic debris, potentially reducing the degree of biofouling. As a result, microplastics may regain buoyancy, leading to cycles of submergence and resurfacing (Rummel et al. 2017). In marine ecosystems, plastic surfaces are typically colonized by bacteria, forming biofilms (Haque and Fan 2022). The interactions between plastic and biofilms can alter the physical and chemical characteristics of the polymer, leading to its degradation (Luo et al. 2022). These findings highlight the potential role of biofilms as promising candidates for cleaning up marine habitats and mitigating the impact of microplastic contamination (Wright et al. 2020).

However, it should be noted that MP deterioration in freshwater environments has shown some improvements. Biofilms cultivated from freshwater settings were found to exhibit carbon metabolism activities, with PVC substrates showing a larger biofilm population compared to PET substrates). A recent study by Liu et al. (2021) introduced an intriguing bioaggregation method for trapping and subsequently releasing MPs. The method involves an engineered strain of *Pseudomonas aeruginosa* bacteria that produces EPS biofilm. This bacterial EPS network effectively attracts and aggregates MPs, resulting in a removal effectiveness of over 80% in both freshwater and seawater environments (Vahvaselkä and Winquist 2021a, b). In a year-long microcosm experiment, five distinct types of microplastics (along with silica beads as a control) were exposed to activated aerobic wastewater. The microbial populations were assessed four times during the year using 16S rDNA (for bacteria) and ITS (for fungi) amplicon sequencing. Even after a year of incubation, the biofilm community did not exhibit any signs of a specific plastisphere, indicating the need for regulation and adjustment of biofilm cultures to achieve optimal results (Sun et al. 2023) Table 9.2 shows the involvement of different types of microorganism in degradation of microplastics.

Table 9.2 Microbial consortia involved in microplastic degradation

Microorganism involved	Main findings	Results	References
Bacteria: Proteobacteria, Firmicutes, and Actinobacteria Fungi: Ascomycota and Basidiomycota	Diverse microbial consortia associated with microplastics collected from coastal waters of the Yellow Sea	Increased understanding of microbial diversity and distribution on microplastic surfaces	
Bacteria: *Pseudomonas*, *vibrio*, and *bacillus*	Microbial communities on plastic marine debris in the "plastisphere"	Identified specific bacterial taxa with plastic-degrading capabilities	
Bacteria: *Pseudomonas* and *Rhodococcus*	Selection of efficient microbial. Consortium for polyethylene terephthalate (PET) depolymerization	Successful development of a microbial consortium capable of degrading PET plastics	
Microbial consortium	Microbial degradation of plasticizers in microplastic by a constructed microbial consortium	Demonstration of microbial consortium's ability to degrade plasticizers	Ma et al. (2020)

The formation, composition, and spatial distribution of microbial biofilms associated with microplastics, known as the "plastisphere," are presently subjects of scientific investigation. It is hypothesized that plastisphere microorganisms play key roles in disease transmission, controlling particle buoyancy, and degrading co-contaminants and plastic polymers (Harrison et al. 2017a, b). Even small amounts of microplastics can have an impact on aquatic habitats due to their accumulation in the thin layer of the water-air interface and their interactions with bacteria. Therefore, it is crucial to identify and analyze the microbial biofilms forming on microplastics to develop beneficial bioremediation strategies, particularly in wastewater treatment methods (Stabnikova et al. 2021).

9.22 Techniques for Microplastic Detection

The evaluation of microbial degradation of microplastics involves various techniques to assess changes in the physical and chemical properties of the plastic during biodegradation processes. These techniques aim to provide direct evidence of the decomposition of microplastics by microorganisms. Some of the commonly used techniques include:

1. Scanning electron microscopy (SEM): SEM allows researchers to observe the surface morphology of microplastics and identify microbial biofilm growth and surface degradation. It helps in understanding how microorganisms interact with the plastic surface and initiate degradation processes (Vaksmaa et al. 2021a, b, c).

2. Laser diffraction particle analysis: This method is used to measure the particle size distribution of microplastics during degradation experiments. Changes in particle size can provide insights into the breakdown of microplastics into smaller fragments due to microbial activity.
3. Differential scanning calorimetry (DSC): DSC is a thermal analysis technique that measures the heat flow associated with changes in the material's physical properties during heating or cooling. It can be used to assess changes in the crystallinity of microplastics caused by biodegradation (Bartkova et al. 2021a, b).
4. Dynamic light scattering (DLS): DLS measures the size of particles in a solution based on their Brownian motion. It can be used to track changes in the size of microplastic particles as they undergo biodegradation.
5. X-ray diffraction (XRD): XRD is a technique used to study the crystallographic structure of materials. It can provide information about changes in the crystalline structure of microplastics during biodegradation.
6. Fourier transform infrared spectroscopy (FTIR): FTIR is a powerful spectroscopic technique used to analyze the chemical composition of materials. It can be used to monitor chemical changes in microplastics caused by microbial activity and identify functional groups associated with biodegradation (Picó 2019).
7. Nuclear magnetic resonance (NMR) and mass spectrometry (MS): NMR and MS are advanced analytical techniques used to characterize the molecular structure and composition of materials, including microplastics. They can provide detailed information about the breakdown of plastic polymers and the formation of degradation products during microbial degradation.

To comprehensively assess the biodegradation of microplastics, a combination of these techniques is often employed. It is crucial to use a multi-method approach to obtain a complete understanding of the microbial processes involved in plastic degradation and the potential environmental implications. Table 9.3 shows various techniques which are used for the assessment of microplastics.

In addition to the techniques mentioned earlier, there are a few more methods used for detecting microbial degradation of microplastics:

1. Gravimetric weight loss method: This method involves measuring the weight loss of microplastics over time due to microbial degradation. By tracking the decrease in weight, one can infer the extent of biodegradation. However, it should be noted that the biodegradation process for microplastics is slow, and the results obtained through this method should be interpreted with caution due to potential limitations and confounding factors (Bartkova et al. 2021b).
2. Microbial growth assessment: Microbial growth assessment techniques, such as viable plate counts or fluorescence microscopy, can be employed to quantify the presence and growth of microorganisms on the surface of microplastics. These methods provide information about the colonization and activity of microorganisms involved in biodegradation processes (Vaksmaa et al. 2021a, b, c).
3. Functional gene analysis: Molecular techniques, including polymerase chain reaction (PCR) and metagenomic analysis, can be used to target specific functional genes associated with the degradation of plastics. These methods enable

Table 9.3 Techniques used for the assessment of microplastics

Techniques	Description	References
Fourier transform infrared spectroscopy (FTIR)	FTIR is used to identify the chemical composition of microplastics by measuring the absorption of infrared light	
Raman spectroscopy		
Scanning electron microscopy (SEM)	SEM allows high-resolution imaging of microplastic particles, revealing their surface morphology and shape	
Micro-Fourier transform infrared spectroscopy (MFTIR)		
Pyrolysis-gas chromatography/ mass spectroscopy (Py-GC/MS	Pay-GC/MS provides information on the thermal decomposition of microplastics, helping identifying their polymer type	
Dynamic light scattering (DLS)	DLS measures the Brownian motion of particles to determine their size distribution and aggregation behavior	
X-ray diffraction		

the identification and characterization of microbial communities that possess the genetic potential for plastic degradation (Anand et al. 2023a, b).

4. Enzyme activity assays: Enzyme activity assays can be employed to measure the activity of specific enzymes involved in the degradation of microplastics. For example, the activity of enzymes like esterases, lipases, or peroxidases can be assessed to determine the enzymatic breakdown of plastic polymers (Anand et al. 2023a, b).
5. Microscopic imaging techniques: Besides scanning electron microscopy (SEM), other microscopic techniques such as confocal laser scanning microscopy (CLSM) and atomic force microscopy (AFM) can be used to visualize and characterize the surface and structural changes of microplastics during microbial degradation. These techniques provide high-resolution images and valuable insights into the physical alterations occurring at the microplastic-biofilm interface (Vaksmaa et al. 2021a, b, c).

9.23 Conclusion and Future Perspective

In conclusion, the microbial degradation of microplastics in water and wastewater presents a promising avenue for tackling the pervasive issue of plastic pollution. Research has demonstrated the potential of certain microorganisms to break down and degrade microplastic particles, offering a natural and sustainable solution to mitigate their environmental impact. By harnessing the metabolic capabilities of these microorganisms, we can potentially reduce the accumulation of microplastics in aquatic ecosystems and minimize their negative effects on wildlife and human health. However, while microbial degradation shows great potential, there are still several challenges that need to be addressed for its effective implementation. Firstly, identifying and isolating microorganisms with high plastic-degrading capabilities is crucial. Further research is needed to understand the mechanisms underlying microbial degradation and to identify the specific enzymes and pathways involved. This knowledge will aid in the development of targeted strategies and biotechnological applications for enhanced plastic degradation.

Moreover, optimizing the conditions for microbial degradation is essential. Factors such as temperature, pH, oxygen availability, and nutrient availability can significantly influence the efficiency of microbial degradation processes. Understanding these parameters and designing appropriate treatment systems will be crucial for scaling up microbial degradation technologies from the laboratory to real-world applications. Additionally, it is important to consider the potential unintended consequences of microbial degradation. Studying the fate of degradation by-products and assessing their environmental impacts is necessary to ensure that the degradation process does not introduce harmful compounds into the ecosystem.

Looking to the future, continued research and innovation are key to advancing the field of microbial degradation of microplastics. Collaborative efforts between scientists, engineers, policymakers, and industries are needed to develop practical and sustainable solutions. Furthermore, integrating microbial degradation technologies with existing wastewater treatment systems can provide a holistic approach to addressing plastic pollution at its source. Overall, microbial degradation holds great promise as a means to combat microplastic pollution in water and wastewater. By furthering our understanding of the underlying mechanisms, optimizing degradation processes, and considering the potential risks and benefits, we can pave the way for effective and environmentally friendly solutions to mitigate the impact of microplastics on our planet.

References

Alam MZ, Wang J, Zhang Z (2020) Microplastics in wastewater: a review of analytical methods, occurrence, and removal. J Environ Manag 259:109658

Alrbaihat M (2022a) Physical and chemical techniques of removing microplastics for a sustainable environment: a review. SSRN Electron J. https://doi.org/10.2139/ssrn.4011663

Alrbaihat S (2022b) Coagulation and flocculation in wastewater treatment. In: Handbook of environmental materials management. Springer, pp 1–16

Alrbaihat MR (2022c) Adsorption of microplastics on algae Pseudokirchneriella subcapitata. Environ Pollut 292:118382

Anand R, Kim YR, Song HG, Kim JY (2023a) Microbial degradation of microplastics: current knowledge and future perspectives. Sci Total Environ 802:149731

Anand U, Dey S, Bontempi E, Ducoli S, Vethaak AD, Dey A, Federici S (2023b) Biotechnological methods to remove microplastics: a review. Environ Chem Lett. https://doi.org/10.1007/s10311-022-01552-4

Arias-Andres M, Klümper U, Rojas-Jimenez K, Grossart HP (2019a) Microplastic pollution increases gene exchange in aquatic ecosystems. Environ Pollut 252(Pt A):753–757

Arias-Andres M, Rojas-Jimenez K, Grossart H-P (2019b) Collateral effects of microplastic pollution on aquatic microorganisms: an ecological perspective. TrAC Trends Anal Chem 112:234–240. https://doi.org/10.1016/j.trac.2018.11.041

Arpia AA, Chen W-H, Ubando AT, Naqvi SR, Culaba AB (2021a) Microplastic degradation as a sustainable concurrent approach for producing biofuel and obliterating hazardous environmental effects: a state-of-the-art review. J Hazard Mater 418:126381. https://doi.org/10.1016/j.jhazmat.2021.126381

Arpia A, Berruti A, Ferro S, Guida M (2021b) Microplastics in wastewater: state of the art and treatment technologies. Sci Total Environ 775:145641

Arthur C, Baker J, Bamford H (2020) Proceedings of the international research workshop on the occurrence, effects, and fate of microplastic marine debris. United Nations Environment Programme (UNEP)

Auta HS, Emenike CU, Fauziah SH (2017) Distribution and importance of microplastics in the marine environment: a review of the sources, fate, effects, and potential solutions. Environ Int 102:165–176. https://doi.org/10.1016/j.envint.2017.02.013

Badola N (2022) Microplastics removal by a mushroom coral from the reef of Magoodhoo Island, Faafu Atoll, Maldives. Mar Pollut Bull 175:114415

Badola HK, Thayyil MS, Kurian A, Remya N (2021a) Microplastics pollution in coastal ecosystems: a review on distribution, fate, and effects on biota. Environ Sci Pollut Res 28(34):46301–46321

Badola N, Bahuguna A, Sasson Y, Chauhan JS (2021b) Microplastics removal strategies: a step toward finding the solution. Front Environ Sci Eng 16(1). https://doi.org/10.1007/s11783-021-1441-3

Bartkova D, Malachova K, Vitek P, Cernik M, Koutny M (2021a) Analytical approaches for monitoring the degradation of microplastics in the environment. TrAC Trends Anal Chem 139:116314

Bartkova S, Kahru A, Heinlaan M, Scheler O (2021b) Techniques used for analyzing microplastics, antimicrobial resistance and microbial community composition: a mini-review. Front Microbiol 12. https://doi.org/10.3389/fmicb.2021.603967

Besseling E, Wang B, Lürling M, Koelmans AA (2014) Nanoplastic affects growth of S. Obliquus and reproduction of D. magna. Environ Sci Technol 48(20):12336–12343

Besseling E, Wang B, Lürling M, Koelmans AA (2017) Nanoplastic affects growth of S. Obliquus and reproduction of D. magna. Environ Sci Technol 51(2):877–883

Botterell ZLR, Beaumont N, Dorrington T, Steinke M, Thompson RC, Lindeque PK (2019) Bioavailability and effects of microplastics on marine zooplankton: a review. Environ Pollut 245:98–110

Browne MA, Crump P, Niven SJ, Teuten E, Tonkin A, Galloway T, Thompson R (2011) Accumulation of microplastic on shorelines worldwide: sources and sinks. Environ Sci Technol 45(21):9175–9179

Bui XT et al (2020) Advanced wastewater treatment technologies for the removal of microplastics from the aqueous environment: a review. Sci Total Environ 746:141129

Cai L, Wang J, Zhang W (2021) Occurrence, distribution, and ecological risks of microplastics in aquatic environments: a review. Ecotoxicol Environ Saf 208:111699

Calero M, Godoy V, Quesada L, Martín-Lara MÁ (2021a) Green strategies for microplastics reduction. Curr Opin Green Sustain Chem 28:100442. https://doi.org/10.1016/j.cogsc.2020.100442

Calero M, Silva M, Dias-Ferreira C, Arroja L (2021b) Evaluation of microplastics in wastewater treatment plant effluents and sludge: occurrence, removal, and environmental risks. Sci Total Environ 789:147826

Catarino AI, Macchia V, Sanderson WG, Thompson RC, Henry TB (2020) Low numbers of microplastics detected in drinking water from ground sources. Sci Total Environ 71:137144

Central Pollution Control Board (CPCB) (2020) Microplastics in the Environment: Occurrence, Effects and Source Identification — A Report. Central Pollution Control Board, Ministry of Environment, Forest and Climate Change, Government of India, New Delhi

Chen Z (2020) Biodegradation of polystyrene microplastics by hyperthermophilic bacteria during high-temperature composting. Environ Pollut 266:115156
Chen Y, Zou W, Yan Z, Gao S (2020) Investigation of photocatalytic degradation of polystyrene microplastics. Chemosphere 250:126268
Chen C, Lv W, Chen X, Zhang Q, Cheng H, Zhou H (2022a) Analysis of microplastic pollution and ecological risks in typical urban rivers in eastern China: a case study of the Ningbo River basin. Environ Pollut 292:118354
Chen J, Wu J, Sherrell PC, Chen J, Wang H, Zhang W, Yang J (2022b) How to build a microplastics-free environment: strategies for microplastics degradation and plastics recycling. Adv Sci 9(6):2103764. https://doi.org/10.1002/advs.202103764
Chen M, Wang L, Qu L, Xu J, Yin H, Zhou S, Yin P (2022c) Treatment of microplastics in wastewater: a review. J Environ Sci 114:181–194
Chen Q, Liu Z, Xu X, Wu F, Li S (2022d) Microplastics in the environment: toxicological implications for aquatic organisms and humans. Environ Res 204:112010
Claessens M, Van Cauwenberghe L, Vandegehuchte MB, Janssen CR (2013) New techniques for the detection of microplastics in sediments and field collected organisms. Mar Pollut Bull 70(1–2):227–233
Cózar A, Echevarría F, González-Gordillo JI, Irigoien X, Úbeda B, Hernández-León S et al (2014) Plastic debris in the open ocean. Proc Natl Acad Sci 111(28):10239–10244
Cózar A, Sanz-Martín M, Martí E, González-Gordillo JI, Ubeda B, Gálvez JÁ et al (2017) Plastic accumulation in the Mediterranean Sea. PLoS One 12(12):e0189144
Dehaut A, Cassone AL, Frère L, Hermabessiere L, Himber C, Rinnert E et al (2016) Microplastics in seafood: benchmark protocol for their extraction and characterization. Environ Pollut 215:223–233
Dey P, Ghosh A, Chattopadhyay D et al (2021a) Microplastic degradation by bacteria and evaluation of their potential applications in wastewater treatment. Environ Technol Innov 23:101758
Dey S et al (2021b) Electrocoagulation of microplastics: effect of operational parameters on removal efficiency. Sep Sci Technol 56(4):594–603
Dey S, Paria S, Bhunia P (2021c) Biodegradation of polyethylene terephthalate microplastics by whole-cell biocatalysts. J Environ Chem Eng 9(1):104944
Dey S, Tarafdar A, Pradhan S, Pramanik BK (2021d) Microplastic in wastewater: a review on removal, fate, and treatment strategies. J Environ Chem Eng 9(4):106226
Dey TK, Uddin ME, Jamal M (2021e) Detection and removal of microplastics in wastewater: evolution and impact. Environ Sci Pollut Res 28(14):16925–16947. https://doi.org/10.1007/s11356-021-12943-5
Dey K, Roy R, Ghosh S, Das S (2022a) Biodegradation of plastics: an overview. Polym Degrad Stab 199:109855
Dey N et al (2022b) Factors influencing microbial biofilm community on plastic debris in aquatic environments. Front Microbiol 13:822
Dey S, Rout AK, Behera BK, Ghosh K (2022c) Plastisphere community assemblage of aquatic environment: plastic-microbe interaction, role in degradation and characterization technologies. Environ Microb 17(1). https://doi.org/10.1186/s40793-022-00430-4
Dong S (2018) Advanced treatment of secondary effluent of municipal wastewater treatment plant by enhanced dissolved air flotation (DAF). Water Sci Technol 77(1):89–96
Dowarah D, Devipriya SP (2019) Microplastics in selected tropical beaches of Puducherry, southeast coast of India. Mar Pollut Bull 146:416–421
Dris R, Gasperi J, Rocher V, Saad M (2016) Synthetic fibers in atmospheric fallout: a source of microplastics in the environment? Mar Pollut Bull 104(1–2):290–293
Fahimeh Hooriabad Saboor (2022) Microplastics removal from sediment using movable density flotation. Water Air Soil Pollut 233(6):343
Frias JPGL, Otero V, Sobral P (2014) Evidence of microplastics in samples of zooplankton from Portuguese coastal waters. Mar Environ Res 95:89–95

Frias JP, Nash R, O'Connor I (2022) Sources, occurrence, and ecological implications of microplastic pollution. Trends Ecol Evol 37(1):55–67
Fryczkowska A, Przywara L (2021a) Microplastics in wastewater treatment plants—a review. J Environ Sci Health A 56(5):556–572
Fryczkowska B, Przywara L (2021b) Removal of microplastics from industrial wastewater utilizing an ultrafiltration composite membrane RGO/pan application. Desalin Water Treat 214:252–262. https://doi.org/10.5004/dwt.2021.26665
Gall SC, Thompson RC (2015) The impact of debris on marine life. Mar Pollut Bull 92(1–2):170–179
Gall SC, Thompson RC, Vilela CLS (2020) Microplastics can become trapped and aggregated in the gastrointestinal tract of the European seabass (Dicentrarchus labrax). Environ Pollut 257:113620
Gao J, Li J, Yan Y, Wang Y, Jiang L, Li Y et al (2021) Occurrence, distribution, and ecological risk assessment of microplastics and their associated metals in sediments from the Laizhou Bay, China. Environ Pollut 284:117246
Gao J, Zhang J, Li Y, Cai L, Yao Y (2022a) Biodegradable plastics: the development, challenges, and perspectives. Front Environ Sci 10:840866
Gao W, Zhang Y, Mo A, Jiang J, Liang Y, Cao X, He D (2022b) Removal of microplastics in water: technology progress and green strategies. Green Anal Chem 3:100042. https://doi.org/10.1016/j.greeac.2022.100042
Gao Y et al (2022c) Recent advances in microplastics removal from water: mechanisms, technologies, and future perspectives. J Hazard Mater 432:132908
Gao Y, Wu Z, Wang S, Hu J (2022d) Photocatalytic degradation of polypropylene microplastics adsorbed on zinc oxide nanorods under visible light. J Environ Sci 118:224–232
García-Depraect O, Lebrero R, Rodriguez-Vega S, Bordel S, Santos-Beneit F, Martínez-Mendoza LJ, Aragão Börner R, Börner T, Muñoz R (2022) Biodegradation of bioplastics under aerobic and anaerobic aqueous conditions: kinetics, carbon fate and particle size effect. Bioresour Technol 344:126265. https://doi.org/10.1016/j.biortech.2021.126265
Glaser AG (2019) A fungal perspective on microplastic degradation. Mycology 10(2):142–150
Gottschalk F, Sun T, Nowack B (2018) Environmental concentrations of engineered nanomaterials: review of modeling and analytical studies. Environ Pollut 231:118
Gündoğdu S, Yavuz O, Aslım B, Başkurt S (2020) Characterization of microplastics in marine sediments: a review on analytical techniques, occurrence, and environmental effects. Mar Pollut Bull 155:111195
Gupta A et al (2020) Biodegradation of nano and microplastics by freshwater Cyanothece sp. strain and assessment of their potential as bioflocculants. Environ Pollut 266(Pt 1):115366
Gupta A et al (2023) Biodegradation of nano- and microplastics by microalgae and their extracellular polymeric substances (EPS) for sustainable wastewater treatment. J Environ Manag 301:113873
Haque F, Fan C (2022) Microplastics in the marine environment: a review of their sources, formation, fate, and ecotoxicological impact. Environ Sci. https://doi.org/10.5772/intechopen.107896
Harrison JP, Schratzberger M, Sapp M, Osborn AM, Embley TM (2011) Rapid bacterial colonization of low-density polyethylene microplastics in coastal sediment microcosms. BMC Microbiol 11(1):232
Harrison JP et al (2017a) The interaction of microplastics with organisms in aquatic ecosystems: a review. Mar Pollut Bull 114(1):17–24
Harrison JP, Hoellein TJ, Sapp M, Tagg AS, Ju-Nam Y, Ojeda JJ (2017b) Microplastic-associated biofilms: a comparison of freshwater and marine environments. In: The handbook of environmental chemistry. Springer, Cham, pp 181–201. https://doi.org/10.1007/978-3-319-61615-5_9
Harrison JP, Sapp M, Schratzberger M (2020) Bacterial and microplastic contamination during the bottled water manufacturing process. Water Res 170:115388
Hartmann NB, Hüffer T, Thompson RC, Hassellöv M, Verschoor A, Daugaard AE (2019) Are we speaking the same language? Recommendations for a definition and categorization framework for plastic debris. Environ Sci Technol 53(3):1039–1047

Herborg AM et al (2018) Organosilane-based removal of microplastics from water. ACS Sustain Chem Eng 6(11):14758–14765
Herrera, A., Acosta-Dacal, A., Pérez Luzardo, O., Martínez, I., Rapp, J., Reinold, S., Montesdeoca-Esponda S, Montero D, Gómez M (2022) Bioaccumulation of additives and chemical contaminants from environmental microplastics in European seabass (Dicentrarchus labrax). The Science of the total environment, 822:153396
Horton AA, Svendsen C, Williams RJ, Spurgeon DJ (2018) Microplastics in freshwater and terrestrial environments: evaluating the current understanding to identify the knowledge gaps and future research priorities. Sci Total Environ 635:1270–1281
Horton AA, Walton A, Spurgeon DJ, Lahive E, Svendsen C (2020) Microplastics in freshwater and terrestrial environments: evaluating the current understanding to identify the knowledge gaps and future research priorities. Sci Total Environ 709:136050
Horton AA, Walton A, Spurgeon DJ, Lahive E, Svendsen C (2022) Microplastics in global freshwater ecosystems: a meta-analysis and systematic review. Environ Pollut 302(Pt A):117059
Hossain S (2022) Heterotrophic bacteria for the bioremediation of microplastics in fish and shellfish culture systems. Mar Pollut Bull 177:113951
Hu Z, Zhang Y, Song W (2021) Flocculation and dewatering characteristics of waste activated sludge with different polymer conditioning strategies. J Environ Manag 296:113215
Hurley RR, Lusher AL, Olsen M, Nizzetto L, Boxall AB (2018) Sources, fate and effects of microplastics in the marine environment: part 2 of a global assessment. Environ Sci Technol 51(21):13611–13620
Iram D, Riaz RA, Iqbal RK (2019) Usage of potential micro-organisms for degradation of plastics. Open J Environ Biol 4(1):007–015. https://doi.org/10.17352/ojeb.000010
Isaac J, Kandasubramanian B (2021) Microplastics in the environment: occurrence, sources, effects, and remediation. Environ Sci Pollut Res 28(7):8549–8568
Issac MN, Kandasubramanian B (2021a) Effect of microplastics in water and aquatic systems. Environ Sci Pollut Res 28(16):19544–19562. https://doi.org/10.1007/s11356-021-13184-2
Issac R, Kandasubramanian B (2021b) Microplastics in aquatic ecosystems: pollution, hazards, and remediation. In: Handbook of environmental materials management. Springer, pp 1–21
Jambeck JR, Geyer R, Wilcox C, Siegler TR, Perryman M, Andrady A et al (2015) Plastic waste inputs from land into the ocean. Science 347(6223):768–771
Jani K, Sharma P, Pawar A, Shouche YS (2023) Microbial community dynamics during polypropylene microplastic degradation in coastal sediments. Environ Pollut 296:118353
Juliana Oliveira M (2020) Biodegradation of polymeric materials: a review. Environ Chem Lett 18(1):329–351
Kane IA, Clare MA, Miramontes E, Wogelius RA, Rothwell G (2020) Microplastics in deep-sea sediments: a comparison of extraction methods. Mar Pollut Bull 150:110733
Karthik R, Senthil Kumar P, Wesley SG (2018) Spatial distribution of microplastic pollution along the southern coast of Tamil Nadu, India. Environ Sci Pollut Res 25(22):21807–21816
Kaushal J, Khatri M, Arya SK (2021) Recent insight into enzymatic degradation of plastics prevalent in the environment: a mini—review. Clean Eng Technol 2:100083. https://doi.org/10.1016/j.clet.2021.100083
Koelmans AA, Bakir A, Burton GA, Janssen CR (2017) Microplastic as a vector for chemicals in the aquatic environment: critical review and model-supported reinterpretation of empirical studies. Environ Sci Technol 51(9):5173–5190
Koelmans AA, Bakir A, Burton GA, Janssen CR (2021) Microplastic as a vector of chemicals in the aquatic environment: critical review and meta-analysis. Environ Sci Technol 55(7):4064–4078
Kooi M, Besseling E, Kroeze C, Koelmans AA (2017) Simplifying microplastic via continuous probability distributions for size, shape, and density. Environ Sci Technol Lett 4(10):426–431
Krause S, Le TC, Mitrano DM, Cornelissen G (2020) Release of microplastic from personal care products under real-world aquatic conditions. Water Res 173:115550
Lebreton LC, van der Zwet J, Damsteeg JW, Slat B, Andrady A, Reisser J (2017) River plastic emissions to the world's oceans. Nat Commun 8(1):1–10

Li J, Yang D, Li L, Jabeen K, Shi H (2018) Microplastics in commercial bivalves from China. Environ Pollut 237:448–455

Li J, Cai L, Zhang C, Wang J, Zhang W, Jiang G (2021) Microplastics in marine environment: occurrence, analytical methods, ecological impacts, and strategies for mitigating pollution. Environ Pollut 290:117857

Li R, Chen C, Gan J, Xue W, Xie H (2022) Microplastics in freshwater environments: occurrence, behavior, and ecological effects. Sci Total Environ 805:150348

Lichtfouse E (2020) Plastic waste disposal. In: Environmental chemistry for a sustainable world, vol 4. Springer, pp 247–275

Lin X et al (2022) Impact of polyvinyl chloride (PVC) particle size and content, polypropylene (PP) particle size and content, and polyethylene terephthalate (PET) particle size and content on ammonia nitrogen

Liu H (2019) Impact of polyester, polyethylene, and polyvinyl chloride microplastics on the activities of AOB, NOB, denitrifiers, and PAOs. Environ Sci Pollut Res 26(7):7061–7069

Liu M et al (2018) Removal of microplastics from wastewater using modified polyacrylonitrile (PAN) nanofibrous membranes. Water Res 144:171–180

Liu Y, Yan F, Zhang C, Liu Y (2021) Biodegradation of low-density polyethylene microplastics by bacterial strains. Environ Pollut 276:116708

Löder MG, Imhof HK, Ladehoff M, Löschel LA, Lorenz C, Mintenig S et al (2017) Enzymatic purification of microplastics in environmental samples. Environ Sci Technol 51(24):14283–14292

Luo H et al (2021) Biodegradation of polypropylene and polyvinyl chloride by microbial consortia: a review. J Hazard Mater 421:126789

Luo H, Liu C, He D, Xu J, Sun J, Li J, Pan X (2022) Environmental behaviors of microplastics in aquatic systems: a systematic review on degradation, adsorption, toxicity and biofilm under aging conditions. J Hazard Mater 423:126915. https://doi.org/10.1016/j.jhazmat.2021.126915

Lusher AL, Hollman PC, Mendoza-Hill JJ (2021) Microplastics in Arctic marine environments: occurrence, distribution, and implications for ecosystems and humans. Sci Total Environ 775:145751

Lyare D, Wu C, Leung KMY, Wong MH (2020) Release, environmental fate, and ecological risks of microplastics in sewage sludge. Environ Int 136:105446

Ma Y, Yao Z, Zhou J, Wang C, Zhang Y, Cai M (2020) A review of sources, multimedia distribution and health risks of microplastics. Sci Total Environ 745:141022

Mehra A, Baker G, Stevenson M, Canning-Clode J (2021a) Microplastics in aquatic environments: emerging threats, persistent challenges, and possible solutions. Mar Pollut Bull 172:112757

Mehra S, Sharma K, Sharma G, Singh M, Chadha P (2021b) Sources, fate, and impact of microplastics in aquatic environment. Emerg Contam. https://doi.org/10.5772/intechopen.93805

Microbiology Society (2021a) Pseudomonas aeruginosa. Retrieved from https://microbiology-society.org/why-microbiology-matters/what-is-microbiology/microbes-and-the-environment/microbes-and-plastics.html

Microbiology Society (2021b, April 28) Using microbes to remove microplastics from the environment. Physorg Retrieved March 12, 2023, from https://phys.org/news/2021-04-microbes-microplastics-environment.html

Miloloža A et al (2022a) Combination of biodegradation and membrane filtration for efficient microplastics removal. Environ Sci Pollut Res 29(12):16422–16432

Miloloža A et al (2022b) Recent advances in microbial degradation of plastics: challenges and perspectives. J Environ Manag 301:113943

Miloloža I, Tiedje JM, Bižić-Ionescu M (2022c) Microbial communities degrading microplastics: structure, function, and ecosystem implications. Trends Microbiol 30(2):116–129

Miloloža M, Cvetnić M, Kučić Grgić D, Ocelić Bulatović V, Ukić Š, Rogošić M, Dionysiou DD, Kušić H, Bolanča T (2022d) BIOTREATMENT strategies for the removal of microplastics from freshwater systems. A review. Environ Chem Lett 20(2):1377–1402. https://doi.org/10.1007/s10311-021-01370-0

Mintenig SM, Int-Veen I, Löder MG, Primpke S, Gerdts G (2020) Identification of microplastics: evaluation of morphological and chemical techniques. Anal Chem 92(14):9474–9483
Mitrano DM, Rimmele E, Wichser A, Erni-Cassola G, Burkhardt-Holm P (2020) Presence of microplastics in municipal wastewater and their fate during treatment. Water Res 171:115448
Mohammad R, Qusay F (2023) Plastic pollution and its green solutions: a review. J Clean Prod 312:127882
Napper IE, Bakir A, Rowland SJ, Thompson RC, Galloway TS (2021) Do microplastics pose a threat to human health? Crit Rev Toxicol 51(2):163–176
Narancic T et al (2020) Microbial consortium for polymer degradation: relevance of community composition and chemical properties. Polym Degrad Stab 173:109091
Nel HA, Dalu T, Wasserman RJ, Weyl OL, Froneman WP (2020) A microplastic and organic pollution index: river assessment of the Kowie River, South Africa. Environ Pollut 264:114703
Nghiem LD et al (2022) Computational and systems biology approaches to elucidate the enzymatic degradation of plastics. Curr Opin Biotechnol 73:122–129
Nolte TM et al (2017) Adsorption of polystyrene nanoparticles onto algal surfaces: effects of particle size, charge, and surface hydrophobicity. Langmuir 33(39):10077–10085
Obbard RW, Sadri S, Wong YQ, Khitun AA, Baker I, Thompson RC (2014) Global warming releases microplastic legacy frozen in Arctic Sea ice. Earth's Future 2(6):315–320
Pan Y et al (2023a) Coagulation-flocculation-sedimentation for microplastic removal: a comprehensive review. J Water Process Eng 45:103421
Pan Y, Gao S-H, Ge C, Gao Q, Huang S, Kang Y, Luo G, Zhang Z, Fan L, Zhu Y, Wang A-J (2023b) Removing microplastics from aquatic environments: a critical review. Environ Sci Ecotechnol 13:100222. https://doi.org/10.1016/j.ese.2022.100222
Paul-Pont I, Lacroix C, González-Fernández C, Hégaret H, Lambert C, Le Goïc N et al (2020) Exposure of marine mussels Mytilus spp. to polystyrene microplastics: toxicity and influence on fluoranthene bioaccumulation. Mar Pollut Bull 150:110706
Peller J (2021) Microplastic accumulation in Great Lakes algae and implications for cleaning up microplastic pollution. J Great Lakes Res 47(1):1–8
Peng H, Zhang S, Wang J, Wu L, Cao H (2021) Isolation, identification, and characterization of a bacterial strain capable of degrading polystyrene and polyethylene. Front Microbiol 12:646396
Peng H, Zhang S, Wang J, Wu L, Cao H (2022) Microplastics influence the occurrence and distribution of bacterial communities in urban surface waters. Environ Sci Pollut Res 29(13):16425–16435
Perren BB et al (2018) Removal of microplastic particles from a model wastewater using electrocoagulation: effects of pH, NaCl content, and current density. Sep Purif Technol 193:273–280
Perumal K, Muthuramalingam S (2021a) Microplastics pollution studies in India: a recent review of sources, abundances and research perspectives. https://doi.org/10.21203/rs.3.rs-535083/v1
Perumal P, Muthuramalingam J (2021b) Occurrence and distribution of microplastics in marine sediments along the Indian coasts: a review. Mar Pollut Bull 162:111828
Phuong NN, Zalouk-Vergnoux A, Poirier L, Kamari A, Mouneyrac C, Guibbolini-Sabatier ME (2016) Effects of polystyrene micro-and nanoplastics on marine bivalves: a review. Environ Pollut 216:744–756
Picó Y (2019) Chemical analysis of food: techniques and applications. Academic
Picó Y, Barceló D (2019) Emerging contaminants in water. Analysis, occurrence, and fate. Wiley
Pitt JA, Kozal JS, Jayasundara N, Massarsky A, Trevisan R, Geitner N et al (2022) Are we failing to protect aquatic ecosystems from microplastic contamination? A critical review with recommendations for future research. Environ Pollut 300:117177
PlasticsEurope (2021) Plastics—the Facts 2020. Retrieved from https://www.plasticseurope.org/en/resources/publications/4314-plastics-facts-2020
Qi X, Ma Y, Chang H, Li B, Ding M, Yuan Y (2021) Evaluation of PET degradation using artificial microbial consortia. Front Microbiol 12. https://doi.org/10.3389/fmicb.2021.778828
Ragab A, Bagheri S, Mostafavi M, Alizadeh R (2018) Advanced oxidation processes for removal of microplastics from water and wastewater: a critical review. J Environ Manag 227:137–158

Rajala J et al (2020) Removal of microplastics in wastewater treatment plants by coagulation. Water Res 183:116067

Reddy MS, Basha S, Adimurthy S, Ramachandraiah G (2006) Microplastic pollution in the Bay of Bengal, India. Mar Pollut Bull 52(7):818–820

Rochman CM, Browne MA, Halpern BS, Hentschel BT, Hoh E, Karapanagioti HK, Thompson RC (2019) Policy: classify plastic waste as hazardous. Nature 494(7436):169–171

Rochman CM, Kross SM, Armstrong JB, Bogan MT, Darling ES, Green SJ et al (2020) Scientific evidence supports a ban on microbeads. Environ Sci Technol 54(9):5504–5509

Rosenfeldt RR, Seidensticker S, Scharf S, MacLeod M, Packman AI (2019) Microplastic retention in river sediments: a probabilistic modeling approach to analyze microplastic sinks and sources. Environ Sci Technol 53(4):2015–2023

Roy PK, Jabeen K, Selvaraj S, Chandrasekaran N (2022a) Impact of microplastics on aquatic biota: recent advances and future challenges. Environ Pollut 292:118457

Roy P, Mohanty AK, Misra M (2022b) Microplastics in ecosystems: their implications and mitigation pathways. Environ Sci Adv 1(1):9–29. https://doi.org/10.1039/d1va00012h

Rummel CD, Jahnke A, Gorokhova E, Kühnel D, Schmitt-Jansen M (2017) Impacts of biofilm formation on the fate and potential effects of microplastic in the aquatic environment. Environ Sci Technol Lett 4(7):258–267. https://doi.org/10.1021/acs.estlett.7b00164

Sadia A, Tariq S, Ali L (2022) Microplastics in wastewater treatment plants: sources, fates, and remediation. Sci Total Environ 806:150563

Schwarz M, Kettner MT, Gajewski J, Weber FA (2021) Evaluation of coagulation/flocculation processes for the removal of microplastics from synthetic laundry wastewater. Sep Purif Technol 280:119660

Schymanski D, Goldbeck C, Humpf HU, Fürst P (2018) Analysis of microplastics in water by micro-Raman spectroscopy: release of plastic particles from different packaging into mineral water. Water Res 129:154–162

Setälä O, Norkko A, Lehtiniemi M, Lindström K, Vedenpää M (2018) Feeding type affects microplastic ingestion in a coastal invertebrate community. Mar Pollut Bull 135:61–66

Sharma A, Singh A, Aseri V, Nagar V, Chopade RL, Pandit PP (2022) Eradication of microplastics in wastewater treatment: overview. Biointerface Res Appl Chem 13(3):223. 10.33263/briac133.223

Song YK, Hong SH, Jang M, Kang JH, Kwon OY, Han GM (2021) Microplastic identification and quantification in marine sediment: a comparison of microscopic and spectroscopic methods. Environ Pollut 274:116575

Sooriyakumar P (2022) Microbial biofilm and the degradation of microplastics: a review. Environ Pollut 296:118258

Stabnikova O, Stabnikov V, Marinin A, Klavins M, Klavins L, Vaseashta A (2021) Microbial life on the surface of microplastics in natural waters. Appl Sci 11(24):11692. https://doi.org/10.3390/app112411692

Statista (2021) Global plastic production from 1950 to 2020. Retrieved from https://www.statista.com/statistics/282732/global-production-of-plastics-since-1950/

Su L, Wang L, Zhang W, Wu C (2021) Occurrence, distribution, and ecological risks of microplastics in the South China Sea: a review. Sci Total Environ 767:144999

Su L, Li J, Zhang W, Zhao L, Lu Y (2022a) Isolation and characterization of polyethylene-degrading bacteria from marine environments. Front Microbiol 13:841

Su L, Wang Q, Xiao Y, Wu J, Liu X, Hu N (2022b) Response of bacterial community structure to microplastic pollution in estuarine sediments. Sci Total Environ 805:150302

Sudharshan Reddy A (2022) Microplastic pollution: sources, distribution, and remediation. In: Microplastics in aquatic environments. Elsevier, pp 211–228

Sun X-L, Xiang H, Xiong H-Q, Fang Y-C, Wang Y (2023) Bioremediation of microplastics in freshwater environments: a systematic review of biofilm culture, degradation mechanisms, and analytical methods. Sci Total Environ 863:160953. https://doi.org/10.1016/j.scitotenv.2022.160953

Syed JH, Malik RN, Li J (2021) Occurrence, sources, and ecotoxicological effects of microplastics on freshwater ecosystems: a review. Environ Sci Pollut Res 28(9):10775–10796

Tagg AS, Sapp M, Harrison JP, Ojeda JJ, Liboiron M, Thiel M, Provencher JF (2021) A global perspective on microplastic pollution in aquatic ecosystems. Environ Sci Technol 55(8):47

Thompson RC (2021) Microplastics in the marine environment: sources, consequences, and solutions. Marine Nature 12(1):18

Tursi A, Baratta M, Easton T, Chatzisymeon E, Chidichimo F, De Biase M, De Filpo G (2022a) Microplastics in aquatic systems, a comprehensive review: origination, accumulation, impact, and removal technologies. RSC Adv 12(44):28318–28340. https://doi.org/10.1039/d2ra04713f

Tursi A et al (2022b) Microplastics in urban wastewater treatment plants: a review on occurrence, removal, and monitoring strategies. J Environ Manag 306:114903

Urbanek AK, Rymowicz W, Mirończuk AM (2018a) Degradation of polystyrene and selected plastics by cold-adapted bacteria from Siberian and Arctic environments. Polym Adv Technol 29(1):98–104

Urbanek AK, Rymowicz W, Mirończuk AM (2018b) Degradation of plastics and plastic-degrading bacteria in cold marine habitats. Appl Microbiol Biotechnol 102(18):7669–7678. https://doi.org/10.1007/s00253-018-9195-y

Vahvaselkä M, Winquist E (2021a) Existing and emerging technologies for microplasticsremoval: review report of the FanpLESStic-sea project. Natural resources and bioeconomy studies 81/2021. Natural Resources Institute Finland Helsinki 34 p

Vahvaselkä M, Winquist E (2021b) Evaluation of microplastic removal efficiency in wastewater treatment plants utilizing flow cytometry. J Hazard Mater 416:125914

Vaid M, Mehra K, Gupta A (2021a) Microplastics as contaminants in indian environment: a review. Environ Sci Pollut Res 28(48):68025–68052. https://doi.org/10.1007/s11356-021-16827-6

Vaid R, Singh N, Garg R, Dhir A, Kumar A, Kumar S (2021b) Microplastics pollution in Indian freshwaters: occurrence, sources, and potential risks. Environ Pollut 287:117374

Vaksmaa A, Hernando-Morales V, Zeghal E, Niemann H (2021a) Microbial degradation of marine plastics: current state and future prospects. In: Biotechnology for sustainable environment. Springer, pp 111–154. https://doi.org/10.1007/978-981-16-1955-7_5

Vaksmaa A, Ivanova I, Kers J (2021b) Biofilm formation on microplastic particles and its impact on microplastic transport in the environment. In: Microplastic contamination in aquatic environments. Springer, Cham, pp 203–224

Vaksmaa A, Põllumaa L, Kuznetsova T, Ivanova K, Laht M, Vasileva T et al (2021c) Microplastics shape the composition and function of particle-attached and free-living bacteria in coastal waters. Environ Pollut 283:117070

van Sebille E, Aliani S, Law KL, Maximenko N, Alsina JM, Bagaev A et al (2020) The physical oceanography of the transport of floating marine debris. Environ Res Lett 15(2):023003

Venkatesan AK, Somasundaran P, Ahmad AL (2022a) Microbial degradation of plastics: a review. Bioresour Technol 346:126595

Venkatesan R, Santhamoorthy M, Alagumalai K, Haldhar R, Raorane CJ, Raj V, Kim S-C (2022b) Novel approach in biodegradation of synthetic thermoplastic polymers: an overview. Polymers 14(20):4271. https://doi.org/10.3390/polym14204271

Vimala R, Mathew L (2016a) Biodegradation of low density polyethylene (LDPE) by mixed culture of Lysinibacillus xylanilyticus and aspergillus Niger. Int Biodeterior Biodegradation 114:216–225

Vimala PP, Mathew L (2016b) Biodegradation of polyethylene using Bacillus subtilis. Procedia Technol 24:232–239

Wang J, Song YK, Schmid M (2022) Microplastic pollution in freshwater systems: sources, sinks, and solutions. Sci Total Environ 822:153774

Westphalen G, Abdelrasoul A (2018a) Microplastics in aquatic environments: methodologies and trends—a review. Environ Sci Pollut Res 25(11):10192–10206

Westphalen H, Abdelrasoul A (2018b) Challenges and treatment of microplastics in water. In: Water challenges of an urbanizing world. InTech, Rijeka. https://doi.org/10.5772/intechopen.71494

Wright SL, Thompson RC, Galloway TS (2013) The physical impacts of microplastics on marine organisms: a review. Environ Pollut 178:483–492

Wright RJ, Erni-Cassola G, Zadjelovic V, Latva M, Christie-Oleza JA (2020) Marine plastic debris: a new surface for microbial colonization. Environ Sci Technol 54(19):11657–11672. https://doi.org/10.1021/acs.est.0c02305

Yan M et al (2022) Characterization of a bacterial consortium capable of degrading polyethylene terephthalate and its potential for plastic waste management. Environ Sci Pollut Res 29(8):10351–10364

Yang D, Shi H, Li L, Li J, Jabeen K, Kolandhasamy P (2019) Microplastic pollution in table salts from China. Environ Sci Technol 53(19):11702–11711

Yang X, Zhao X, Zhong X, Xiong W (2020a) Enhanced sludge dewatering and phosphorus recovery using ferric chloride combined with titanium tetrachloride as dual coagulants. Environ Sci Pollut Res 27(16):19319–19327

Yang Y et al (2020b) Microplastic-biota interactions: dynamic microbial colonization on microplastics. Front Microbiol 11:572751

Yang Y, Liu W, Zhang Z, Grossart H-P, Gadd GM (2020c) Microplastics provide new microbial niches in aquatic environments. Appl Microbiol Biotechnol 104(15):6501–6511. https://doi.org/10.1007/s00253-020-10704-x

Yoshida, S., Hiraga, K., Takehana, T., Taniguchi, I., Yamaji, H., Maeda, Y., Toyohara, K., Miyamoto, K., Kimura, Y., & Oda, K (2016) A bacterium that degrades and assimilates poly(ethylene terephthalate) Science (New York, N.Y), 351(6278), 1196–1199. https://doi.org/10.1126/science.aad6359

Zhang J, Xu L, Jiang Z, Zheng J (2020) Photocatalytic degradation of polystyrene microplastics using TiO2 nanoparticle coatings under UV light irradiation. J Hazard Mater 394:122529

Zhao Z, Hu J, Qin X, Zeng J, Cui J, Yang S et al (2021) Review on biological removal of microplastics: new perspectives and challenges. Water Res 197:117089

Ziajahromi S, Neale PA, Rintoul L, Leusch FD, Wijekoon KC (2017) Occurrence and fate of microplastic particles and fibers in the metropolitan wastewater treatment plant in Melbourne, Australia. J Water Health 15(1):143–153

Chapter 10
Innovative Nature-Based Solutions for Hazardous Effluent Treatment

Garima Upreti, S. Mary Celin, and Smita S. Kumar

Abstract Environmental pollution from anthropogenic activities is a global concern requiring innovative treatment solutions. This chapter explores the application of nature-based technologies for treating hazardous contaminants in wastewater. By harnessing the inherent capabilities of nature, strategies, like constructed wetlands, phytoremediation, biofiltration, and bioremediation, offer effective pollutant mitigation. Emerging approaches, such as floating treatment wetlands, algal biotechnology, and sustainable aquaculture systems, are also examined for their potential to redefine wastewater treatment paradigms. Through real-world case studies and the latest research, this chapter illustrates the practical applications and benefits of these methods, highlighting their environmental advantages, cost-effectiveness, and adaptability across diverse contexts. The focus on nature-based technologies emphasizes their role in fostering ecological resilience and sustainability, making them vital tools for mitigating the environmental and health risks associated with hazardous wastewater. This chapter serves as a valuable resource for those seeking holistic and sustainable wastewater treatment solutions.

Keywords Hazardous wastewater treatment · Constructed wetlands · Phytoremediation · Biofiltration · Bioremediation · Floating treatment wetlands · Algal biotechnology · Sustainable aquaculture · Ecological resilience

G. Upreti · S. Mary Celin (✉)
Centre for Fire Explosive and Environment Safety (CFEES), Defence Research and Development Organization, Delhi, India
e-mail: marycelin.cfees@gov.in

S. S. Kumar
Department of Environmental Studies, Central University of Haryana, Mahendragarh, Haryana, India

S. S. Kumar, V. Kumar (eds.), *Nature-Based Solutions for Decentralised Liquid Waste Management*, Environmental Challenges and Solutions,
https://doi.org/10.1007/978-3-032-08362-3_10

10.1 Introduction

Since the last decade with the increase in population, the consumption of water has also increased twice at the global scale and so has the discharge of wastewater into natural water bodies (UN-FAO 2023). The release of wastewater from different industries is known as effluents. The effluents contain hazardous chemicals consisting of both organic and inorganic compounds, microbes, etc. Hazardous effluents are released during production, machining, and treatment and waste management processes from industries like tannery and leather, metals and electroplating, paint, pharmaceutical, chemicals, dye, printing, petroleum, paper and explosives manufacturing, etc. These effluents contain spent solvents and wastes like electroplating and other metal finishing wastes, dioxin-bearing wastes, chlorinated aliphatic hydrocarbon production wastes, wood preserving wastes, petroleum refinery wastewater treatment sludges, and multisource leachates. Effluents from hospitals and pharmaceutical industries contain various contaminants including ibuprofen; phenols; microbial residues of *Acinetobacter, Enterococcus*, and *Pseudomonas* species; and radioisotopes such as technetium-99 m (Tc-99 m), iodine-131 (I-131), iodine-125 (I-125), iodine-123 (I-123), fluorine-18 (F-18), tritium (H-3), and carbon-14 (C-14) (Elisabetta et al. 2016). Other contaminants present in hazardous effluents are heavy metals such as lead, mercury, cadmium, arsenic, and chromium; pathogens like bacteria, viruses, and parasites; excessive amounts of nutrients like nitrogen and phosphorus from agricultural runoff; hydrocarbons, oils, and greases; radioactive materials; and suspended solids, such as sediment and particulate matter. In recent years, hazardous effluents containing emerging contaminants have gained attention due to their potential environmental and health risks, even though they may not have been extensively regulated or studied in the past. These contaminants often come from a variety of sources, including industrial processes, pharmaceuticals, personal care products, and agricultural practices. Following is the list of emerging contaminants of concern:

- Pharmaceuticals and personal care products (PPCPs): These include prescription and over-the-counter drugs, as well as ingredients found in personal care products like cosmetics and lotions. PPCPs can enter effluents through human use and excretion, and they may persist in the environment, potentially affecting aquatic life and water quality.
- Endocrine-disrupting chemicals (EDCs): Substances like bisphenol A (BPA) and phthalates are examples of EDCs that can be found in effluents. EDCs can interfere with the endocrine system, potentially disrupting hormonal functions in humans and wildlife.
- Per- and polyfluoroalkyl substances (PFAS): PFAS are a group of man-made chemicals used in a wide range of products, including firefighting foam, nonstick cookware, and waterproof fabrics. They are known for their persistence in the environment and potential health risks. PFAS contamination in effluents has raised concerns about water quality.

- Microplastics: Tiny plastic particles, often less than 5 mm in size, have become a concern in effluents due to their widespread presence in the environment and potential harm to aquatic organisms that ingest them.
- Nanomaterials: Engineered nanoparticles used in various industries, such as electronics and cosmetics, can enter effluents and raise questions about their potential ecological and health impacts.
- Pharmaceuticals of emerging concern (PECs): PECs are pharmaceuticals that have not traditionally been regulated, but there is growing concern about their presence in effluents and their potential effects on aquatic ecosystems and even drinking water supplies.
- New and designer drugs: Illicit drugs and novel psychoactive substances can end up in wastewater effluents, posing challenges for treatment plants and potential environmental and public health concerns.
- Flame retardants: Certain flame-retardant chemicals used in textiles, electronics, and construction materials can leach into effluents, potentially impacting aquatic ecosystems and human health.
- Pesticide metabolites: Some pesticide metabolites, which are breakdown products of pesticides, can persist in the environment and be found in effluents, raising concerns about their ecological impact.

Contaminants in hazardous effluents enter the environment from industries, treatment plants, agriculture, mining, oil and chemical spills, landfills, and nuclear facilities and pave their ways through various routes in air, water, and soil ecosystems. Contaminants like VOCs, VICs, ammonia, hydrogen sulfide, PM (soot and dust), and gasses (chlorine, methane, and sulfur dioxide) may get released into the air during the treatment by processes, viz., volatilization, aerosolization, spray drift, stack and landfill gas emission, and evaporation ponds and lagoons (Bianciardi et al. 2022). Contaminants like heavy metals, pesticides and residues, herbicides, PAHs, POPs, PFAS, radioactive materials, salts, minerals, acids, alkali, and microplastics may enter the soil matrix due to processes, viz., surface runoff, leaching, erosion, sedimentation, infiltration through sewage system or underground storage tank, and biological uptake. Similarly, these contaminants can also enter water bodies due to leaching, point source pollution, and direct/indirect discharge to water systems.

There are region-specific and country-specific limits/guidelines for wastewater discharge into streams, and criteria are set for reusing the treated water for irrigation, toilet flushing, etc. (WHO, BIS). But with the complex mixture of contaminants in hazardous effluents, increasing wastewater discharge, high BOD (approx 3000–4000 mg/L leather industry), and high COD (5000–8000 mg/L explosive manufacturing industries), it becomes challenging to meet the treatment goals. These effluents tend to pollute the environment, contaminate natural resources, and impact human health and environment. The contaminants affect aquatic life through the process of bioaccumulation and increase nutrient concentration resulting in algal bloom. Along with aquatic ecosystems, the interconnected coastal ecosystems are also affected by the process of biomagnification. These processes affect the health of both flora and fauna. Various health-related ailments have been reported on exposure to contaminants like heavy metals, radioactive substances, nitro-organic compounds, and VOCs (Ademuyiwa et al. 2007).

There are challenges faced in treating these contaminants due to their recalcitrant nature leading to ecosystem imbalance. It has also been found that wastewater characteristics can vary seasonally and with changes in flow rates, which can impact the performance of conventional treatment systems that may not be designed to handle such variability. Conventional treatment methods also have limited ability to remove microplastics and nanoparticles, which can persist in effluents and accumulate in the environment. Present treatment techniques based on physical and chemical methods for hazardous waste water are prone to create more waste. These challenges can be overcome by exploring nature-based treatment (NBT) technologies which draw inspiration from the natural ecosystem. The NBTs aim to reduce the environmental footprint. These techniques are being explored by various research groups and are being developed and tested at pilot scale. An attempt has been made in this chapter to provide an overview on the types, performance, and regulatory regime of nature-based technologies; few relevant case studies are discussed. Environmental and economic benefits are presented along with challenges. Future research insights of NBTs are also highlighted.

10.2 Nature-Based Treatment Methods for Hazardous Effluents

Nature-based solutions (NBS) for wastewater treatment focus on creating engineered systems that emulate the processes and benefits of natural ecosystems while relying minimally on mechanical components. These systems utilize elements such as vegetation, soil, porous materials, microbial communities, and other naturally occurring processes to effectively eliminate contaminants from wastewater. These contaminants may include suspended solids, organic matter, nutrients like nitrogen and phosphorus, and harmful pathogens (Kadlec and Wallace 2009). The prominent NBTs being studied are detailed below:

10.2.1 Constructed Wetlands

Constructed wetlands mimic natural wetland ecosystems. They use aquatic plants, bacteria, and substrate to remove pollutants through physical, biochemical, and biological processes. They are effective in removing heavy metals, organic compounds, and nutrients from industrial wastewater. Constructed wetland consists of a shallow long stream bed in which hazardous effluents flow and is diverted to the sedimentation pond/area. Algae and species like *Potamogeton*, *Typha*, etc., are planted. Plant selection may vary depending upon contaminant type. Natural conditions developed along with flora support the treatment process. The important parameters impacting the efficiency of a constructed wetland treatment are plant species, contaminant load, hydraulic design, and substratum (Liao and Chang 2004). The constructed wetland can be developed with horizontal and vertical flow of wastewater. Though horizontal subsurface flow has higher potential, the combined flow model is reported to be more effective (Laaffat et al. 2015). Apart from heavy metals, constructed wetlands are implemented for treatment of effluents from textile, tannery, oil and gas coking, paper and pulp mills, etc. (Wallace et al. 2015). Modeling studies have been proven beneficial in improving the effectiveness of the vertical flow treatment system (Mohammed 2017). Table 10.1 provides detailed overview of the removal efficiency of various contaminants using different types of constructed wetlands:

Table 10.1 highlights the effectiveness of treatment systems in treating pollutants like pharmaceuticals (e.g., diclofenac, ibuprofen), industrial chemicals (e.g., benzene, MTBE), and waste from swine and petrochemical sources. Constructed wetlands utilizing plants like *Typha* and *Phragmites*, along with materials like gravel, soil, and oyster shells, show varied removal rates, with efficiencies ranging from 43.11% to 100%. The data demonstrates the potential of nature-based technologies in wastewater treatment across diverse contexts.

Table 10.1 Comparative removal efficiency of different constructed wetlands

Sr. no.	Waste/contaminant	Constructed wetland treatment type	Removal efficiency (%)	References
1.	Diclofenac (DCF)	Wastewater stabilization pond (WSP) and sludge waste	75.2	Nas et al. (2021)
2.	Ibuprofen (IBU)	Constructed wetland (CW) and *Typha* and *Phragmites*	100	Zhang et al. (2016)
3.	Iohexol (IOH)	Constructed wetland (CW) and *Typha* and *Phragmites*	80	Zhang et al. (2016)
4.	Benzene	Gravel bed constructed wetland and common reed (*Phragmites*)	76	Seeger et al. (2011)
5.	MTBE	Plant root mat system and common reed (*Phragmites*)	83	Seeger et al. (2011)
6.	Nitrogen	Constructed wetland (CW) and *Typha angustifolia*	65.3	Klomjek and Nitisoravut (2005)
7.	BOD	Constructed wetland (CW) and *Asian crabgrass*	67.9	Klomjek and Nitisoravut (2005)
8.	Textile dye waste COD removal	Constructed wetland (CW), soil and rice husk, and *Persicaria barbata*	43.11	Saba et al. (2014)
9.	Swine waste total nitrogen and phosphorus	Vertical upflow constructed wetland (CW), oyster shell, brick, red soil, and *Phragmites australis*	63.3 and 92.29	Huang et al. (2015)
10.	Petrochemical waste COD removal	Horizontal flow constructed wetland (HFCW), gravels and sand layer, and *Phragmites* spp.	89.9	Sudarsan et al. (2018)

10.2.2 Phytoremediation

Phytoremediation utilizes specific plant species such as *P. stratiotes*, *Pistia stratiotes*, *Eichhornia* spp., *Lemna* spp., and other plants belonging to Gramineae, Pontederiaceae, Ceratophyllaceae, Typhaceae, and Haloragaceae family and their suitability for different contaminants to absorb, accumulate, or break down contaminants (Mustafa and Hayder 2020). These plants have the ability to metabolize or sequester hazardous substances from effluents, particularly useful for addressing heavy metals and organic pollutants. The system consists of contaminant-specific plants accompanied by native aquatic flora and includes processes, viz., phytofiltration, phytodegradation, rhizofiltration, and phytotransformation (Khan et al. 2022a, b). The process has been successful in treating the effluents containing dyes, inorganic and organic compounds, oils, pesticides, heavy metals, hydrocarbons, and pharmaceuticals. The enzyme present in the roots, leaves, and stem of the plants plays a crucial role in treatment, and hence it is being used for developing further innovative treatment methods.

10.2.3 Biofiltration

Biofiltration systems employ algae, bacteria, plants, protozoa, viruses, yeast, and mixed microbes to degrade organic compounds in contaminated water. This bioremediation technique uses microorganisms to break down hazardous chemicals, converting them into less harmful substances. It has been found effective for treating organic pollutants like petroleum VOCs, heavy metals, hydrocarbons, and pesticides (Pachaiappan et al. 2022). Biofiltration mechanism involves addition of biological creatures to pollutants which in the presence of oxygen reduces the contaminants to form biomass, water, and carbon dioxide (Bressani-Ribeiro et al. 2018). It can take place in three ways: (i) self-attachment of bacteria to pollutants, (ii) immobilization of bacteria to pollutants, or (iii) by microencapsulation process. Parameters affecting the treatment process are moisture content, temperature, pH, nutrient content, biofilter bed, and microbes (Pachaiappan et al. 2022). Effluents containing VOCs such as toluene, benzene, ethylbenzene, xylene, styrene, a-pinene, etc., have been reported to be treated using biofiltration methods. Biofilters can be incorporated in different ways, viz., rotary drum biofilters, submerged aerated biofilter, macrophyte biofilters, and stormwater biofilters. Biofilters are reported to remove variable concentration of mixed contaminants in effluents, be it the high concentration of dimethyl sulfide or low concentration (<200 ppmv) of nitrous oxide with approximately 95% efficiency (Yoon et al. 2017).

10.2.4 Constructed Solar Stills

Solar stills use solar energy to evaporate water from contaminated sources, leaving behind concentrated pollutants, to concentrate dilute hazardous waste, and to remove organic and inorganic substances through a photooxidation process (Funken et al. 1999). This technology is especially useful for treating saline or brackish water and can be combined with other treatment methods. Solar stills consist of major components, viz., glass cover, collector plate, water container feed, and distillate output (Ranjan and Kaushik 2013). Depending upon the altitude (availability of solar radiation), effluent load, and resources, different types of solar stills can be incorporated such as single slope, double slope, water film solar type, multiple effect diffusion, basin type, double basin, triple basin, hybrid, regenerative solar still, etc. (Taha et al. 2022). Efficiency of solar still depends on factors, viz., temperature, area of radiation collection, material (interior and exterior), wind speed, solar radiation, and inclination angle (Velmurugan et al. 2008). Solar stills are adopted for their low maintenance, design flexibility, and easy implementation features.

10.2.5 Mangrove Swamps and Coastal Vegetation

Coastal vegetation, including mangrove swamps, can naturally filter and treat wastewater from industrial and agricultural sources. These are found useful in treating industrial wastewater, stormwater runoff, and agricultural runoff. The extensive root system of this nature-based method provides higher surface area for absorption of contaminants; hence, it becomes easier to filter, trap, and absorb contaminants from the wastewater. Coastal vegetation helps in removing excess nutrients like nitrogen and phosphorus from effluents. The root systems of mangroves and coastal plants create microenvironments with anaerobic conditions that promote the growth of beneficial microorganisms. These microbes help in breakdown of complex organic contaminants. Coastal vegetation, especially mangroves, can stabilize the sediment in coastal areas, reducing erosion and preventing the release of contaminants that may be bound to sediment particles. These NBS also act as natural buffer zones, helping to absorb and dissipate the energy of incoming tidal surges and stormwater, which can contain pollutants. This reduces the impact of such events on the surrounding ecosystems. Different factors affect the working of mangrove swamps and coastal vegetation, viz., presence of specific contaminants, local environmental conditions, and the types of vegetation that are naturally adapted to the area. Proper planning, monitoring, and maintenance are crucial to ensure success. Impact of mangrove ecosystems with controlled wastewater discharge was monitored in a study, which showed decrease in crab density of coastal area suggesting the control discharge of treated water (Dimitri et al. 2019).

10.3 Case Studies on Treatment of Hazardous Contaminants by NBTs

Hazardous effluents contain majorly two types of contaminants: organic and inorganic contaminants. Organic contaminants include pesticides, herbicides, dyes, PAH, and pharmaceuticals. Inorganic contaminants include heavy metals, metalloids, inorganic nonmetallic compounds, salts, cyanide, acids and base, radioactive elements, and organometallic compounds. Some of the nature-based technologies for treating organic contaminants are discussed below in brief:

10.3.1 Organic Compounds in Hazardous Effluents

10.3.1.1 Pesticides/Herbicides/Insecticide

The aquatic floating plant species duckweed was found to be effective in treating dimethomorph-contaminated water in a lab-scale study. Concentrations in the range from 25, 50, 75, 150, 300, and 600 μg/L of dimethomorph were tested with duckweed plants for 4 days. Among the species *Lemna minor* and *Spirodela polyrhiza*, the efficient removal was made by *Lemna minor*. Approximately 9, 14, 20, 33, 60, and 115 μg/L dimethomorph was removed by the end of the treatment, respectively (Dosnon-Olette et al. 2010). The use of biopolymers such as starch, chitin, chitosan, alginates, cyclodextrins, and cellulose has also been tested for the treatment of pesticides and herbicides (Sanchez et al. 2020). The use of chitosan and gold nanoparticles has been evaluated to remove pesticide from wastewater. The nature-based adsorbent increased the capacity of beads from 11 to 58 μmole/g, resulting in enhanced removal of pesticides (Dwivedi et al. 2014). Biopolymer layer made with alginate and chitosan has been tested for adsorption of herbicides diquat, difenzoquat, and clomazone. Concentration from 5 to 200 μM of each herbicide was tested for adsorption with biopolymer, and no adsorption for concentrations of clomazone was observed. In case of diquat and difenzoquat, among all concentrations, only 50 μM concentration showed adsorption up to 95% and 62%, respectively (Agostini et al. 2013).

10.3.1.2 Dyes

Experiments have demonstrated the enhanced removal of toxic dyes and contaminants using plant species such as *Salvinia molesta*, *Aster amellus*, *Glandularia pulchella*, and *Zinnia angustifolia* plants (Priyadarshani et al. 2021). The effectiveness of macrophytes in the phytoremediation of dye wastewater was investigated. It was found that among the macrophytes studied, *Salvinia molesta* exhibited the most significant reduction efficiency for phenol and ammoniacal nitrogen when compared to other macrophytes, including *Lemna minor*, *Eichhornia crassipes*, *Pistia stratiotes*, and *Azolla pinnata* (Ugya et al. 2019). A similar study was also performed to assess the recuperative capabilities of *Pistia stratiotes*, *Eichhornia crassipes*, and an alga (*Oedogonium* sp.) in textile wastewater with elevated levels of COD, BOD, dyes, and heavy metals (Cd, Cu, Fe, and Pb) over a 7-day period. The research revealed that *P. stratiotes* exhibited a notably higher accumulation of Pb and Fe compared to *E. crassipes* aquatic plants (Tabinda 2019). Dye removal from wastewater has been experimented with enzymatic bioremediation using laccase, azoreductase, photocatalytic degradation, and bioelectrochemical system (Shindhal et al. 2020). Treatment of malachite green present in wastewater was reduced up to 96.8% (200 mg/L) using *Aeromonas hydrophila LZMG14* in 12 h (Jing et al. 2020). The use of extracted yeast in studies reports the degradation and decolorization of effluent during treatment (Akansha et al. 2019). Degradation of dye effluents containing methylene blue has been tested with algal species like *Oscillatoria*, *Chlorella*, *C. pyrenoidosa*, and *Spirulina maxima*. Algal degradation was found efficient in degrading 94–98% of contaminants, resulting in intermediates like water and CO_2 (Lebron et al. 2018).

10.3.1.3 PAHs

A horizontal subsurface flow constructed wetland (HSFCW) was tested for treating the wastewater containing PAHs (phenanthrene, pyrene, and benzo-a-pyrene) using vetiver and *Phragmites* (Alsghayer et al. 2020). Highest concentrations of PAHs were found in the root and shoot system of constructed wetland plants. *Phragmites* sp. reported the removal of phenanthrene, pyrene, and benzo-a-pyrene up to 83%, 71%, and 81%, respectively, while vetiver was able to remove 67%, 66%, and 73%, respectively. A study explored the effectiveness of various natural and cost-effective adsorbents, including sugarcane bagasse, green coconut shells, chitin, and chitosan for the removal of polycyclic aromatic hydrocarbons (PAHs) from wastewater in the petrochemical industry. The experiments involved the adsorption of PAH mixtures at concentrations ranging from 5.0 to 15.0 mg/L, with the process taking place under ambient conditions at approximately 28 °C and at a pH of 7.5. Research revealed that the adsorption behavior of PAHs conformed to the Freundlich model, signifying a multilayer adsorption process on heterogeneous surfaces. In terms of adsorption capacity, the natural adsorbents were ranked in the following order, from the most effective to the least: green coconut shells > sugarcane bagasse > chitin > chitosan. Notably, the adsorption characteristics of green coconut shells were found to be comparable to those of established conventional adsorbents like Amberlite T (Crisafully et al. 2021). A 5 L batch bioreactor (working capacity 3 L) was developed to remediate naphthalene, phenanthrene, and fluoranthene (200 mg/L each) mixture in wastewater, and the study reported 80–91% removal of PAHs at the end of the treatment (Goswami et al. 2018). The bacteria *Rhodococcus opacus* was found to be efficient in degrading the contaminants.

10.3.1.4 Pharmaceuticals

In a comparative study, diclofenac (DCF)-, carbamazepine (CBZ)-, 17β-estradiol (17β-E2)-, and 17α-ethynylestradiol (EE2)-containing wastewater was treated using wastewater stabilization pond (WSP) and constructed wetland (CW) method (Nas et al. 2021). The study reported significant (75.2%) removal of diclofenac using wastewater stabilization ponds. Studies on the utilization of the ozonation process as a pretreatment or integrated treatment approach for the degradation of pharmaceuticals in effluents have demonstrated its effectiveness in breaking down complex organic molecules, enhancing biodegradability, and improving the overall efficiency of subsequent biological or physicochemical treatment processes (Kharel et al. 2021). In a constructed wetland system, *Typha* and *Phragmites* sps. were used to treat pharmaceutical (ibuprofen [IBU] and iohexol [IOH])-containing wastewater. At the end of the treatment, ibuprofen (IBU) was completely removed, while iohexol (IOH) could be removed up to 80% (Zhang et al. 2016). Other examples include horizontal subsurface flow constructed wetlands (HSSF) and ozonation for the removal of trace pharmaceuticals in domestic wastewaters. In a similar such constructed wetland study, *Cyperus ligularis* was used with different ozone doses, viz., 10, 15, and 21 mg/L to treat ibuprofen- and naproxen-containing wastewater. Removal efficiency of 97% was achieved at the end, and 15 mg/L of ozone dose was found to be optimum for maximum removal (Lancheros et al. 2019). Figure 10.1 illustrates schematic diagram of constructed wetland for removal of pharmaceutical contaminants.

The figure here illustrates the layout and components of a constructed wetland designed for removing pharmaceuticals from wastewater. It shows the flow of wastewater through different sections, including plant species used (e.g., *Typha*, *Phragmites*), and indicates the points of contaminant removal (e.g., diclofenac, ibuprofen).

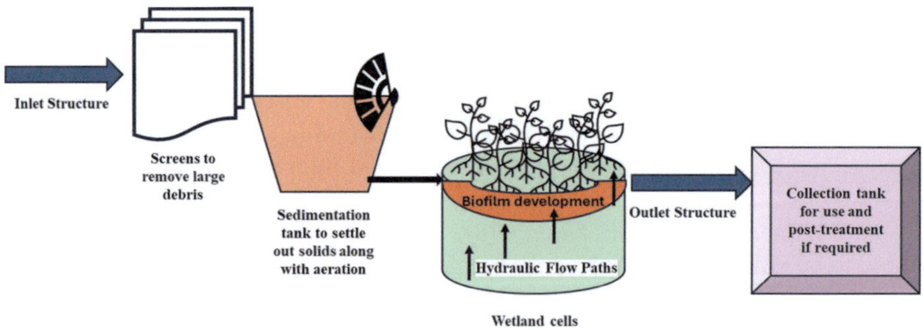

Fig. 10.1 Schematic diagram of a constructed wetland for pharmaceutical removal

10.3.1.5 VOCs

Strains of *Zoogloea resiniphila HJ1* and *Methylobacterium rhodesianum H13* are reported to be efficient in removing toluene, dichloromethane, and o-xylene up to 96.5% from the effluents of pharmaceutical factory, Shaoxing, China (Hu et al. 2016). In a pilot-scale study, removal of VOC (benzene; MTBE) was tested in two types of constructed wetlands, (i) planted gravel filters and (ii) plant root mat systems planted with common reed (*Phragmites australis*) (Seeger et al. 2011). The systems showed degradation in the rhizosphere, and plants were able to uptake benzene and MTBE up to 76% and 13% in the gravel bed constructed wetland and 83% and 11% in the root mat system. The gravel bed constructed wetland exhibited minimal volatilization flux for benzene and MTBE, measuring less than 5%. On the other hand, in the root mat system, the close interaction between the liquid and gas phases promoted significant (up to 24%) volatilization of MTBE (Seeger et al. 2011).

10.3.1.6 Oil Spill

In an investigation focused on the distillation of olive mill wastewater, a solar still was employed over a period of 9 days. This approach led to a reduction of 80% in COD (chemical oxygen demand) and a noteworthy 90% decrease in TKN (total Kjeldahl nitrogen). The outcomes of this study revealed that the distillate produced exhibited an absence of solid particles and there were no detectable odorous emissions (Potoglou et al. 2004).

10.3.2 Inorganic Compounds in Hazardous Effluent

10.3.2.1 Heavy Metals

Research groups worldwide have reported successful nature-based approaches for the removal of heavy metals from wastewater (Saeed and Khan 2019). In one particular study, the utilization of coconut waste, *Miscanthus* (silver grass), and black oak bark demonstrated the capability to effectively extract metals such as lead, cadmium, and mercury (Alalwan et al. 2020) from wastewater. These materials act as biofilters and have demonstrated 90% removal rate for heavy metals such as copper, cadmium, lead, and zinc (Blecken et al. 2011). Furthermore, the incorporation of solar stills in another study was efficient in reducing heavy metal levels meeting safe discharge limits (Ojo 2022). Mangrove swamps and coastal vegetation have also been successfully employed for removing heavy metals from effluents (Nguyen et al. 2020).

10.4 Advancements in Nature-Based Technologies

In recent times, several emerging nature-based technologies have evolved to enhance system performance in scenarios involving high water flow and substantial water volumes. These methods are presently tailored to address only a limited set of contaminants like micropollutants, nutrients, heavy metals, oil spills, and pathogens, and further testing is required to evaluate their effectiveness against additional hazardous effluent pollutants. Below, we provide a brief overview of some of these methodologies:

10.4.1 Floating Treatment Wetlands

Floating treatment wetlands consist of floating vegetation mats on the surface of water. These mats host plants with extensive root systems that absorb and metabolize pollutants. They are versatile and can be deployed in various water bodies, including ponds, lakes, and stormwater retention basins (Quyen et al. 2023). The efficiency of FTW depends on factor, viz., frame of raft/platform, depth of water body, design, seasonal variation, support substrate/media/culture, and temperature. Studies have reported significant removal (89–90%) of ibuprofen and caffeine using FTWs with *Eichhornia crassipes* sp. (Oliveira et al. 2019). The biomass generated after treatment can be used as compost, biogas generation, and biorefinery feedstock (Rodriguez et al. 2021). The technique is being tested at pilot scale for the removal of organic matter, micropollutants, nutrients, heavy metals, oil spills, and pathogens (Stanley et al. 2022).

10.4.2 Algal Biotechnology

Algae can be harnessed to treat wastewater while producing biomass for various applications. Algal biotechnology systems cultivate specific algae strains that can absorb nutrients and heavy metals from wastewater. This approach combines wastewater treatment with the potential for biofuel production (Hashmi et al. 2023). Efficient implementation of algal biotechnology depends on the parameters, viz., light, temperature, nutrients, and pH (Gondi et al. 2021). Algae *Chlorella vulgaris* and *Microspora* sp. were evaluated and are reported to be effective in removing metallic ions from oil mill effluent and absorbing methylene blue dye from textile effluent, respectively (Khan et al. 2022a, b). Algal biotechnology can also be combined with wastewater treatment methods for obtaining dual benefits. The utilization of bioaccumulation and biosorption processes by algae has the potential to effectively treat contaminants. Consequently, there is a need for the development of improved techniques for harvesting algae (Gondi et al. 2021). In a study involving a fed-batch reactor, a combination of algae species, namely, *Chlorella* and *Scenedesmus* sp., was employed for the treatment of textile effluent (Kumar et al. 2018). Recently, algal-based remediation technology has been applied to address the removal of emerging pollutants from effluent sources (Sarker et al. 2023). Different nature-based technologies used for the removal of heavy metals from the environment are tabulated in Table 10.2.

Table 10.2 summarizes different nature-based technologies used for the removal of heavy metals from the environment. Techniques like biofiltration, constructed wetlands, and phytoremediation employ natural materials such as coconut waste, water hyacinth, and tea waste biochar to target metals like lead, cadmium, mercury, chromium, and others. Removal efficiencies vary, with some methods achieving up to 99.3% removal. The table highlights the sustainable potential of these methods in treating heavy metal contamination.

Table 10.2 Nature-based technologies for heavy metal removal

Sr. no.	Technology	Materials/plants used	Heavy metals targeted	Removal rate (%)	References
1.	Biofiltration	Coconut waste, *Miscanthus*	Lead, cadmium, and mercury	90	Alalwan et al. (2020)
2.	Constructed wetlands	Water hyacinth	Various heavy metals	–	Liao and Chang (2004)
3.	Mangrove swamps	Mangrove ecosystems	Lead, cadmium, and mercury	–	Nguyen et al. (2020)
4.	Solar stills	–	General reduction of metals	–	Ojo (2022)
5.	Phytoremediation	Tea waste biochar	Chromium removal	99.3%	Khalil et al. (2020)
6.	Phytoremediation	Water hyacinth	Chromium removal	80%	Saha et al. (2017)
7.	Bioadsorbents	Coconut husk	Iron, copper, and lead	90%	Abdulrasaq and Osinfade (2010)
8.	Flocculation	Biopolymeric flocculant produced by *Terrabacter* sp.	Iron, aluminum, and zinc	77.7%	Agunbiade et al. (2019)
9.	Coagulant	*Opuntia ficus*	Iron and chromium	90%	Vargas-Solano et al. (2022)
10.	Spiral bound RO membrane	Ni(II), Pb(II), and Cu(II)	Nickel, lead, and copper	98.5%	Algureiri and Abdulmajeed (2016)

10.4.3 Advanced Phytoremediation Techniques

Researchers have been developing and optimizing the use of specific plant species and engineered plant systems to enhance phytoremediation efficiency for hazardous effluent contaminants. Genetic modifications and selection of hyperaccumulators have shown promise in improving contaminant uptake and tolerance. Genetic modification is done by isolating the gene responsible for degradation of contaminants and introducing it in plants capable of tolerating contaminants (Nedjimi 2021). The objective of this is to prepare plants to tolerate high concentration of contaminants and be able to degrade or accumulate contaminants. Hyperaccumulators assimilate higher concentrations of contaminants during their normal growth period. Some major classes of hyperaccumulators are Brassicaceae, Asteraceae, Amaranthaceae, Cyperaceae, and Fabaceae (Peer et al. 2003). Transgenic plants engineered for enhancing the remediation are reported to treat heavy metals and organic pollutants present in groundwater and can be incorporated for effluent treatment also. The use of nanoparticles, such as iron nanoparticles, in combination with plants, has demonstrated potential for enhancing the remediation of heavy metals and organic pollutants. Nanoparticle-assisted phytoremediation helps in stimulating plant growth, increases interaction by adsorption/redox reaction, and facilitates enhanced remediation. Nanotechnology can improve the efficiency of contaminant uptake and transformation.

10.4.4 Hybrid Systems

Integrating different nature-based technologies such as constructed wetlands, phytoremediation, and biofiltration into hybrid systems has gained attention. These integrated approaches aim to maximize contaminant removal and system resilience. Porous media function based on the adsorption principle, where it captures and retains contaminants from hazardous effluents. Subsequently, the treated water is released, having had the contaminants removed. The efficacy of this approach depends on factors such as the concentration of contaminants and the choice of adsorbent media, which may include activated carbon, biosorbents, and biomass adsorbents (Emad et al. 2018). Natural porous medium, zeolite, was used to remove heavy metals from oil mill effluent. It was found that heavy metals like Zn^{2+}, Mn^{2+}, and Fe^{3+} were removed up to 50%, 50%, and 60%, respectively (Fenglian and Wang 2011). Recirculating aquaculture system (RAS) is a closed-loop aquaculture system that recirculates and treats water within the system, reducing the need for large water exchanges. Advanced filtration technologies, such as mechanical and biological, are used to remove solids and treat water quality such as sand filters, microscreen drum filters, foam separator, and protein skimmer. Recirculating aquaculture system often incorporates biofilters, which rely on beneficial bacteria to convert ammonia (a fish waste product) into less harmful compounds. In a study, *E. coli* cells were genetically modified to incorporate genes for capturing heavy metals from wastewater and were combined with metallic nanoparticles. These co-assembled structures demonstrated an impressive removal efficiency exceeding 90% even while dealing with very low concentrations of heavy metals. Notably, the removal efficiency of the *E. coli*-magnetic nanoparticle co-assembly outperformed that of magnetic nanoparticles alone. As a result, this approach underscores the impact of utilizing genetically modified *E. coli* in combination with metallic nanoparticles to efficiently eliminate heavy metals from wastewater (Zhu et al. 2020).

In a study, performance of a treatment system was observed over a year. It included a treated wastewater pond (TWP) that exclusively receives 100% of the effluent from a wastewater treatment plant and an aquaculture pond (AP) supplied with treated water from the TWP for fish production. In the influent wastewater directed to the TWP, substances like telmisartan, carbamazepine, diclofenac, and venlafaxine were surpassed. After the treatment, it was found that the TWP-AP system appears to be a valuable tertiary treatment step, effectively reducing PhACs and diminishing water quality hazards related to ecotoxicology and antibiotic resistance before the water is repurposed for aquaculture (Fedorova et al. 2022).

10.5 Performance Evaluation of NBT

Monitoring the efficiency of nature-based technology for the treatment of hazardous effluents is crucial to guarantee their effective operation and to adhere to environmental regulation. Key performance indicators (KPIs) and monitoring techniques play a crucial role in evaluating their performance.

10.5.1 Key Performance Indicators (KPIs)

The collection of indicators includes treatment efficiency, hydraulic retention time (HRT), reduction in pollution load, and ecosystem services enhancement. HRT is a critical KPI for some nature-based systems like constructed wetlands. It measures the average time water spends in the treatment system, affecting treatment efficiency. These KPIs are related to ecosystem services, such as habitat creation, carbon sequestration, and water purification, and assess the broader environmental benefits of the system. Comparative studies against conventional treatment methods may also be conducted comparing the treated effluent's quality to regulatory standards or benchmarks. Assessments of the ecological health of the treatment area, including the presence of indicator species, biodiversity, and habitat quality, indicate the overall effectiveness of the nature-based system.

10.5.2 Monitoring Techniques

Monitoring of the entire treatment process is required for life cycle assessment and overall working of the treatment process. The monitoring techniques are regular water sampling and analysis, vegetation health monitoring, biomonitoring, remote sensing, and modeling and simulation. For systems involving vegetation, monitoring the health and growth of plants can indicate the system's ability to absorb contaminants and maintain its ecological functions. Using bioindicators or biomarkers, such as specific plant or microbial species, can provide insights into the system's response to contaminants. Remote sensing techniques, including satellite imagery and aerial photography, can be used to assess changes in land cover and vegetation health within the treatment area. Mathematical models and computer simulations can help predict the system's performance under different conditions and provide insights into its long-term effectiveness. By employing these assessment methods, KPIs, and monitoring techniques, researchers and practitioners can comprehensively evaluate the effectiveness of nature-based technologies for treating hazardous effluents, ensuring their long-term success and environmental benefits.

10.6 Environmental and Economic Benefits

NBTs have various environmental and economic benefits. The environmental benefits highlight the contributions of nature-based treatment in achieving more sustainable and ecologically friendly solutions while safeguarding the health of ecosystems and human populations. NBTs are effective in removing or reducing a wide range of contaminants, including heavy metals, organic pollutants, and nutrients. By doing so, they prevent these pollutants from entering natural ecosystems, thereby reducing the environmental impact. This also improves water quality and natural habitat conditions and ensures industrial compliance with environmental standards.

Many nature-based treatment systems, such as wetlands and riparian buffers, serve as habitats for diverse plant and animal species. Implementing these systems can contribute to the restoration of local ecosystems, support biodiversity, act as carbon sinks, and store organic matter in soils and vegetation. NBTs are also effective in managing stormwater runoff, reducing the risk of flooding and erosion in urban areas. This helps maintain the integrity of local waterways and prevents soil erosion. Green roofs and urban wetlands contribute to greener and more attractive urban spaces enhancing the aesthetic appeal.

Analyzing the economic aspects of nature-based technologies for hazardous effluents is crucial to understanding their cost-effectiveness and financial feasibility (Agaton and Guila 2023). The economic aspects involve a complex interplay of upfront capital costs, lower operational and maintenance expenses, energy efficiency, and long-term sustainability. Implementing nature-based treatment systems often involves significant upfront costs, including land acquisition, construction, and system design. The capital investment can vary widely depending on the specific technology, site characteristics, and scale. When comparing nature-based technologies to traditional treatment methods, such as chemical treatment or mechanical filtration, the initial capital costs may be higher. However, long-term operational and maintenance costs can offset these higher upfront expenses. Maintenance costs primarily involve routine tasks such as plant pruning, debris removal, and sediment management. These costs are generally lower than the ongoing expenses associated with mechanical systems. The utilization of land for nature-based systems can represent an opportunity cost, especially in urban areas where land is valuable. However, this can be balanced by the environmental and aesthetic value added to the site. It is essential to consider the environmental benefits and ecosystem services provided, when conducting cost-benefit analyses. These benefits may include habitat creation, carbon sequestration, and improved water quality, which can have long-term economic value. Economists use benefit transfer methods to estimate the economic value of ecosystem services provided by nature-based systems. This helps in assessing the overall economic impact. While nature-based systems may have higher upfront costs, their long-term sustainability can lead to economic advantages. Reduced reliance on nonrenewable resources and energy savings can translate into cost savings over time. While the initial investment may be higher, the economic benefits of reduced ongoing costs, lower energy consumption, and the value of ecosystem services often make nature-based systems economically competitive and environmentally attractive options for hazardous wastewater treatment.

10.7 Challenges and Limitations

Understanding the challenges in implementing nature-based technologies is essential for effective implementation and management. The key challenges and limitations of nature-based solutions for treating hazardous effluents are site specificity, land and space requirements, climate sensitivity, maintenance, and risk of contaminant release. Many nature-based solutions are highly site-specific; they may not be suitable for all locations. Factors such as soil type, climate, and available land can limit their applicability. Systems, like constructed wetlands and phytoremediation, can require significant land or space, which may not be readily available in urban or densely populated areas. Climate conditions can significantly impact the performance of nature-based solutions. Extreme weather events, temperature fluctuations, and seasonal variations can affect treatment efficiency. Frequent and consistent maintenance is often necessary to uphold the efficiency of nature-based systems. Improper maintenance can lead to system degradation and reduced treatment efficiency. These treatment processes can be slower compared to some conventional treatment methods. This can be a limitation when rapid treatment is required. Treatment performance may vary seasonally, with some systems being less effective during certain times of the year. For example, cold winter temperatures can affect plant-based treatment systems. In some cases, there may be a risk of contaminant release back into the environment if the treatment system is not properly designed or maintained. The drainage leak could release the contaminants into soil and further pollution; hence, public may have concerns or misconceptions about the safety and effectiveness of the method. While nature-based solutions can be cost-effective over the long term, the initial implementation costs, including site preparation and system construction, can be significant. There is a lack of standardized guidelines for designing and operating nature-based treatment systems, making it challenging to ensure consistent performance. Despite these challenges and limitations, nature-based solutions remain valuable tools for treating hazardous effluents, especially when sustainability, ecosystem restoration, and long-term environmental benefits are important considerations (Cross et al. 2021). Effective planning, design, and ongoing management are crucial to address these challenges and maximize the benefits of nature-based treatment systems.

10.8 Regulatory Framework

Relevant regulations and policies governing hazardous wastewater treatment vary by region and country. However, several common principles and frameworks guide the management of hazardous wastewater globally. Nature-based technologies can align with these regulations and policies in various ways:

The Clean Water Act (CWA) in the United States, the Water (Prevention and Control of pollution) Act in India, and the European Water Framework Directive (WFD) in Europe: These acts regulate the discharge of pollutants into water, including hazardous wastewater, and establish water quality standards and require permits for discharges. Nature-based systems, such as constructed wetlands and riparian buffers, align with the acts by naturally filtering and treating wastewater before it enters water bodies. They can be incorporated into compliance strategies to meet discharge standards.

The Basel Convention on the Control of Transboundary Movements of Hazardous Wastes and Their Disposal: The Basel Convention addresses the management and disposal of hazardous wastes, including those generated by industrial processes. NBTs can complement hazardous waste management efforts by treating effluents on-site or before disposal, reducing the need for long-distance transport and disposal of hazardous materials.

The United Nations Sustainable Development Goals (SDGs): The SDGs include Goal 6, which aims to ensure the availability and sustainable management of water and sanitation for all. NBTs contribute to achieving SDG 6 by improving water quality, enhancing ecosystems, and promoting sustainable water management practices. Many countries have their own national environmental regulations and standards and pollution prevention principles for hazardous wastewater treatment, and NBTs can contribute to reduce pollutant discharges, improve water quality, and promote the restoration of ecosystems, all of which are central goals of many environmental regulations worldwide.

10.9 Future Prospects and Research Directions

There are few promising and critical technologies which can be further explored for addressing pressing environmental challenges during treatment of hazardous effluents using NBTs. Some NB systems are discussed below:

Microbial fuel cells (MFCs): Microbial fuel cells utilizing microorganisms to generate electricity while treating wastewater have been explored. These systems offer a dual benefit by harnessing energy and effectively removing contaminants.

Artificial intelligence (AI) and machine learning: AI and machine learning algorithms have been applied to optimize the operation of nature-based treatment systems. They can help predict system performance, detect anomalies, and optimize resource allocation.

Wastewater treatment in urban environments: Research has focused on integrating nature-based treatment technologies into urban environments, such as green roofs, green walls, and urban wetlands. These innovations aim to mitigate the impact of urban pollution on local water bodies.

Monitoring and sensor technologies: Advances in sensor technologies and real-time monitoring systems have enabled better tracking of water quality parameters and system performance. These tools are crucial for ensuring the effectiveness of nature-based treatment solutions.

Life cycle assessment (LCA): Researchers have conducted life cycle assessments to evaluate the environmental and economic sustainability of nature-based treatment systems compared to traditional treatment methods. LCA findings can guide decision-makers in choosing the most sustainable solutions.

Removal of emerging contaminants: Emerging contaminants, such as pharmaceuticals and personal care products, have received increased attention. Research has focused on understanding how nature-based treatment systems can address these emerging challenges.

Ecosystem services and biodiversity enhancement: Studies have explored the broader ecological benefits of nature-based treatment systems, including their role in enhancing biodiversity, providing habitat for wildlife, and promoting ecosystem services like carbon sequestration.

10.10 Recommendations

The recommendations aim to guide actions at various levels, from policy development to on-the-ground implementation, to maximize the potential of nature-based technologies for hazardous wastewater treatment while fostering sustainability and environmental protection.

The specific recommendations are:

Promote research and development in the field of nature-based technologies for hazardous wastewater treatment to assess the effectiveness, scalability, and adaptability of these systems to different contaminants and environmental conditions. Develop or update regulatory frameworks and guidelines that explicitly recognize and support the use of nature-based technologies in hazardous wastewater treatment. Ensure alignment with international standards and best practices. Integrate nature-based wastewater treatment into urban planning and design, broader sustainability, and environmental policies, recognizing the potential for these technologies to address multiple environmental and societal goals. Create financial incentives, tax credits, or grants to motivate industries and municipalities to adopt nature-based treatment systems. This can help offset initial implementation costs and promote sustainable practices. Invest in training programs and capacity-building initiatives to educate professionals, engineers, and operators in the design, construction, operation, and maintenance of nature-based treatment systems. Establish robust monitoring and assessment protocols to continuously evaluate the performance of nature-based systems. Continuously collect data to measure contaminant removal rates, water quality improvement, and ecosystem health.

Raise public awareness about the benefits of nature-based solutions for hazardous wastewater treatment. Engage communities in the planning and implementation of such systems to ensure local support. Recognize the importance of flexibility and adaptability in nature-based systems. Design systems that can respond to changing environmental conditions, climate variability, and emerging contaminants. Establish demonstration/pilot-scale projects that showcase the effectiveness of innovative nature-based techniques in diverse settings. These projects can serve as educational tools and inspire wider adoption. Development of platforms for collaboration and knowledge sharing among researchers, practitioners, and policymakers in the field of innovative nature-based techniques at global scale is essential to promote interdisciplinary research and foster collaboration among experts from diverse fields, including environmental science, engineering, biology, and chemistry, to develop holistic and innovative solutions.

10.11 Conclusion and Implications

The nature-based technologies offer promising solutions for hazardous effluent treatment, contributing to environmental sustainability and reducing the impact of contaminants on ecosystems. Their effectiveness may vary depending on specific contaminants and site conditions, making them valuable tools in the toolbox of wastewater treatment strategies. These innovative approaches offer multifaceted solutions that address several critical environmental and societal challenges simultaneously. These are versatile and adaptable to various contexts and contaminants. As the world seeks innovative approaches to address environmental crises, nature-based technologies have demonstrated their effectiveness in efficiently removing various hazardous contaminants from wastewater, including heavy metals, organic pollutants, and nutrients. The natural processes and diverse plant and microbial communities included in the treatment methods contribute to thorough and consistent treatment. These technologies promote sustainability by integrating wastewater treatment with ecosystem restoration. They enhance biodiversity, create valuable habitats, and contribute to the conservation of natural resources. Nature-based systems typically require less energy than conventional treatment methods, reducing the carbon footprint associated with wastewater treatment. This reduced energy demand contributes to lower greenhouse gas emissions and supports efforts to combat climate change. This energy efficiency aligns with global sustainability goals including those related to clean water and sanitation (SDG 6), life on land (SDG 15), and climate action (SDG 13). Certain nature-based solutions, such as wetlands and coastal vegetation, provide natural buffers against climate change impacts such as storm surges and flooding. They enhance the resilience of ecosystems and communities. While they may have higher initial capital costs, nature-based technologies offer long-term cost savings through lower operational and maintenance expenses. Nature-based systems also play a crucial role in managing local water resources by reducing the risk of flooding, improving stormwater management, and recharging aquifers. They contribute to sustainable water use in urban and rural areas and safeguard public health by preventing the release of harmful contaminants into water bodies. Hence, these systems can help industries and municipalities meet strict environmental regulations governing hazardous effluent treatment. They can be tailored to specific needs and challenges, making them suitable for diverse applications. The use of nature-based hazardous effluent treatment methods has significant implications for both environmental sustainability and public health. These implications underscore the vital role that nature-based solutions play in addressing complex environmental and health challenges. Vegetated systems within nature-based treatment methods sequester carbon dioxide (CO_2) from the atmosphere, helping mitigate climate change. By improving water quality and reducing pollution, nature-based methods contribute to the preservation of aquatic ecosystems and the protection of vulnerable species. Systems, such as mangroves and coastal vegetation, provide natural buffers against the impacts of climate change, including storm surges, sea-level rise, and coastal erosion. They enhance the resilience of coastal ecosystems and communities. Some nature-based systems, like green roofs and vegetated walls, help filter and purify the air by removing pollutants, thereby improving air quality in urban areas and reducing respiratory health risks. NBTs not only address environmental sustainability by restoring ecosystems and reducing resource consumption but also have far-reaching public health implications. These methods contribute to cleaner water, improved air quality, disease prevention, and the overall well-being of communities, fostering a healthier and more sustainable future for all.

References

Abdulrasaq OO, Osinfade GB (2010) Removal of copper (ii), iron (iii) and lead (ii) ions from mono-component simulated waste effluent by adsorption on coconut husk. Afr J Environ Sci Technol 4:6

Ademuyiwa O, Ugbaja RN, Rotimi SO, Abam E, Okediran BS, Dosumu OA, Onunkwor BO (2007) Erythrocyte acetylcholinesterase activity as a surrogate indicator of lead-induced neurotoxicity in occupational lead exposure in Abeokuta, Nigeria. Environ Toxicol Pharmacol 24:183–188. https://doi.org/10.1016/j.etap.2007.05.002

Agaton C, Guila PM (2023) Ecosystem services valuation of constructed wetland as a nature-based solution to wastewater treatment. Earth 4:78–92. https://doi.org/10.3390/earth4010006

Agostini M, Mariana C, Daniela V, Fernando F, Leonardo B (2013) Chitosan and alginate biopolymer membranes for remediation of contaminated water with herbicides. J Environ Manag 131C:222–227. https://doi.org/10.1016/j.jenvman.2013.09.028

Agunbiade MO, Pohl C, Heerden EV, Oyekola O, Ashafa A (2019) Evaluation of fresh water actinomycete bioflocculant and its biotechnological applications in wastewaters treatment and removal of heavy metals. J Environ Res Public Health 16:1–15. https://doi.org/10.3390/ijerph16183337

Akansha K, Chakraborty D, Sachan SG (2019) Decolorization and degradation of methyl orange by bacillus stratosphericus SCA1007. Biocatal Agric Biotechnol 18:101044

Alalwan HA, Kadhom MA, Alminshid AH (2020) Removal of heavy metals from wastewater using agricultural byproducts. J Water Supply Res Technol-Aqua 69(2):99–112

Algureiri AH, Abdulmajeed YR (2016) Removal of heavy metals from industrial wastewater by using RO membrane, Iraqi. J Chem Pet Eng 17:125–136

Alsghayer R, Salmiaton A, Mohammad T, Idris A, Ishak CF (2020) Removal efficiencies of constructed wetland planted with Phragmites and Vetiver in treating synthetic wastewater contaminated with high concentration of PAHs. Sustainability 12(8):3357

Bianciardi A, Becattini N, Cascini G (2022) How would nature design and implement nature-based solutions? Nat Based Sol 3:100047. https://doi.org/10.1016/j.nbsj.2022.100047

Blecken GT, Marsalek J, Viklander M (2011) Laboratory study of stormwater biofiltration in low temperatures: total and dissolved metal removals and fates. Water Air Soil Pollut 219(1–4):303–317

Bressani-Ribeiro T, Almeida PG, Volcke EI (2018) Trickling filters following anaerobic sewage treatment: state of the art and perspectives. Environ Sci Water Res Technol 4(11):1721–1738

Cross K, Tondera K, Rizzo A, Andrews L, Pucher B, Istenic D, Karres N, Mcdonald R (2021) Nature-based solutions for wastewater treatment: a series of factsheets and case studies. IWA Publishing. https://doi.org/10.2166/9781789062267

Dimitri T, Rivera I, Georgina L, Sophia M, Yann L, Mathilde L, Sucré J, Elliott H (2019) Wastewater bioremediation by mangrove ecosystems impacts crab ecophysiology: in-situ caging experiment. Aquat Toxicol 218:105358. https://doi.org/10.1016/j.aquatox.2019.105358

Dosnon-Olette R, Couderchet M, El Arfaoui A, Sayen S, Eullaffroy P (2010) Influence of initial pesticide concentrations and plant population density on dimethomorph toxicity and removal by two duckweed species. Sci Total Environ 408(10):2254–2259

Dwivedi C, Gupta A, Chaudhary A, Nandi C (2014) Gold nanoparticle chitosan composite hydrogel beads shows efficient removal of methyl parathion from waste water. RSC Adv 4. https://doi.org/10.1039/C4RA03870C

Elisabetta C, Bonetta S, Bertino C, Lorenzi E, Bonetta S, Gilli G (2016) Hospital effluents management: chemical, physical, microbiological risks and legislation in different countries. J Environ Manag 168:185–199. https://doi.org/10.1016/j.jenvman.2015.11.021

Emad AJ, Fayq HJ, Rawaa FC (2018) Porous Media for Removal of organic and inorganic contaminants. Nano Biomed Eng 10(2):104–116. https://doi.org/10.5101/nbe.v10i2.p104-116

Fedorova G, Grabic R, Grabicova K, Turek J, Van N, Tuyen R, Tomas B, Bryan Z (2022) Water reuse for aquaculture: comparative removal efficacy and aquatic hazard reduction of pharmaceuticals by a pond treatment system during a one year study. J Hazard Mater 421:126712. https://doi.org/10.1016/j.jhazmat.2021.126712

Fenglian FU, Wang Q (2011) Removal of heavy metal ions from wastewaters: a review. J Environ Manag 92(3):407–418

Funken KH, Pohlmann B, Lupfert E, Dominik R (1999) Application of concentrated solar radiation to high temperature detoxification and recycling processes of hazardous wastes. Sol Energy 65:25–31. https://doi.org/10.1016/S0038-092X(98)00089-9

Gondi R, Kavitha S, Ravi YK, Karthik O, Kumar G, Tyagi V, Banu R (2021) Algal-based system for removal of emerging pollutants from wastewater: a review. Bioresour Technol 344:126245. https://doi.org/10.1016/j.biortech.2021.126245

Goswami L, Manikandan N, Dolman B, Pakshirajan K, Pugazhenthi G (2018) Biological treatment of wastewater containing a mixture of polycyclic aromatic hydrocarbons using the oleaginous bacterium Rhodococcus opacus. J Clean Prod 196. https://doi.org/10.1016/j.jclepro.2018.06.070

Hashmi Z, Bilad M, Fahrurrozi Z, Juliana L, Jun W (2023) Recent Progress in microalgae-based technologies for industrial wastewater treatment. Fermentation 9:311. https://doi.org/10.3390/fermentation9030311

Hu J, Zhang L, Chen J (2016) Performance and microbial analysis of a biotrickling filter inoculated by a specific bacteria consortium for removal of a simulated mixture of pharmaceutical volatile organic compounds. Chem Eng J 304:757–765. https://doi.org/10.1016/j.cej.2016.06.078

Huang X, Liu C, Li K, Su J, Zhu G, Liu L (2015) Performance of vertical up-flow constructed wetlands on swine wastewater containing tetracyclines and tet genes. Water Res 70:109–117. https://doi.org/10.1016/j.watres.2014.11.048

Jing J, Kulshreshtha S, Kakade A (2020) Bioaugmentation of membrane bioreactor with Aeromonas hydrophila LZ-MG14 for enhanced malachite green and hexavalent chromium removal in textile wastewater. Int Biodeter Biodeg 150:104939

Kadlec RH, Wallace SD (2009) Treatment wetlands, 2nd edn. CRC Press, Boca Raton

Crisafully R, Milhome MA L, Cavalcante RM, Silveira E R, Keukeleire D, Nascimento RF, (2008) Removal of some polycyclic aromatic hydrocarbons from petrochemical wastewater using low-cost adsorbents of natural origin. Bioresour Technol 99:4515–4519. https://doi.org/10.1016/j.biortech.2007.08.041

Khalil U, Shakoor MB, Ali S, Rizwan M, Alyemeni MN, Wijaya L (2020) Adsorption-reduction performance of tea waste and rice husk biochars for Cr (VI) elimination from wastewater. J Saudi Chem Soc 24(11):799–810. https://doi.org/10.1016/j.jscs.2020.7.01

Khan AA, Gul J, Naqvi SR, Ali I, Farooq W, Liaqat R, Al Mohamadi H, Stepanec L, Juchelkova D (2022a) Recent progress in microalgae-derived biochar for the treatment of textile industry wastewater. Chemosphere 306:135565

Khan A, Khan A, Waris A, Ilyas M, Zamel D (2022b) Phytoremediation of pollutants from wastewater: a concise review. Open Life Sci 17:488–496. https://doi.org/10.1515/biol-2022-0056

Kharel S, Stapf M, Miehe U, Ekblad M, Cimbritz M, Falas P, Nilsson J, Sehlen R, Bregendahl J, Bester K (2021) Removal of pharmaceutical metabolites in wastewater ozonation including their fate in different post-treatments. Sci Total Environ 759. https://doi.org/10.1016/j.scitotenv.2020.143989

Klomjek P, Nitisoravut S (2005) Constructed treatment wetland: a study of eight plant species under saline conditions. Chemosphere 58(5):585–593. https://doi.org/10.1016/j.chemosphere.2004.08

Kumar G, Huy M, Bakonyi P, Bélafi B, Katalin K, Sang H (2018) Evaluation of gradual adaptation of mixed microalgae consortia cultivation using textile wastewater via fed batch operation. Biotechnol Rep 20:e00289. https://doi.org/10.1016/j.btre.2018.e00289

Laaffat J, Ouazzani N, Mandi L (2015) The evaluation of potential purification of a horizontal subsurface flow constructed wetland treating greywater in semi-arid environment. Process Saf Environ Prot 95. https://doi.org/10.1016/j.psep.2015.02.016

Lancheros JC, Madera CA, Caselles A, Torres WA, Vargasramírez XM (2019) Ibuprofen and naproxen removal from domestic wastewater using a horizontal subsurface flow constructed wetland coupled to ozonation. Ecol Eng 135:89–97. https://doi.org/10.1016/j.ecoleng.2019.05.007

Lebron YAR, Moreira VR, Santos LVS (2018) Remediation of methylene blue from aqueous solution by Chlorella pyrenoidosa and Spirulina maxima biosorption: equilibrium, kinetics, thermodynamics and optimization studies. J Environ Chem Eng 6(5):6680–6690

Liao SW, Chang WL (2004) Heavy metal phytoremediation by water hyacinth at constructed wetlands in Taiwan. J Aquat Plant Manag 42:60–68. https://scholars.lib.ntu.edu.tw/bitstream/123456789/86773/1/14.pdf

Mohammed A (2017) Modelling heavy metals transformation in vertical flow constructed wetlands. Ecol Model 354:62–71. https://doi.org/10.1016/j.ecolmodel.2017.03.012

Mustafa H, Hayder G (2020) Recent studies on applications of aquatic weed plants in phytoremediation of wastewater: a review article. Ain Shams Eng J 12. https://doi.org/10.1016/j.asej.2020.05.009

Nas B, Dolu T, Argun M, Yel E, Ates H, Koyuncu S (2021) Comparison of advanced biological treatment and nature-based solutions for the treatment of pharmaceutically active compounds (PhACs): a comprehensive study for wastewater and sewage sludge. Sci Total Environ 779:146344. https://doi.org/10.1016/j.scitotenv.2021.146344

Nedjimi B (2021) Phytoremediation: a sustainable environmental technology for heavy metals decontamination. SN Appl Sci 3:286. https://doi.org/10.1007/s42452-021-04301-4

Nguyen A, Richter O, Le B, Thi N, Phuong K, Dinh K (2020) Long-term heavy metal retention by mangroves and effect on its growth: a field inventory and scenario simulation. Int J Environ Res Public Health 17:9131. https://doi.org/10.3390/ijerph17239131

Ojo O (2022) Use of solar still in heavy metals removal from kitchen wastewater. Eur J Eng Technol Res 7:44–47. https://doi.org/10.24018/ej-eng.2022.7.1.2710

Oliveira M, Atalla A, Frihling B, Emanuel C, Priscila M, Ludovico MF, Fernando J (2019) Ibuprofen and caffeine removal in vertical flow and free-floating macrophyte constructed wetlands with Heliconia rostrata and Eichornia crassipes. Chem Eng J 373. https://doi.org/10.1016/j.cej.2019.05.064

Pachaiappan R, Cornejo-Ponce L, Rajendran R, Manavalan K, Femilaa V, Awad F (2022) A review on biofiltration techniques: recent advancements in the removal of volatile organic compounds and heavy metals in the treatment of polluted water. Bioengineered 13. https://doi.org/10.1080/21655979.2022.2050538

Peer WA, Mamoudian M, Lahner B, Reeves RD, Murphy AS, Salt DE (2003) Identifying model metal hyperaccumulating plants: germplasm analysis of 20 Brassicaceae accessions from a wide geographic area. New Phytol 159:421–430

Potoglou D, Kouzeli KA, Haralambopoulos DA (2004) Solar distillation of olive mill wastewater. Renew Energy 29:569–579

Priyadarshani SD, Babu PS, Manikandan S, Subbaiya R, Govarthanan M, Karmegam N (2021) Phycoremediation of wastewater for pollutant removal: a green approach to environmental protection and long-term remediation. Environ Pollut 290:117989

Quyen V, Hien V, Ntagia E, Amulya K, Nguyen N, Tran P, Ninh N, Lee S, Sac T, Ha L, Pham T, Bui X, Lens PNL (2023) Pilot and full scale applications of floating treatment wetlands for treating diffuse pollution. Sci Total Environ 899:165595. https://doi.org/10.1016/j.scitotenv.2023.165595

Ranjan K, Kaushik S (2013) Energy, exergy and thermo-economic analysis of solar distillation systems: a review. Renew Sust Energ Rev 27:709–723

Rodriguez MA, Bonefeld B, Jensen M, Brix H, Arias C (2021) The use of treatment wetlands plants for protein and cellulose valorization in biorefinery platform. Sci Total Environ 810:152376. https://doi.org/10.1016/j.scitotenv.2021.152376

Saba B, Jabeen M, Mahmood T, Aziz I (2014) Treatment comparison efficiency of microbial amended agro-waste bio-char constructed wetlands for reactive black textile dye. Int Proc Chem Biol Environ Eng 65:13–16

Saeed T, Khan M (2019) Constructed wetlands for industrial wastewater treatment: alternative media, input biodegradation ratio and unstable loading. J Environ Chem Eng. https://doi.org/10.1016/j.jece.2019.103042

Saha P, Shinde O, Sarkar S (2017) Phytoremediation of industrial mines wastewater using water hyacinth. Int J Phytoremediation 19(1):87–96. https://doi.org/10.1080/15226514.2016.1216078

Sanchez LM, Ollier PR, Pereira AF, Leonardo AV (2020) Pesticide removal from industrial effluents using biopolymeric materials. In: Biopolymer membranes and films. Elsevier, pp 359–382. https://doi.org/10.1016/B978-0-12-818134-8.00015-8

Sarker S, Akter T, Parveen S, Uddin M, Mondal A, Sujan S, Asaduzzaman M (2023) Microalgae-based green approach for effective chromium removal from tannery effluent: a review. Arab J Chem 16:105085. https://doi.org/10.1016/j.arabjc.2023.105085

Seeger E, Reiche N, Kuschk P, Borsdorf H, Kaestner M (2011) Performance evaluation using a three compartment mass balance for the removal of volatile organic compounds in pilot scale constructed wetlands. Environ Sci Technol 45:8467–8474. https://doi.org/10.1021/es201536j

Shindhal T, Rakholiya P, Varjani S, Pandey A, Ngo H, Guo W, Ng Y, Taherzadeh M (2020) A critical review on advances in the practices and perspectives for the treatment of dye industry wastewater. Bioengineered 12:70–87. https://doi.org/10.1080/21655979.2020.1863034

Stanley M, Palace V, Grosshans R, Levin D (2022) Floating treatment wetlands for the bioremediation of oil spills: a review. J Environ Manag 317:115416. https://doi.org/10.1016/j.jenvman.2022.115416

Sudarsan JS, Subramani S, Rajan RJ, Shah I, Nithiyanantham S (2018) Simulation of constructed wetland in treating wastewater using fuzzy logic technique. J Phys 1000(1):012137. IOP Publishing

Tabinda AB (2019) Treatment of textile effluents with Pistia stratiotes, Eichhornia crassipes and Oedogonium sp. Int J Phytoremediation 21(10):939–943. https://doi.org/10.1080/15226514.2019.1577354

Taha O, Hussein A, Attia M, Aljibori H, Kolsi L, Togun H, Ali B, Abderrahmane A, Subkrajang K, Jirawattanapanit A, Ahmadi H (2022) Comprehensive review on solar stills—latest developments and overview. Sustainability. https://doi.org/10.3390/su141610136

Ugya AY, Hua X, Ma J (2019) Phytoremediation as a tool for the remediation of wastewater resulting from dyeing activities. Appl Ecol Environ Res 17(2):3723–3735. https://doi.org/10.15666/aeer/1702

UN-FAO. AQUASTAT—FAO's global information system on water and agriculture. Available online: https://www.fao.org/aquastat/en/overview/methodology/water-use (accessed on 29 September 2023)

Vargas-Solano SV, Rodríguez-González F, Martínez-Velarde R, Morales-García SS, Jonathan MP (2022) Removal of heavy metals present in water from the Yautepec River Morelos México, using Opuntia ficus-indica mucilage. Environ Adv 7. https://doi.org/10.1016/j.envadv.2021.100160

Velmurugan V, Kumaran SS, Prabhu VN, Srithar K (2008) Productivity enhancement of stepped solar still: performance analysis. Therm Sci 12:153–163

Wallace S, Brix H, Kuschk P, Kirui W, Masi F, Dong R (2015) Treatment of industrial effluents in constructed wetlands: challenges, operational strategies and overall performance. Environ Pollut 201:107–120. https://doi.org/10.1016/j.envpol.2015.03.006

Yoon H, Song MJ, Yoon S (2017) Design and feasibility analysis of a self-sustaining biofiltration system for removal of low concentration N2O emitted from wastewater treatment plants. Environ Sci Technol 51(18):10736–10745

Zhang Y, Lv T, Carvalho PN, Arias CA, Chen Z, Brix H (2016) Removal of the pharmaceuticals ibuprofen and iohexol by four wetland plant species in hydroponic culture: plant uptake and microbial degradation. Environ Sci Pollut Res 23(3):2890–2898

Zhu N, Zhang B, Yu Q (2020) Genetic engineering-facilitated coassembly of synthetic bacterial cells and magnetic nanoparticles for efficient heavy metal removal. ACS Appl Mater Interfaces 12:22948–22957

Chapter 11
Application of Constructed Wetlands for the Removal of Contaminants from Wastewater

Jyoti Yadav, Renu Meena, Yudhveer Soora, and Smita S. Kumar

Abstract There is a growing need for cost-effective and more efficient alternative options for wastewater treatment. Green technologies continually evolve to offer sustainable solutions to modern societies' growing and intensifying challenges. Constructed wetlands (CWs) are low-cost engineered systems for treating wastewater with proven treatment efficiency. The efficiency of CWs depends upon many factors, including the types of CWs, vegetation, hydraulic load, and the media utilized in the bed. This chapter will explore a wide range of CWs, including different types based on hydrology, macrophytic growth type, and water flow direction. It aims to highlight the applications of CWs and the recent innovations in CW technology like thermal insulation, bioengineering, reed box, the role of substrates, and the methods of improving the removal efficiency of CWs. This chapter indicates that major advances in the technology design, operation, and optimization of CWs using different mechanisms promote more sustainable, eco-friendly techniques for wastewater treatment.

Keywords Constructed wetlands · Wastewater · Pollutants · Substrates · Hybrid constructed wetlands · Design criteria

J. Yadav · R. Meena · S. S. Kumar (✉)
Department of Environmental Studies, Central University of Haryana, Mahendragarh, Haryana, India

Y. Soora
Department of Civil Engineering, Central University of Haryana, Mahendragarh, Haryana, India

S. S. Kumar, V. Kumar (eds.), *Nature-Based Solutions for Decentralised Liquid Waste Management*, Environmental Challenges and Solutions,
https://doi.org/10.1007/978-3-032-08362-3_11

11.1 Introduction

Wastewater disposal has become a severe issue in recent years due to the rise in pollutants and wastewater generation due to increased living standards, climate change, and rapid population growth. Conventional methods for wastewater management have been utilized effectively in various areas worldwide to overcome water pollution. Strategies for managing wastewater, such as the activated sludge process and membrane bioreactors with membrane isolation, are costly and may challenge their application in local communities (Chen et al. 2014). Due to these limitations, there is a tendency for water demand to increase in wastewater treatment. In developing countries, cost-effective and more efficient alternative options are preferred for wastewater purification. Constructed wetlands (CWs) are widely preferred for addressing contaminant management, cost-effectiveness, lower operative requirements, and simple operation protocols (Rai et al. 2013). CWs are engineered structures that utilize natural mechanisms for wastewater treatment. These specialized systems are designed and built for exploiting the natural cleaning mechanisms of wetland vegetation, substrates, and host microorganisms (Cheng et al. 2018). Vegetation provides suitable conditions for the growth of microorganisms that decompose organic materials in wastewater, whereas the soil acts as a filter to remove pollutants. Compared to traditional treatment methods, CWs have several advantages, including lower energy consumption, reduced maintenance requirements, and the potential to provide habitats for wildlife. CWs have effectively treated various wastewaters (Garcia et al. 2023). Since its inception, CWs have been largely based on a methodology that goes back to the ancient practice of purifying water with naturally functioning wetlands. However, recent developments have turned these mechanisms into complex engineering remedies that can tackle present environmental issues. Whether integrated within an urban environment or dispersed across agricultural regions, CWs exhibit versatile design strategies that harness ecological processes to generate tangible outcomes. CWs have demonstrated their usefulness in various applications beyond conventional wastewater treatment. For example, they are utilized in urban stormwater management to mitigate floods and reduce the number of pollutants entering water systems. Additionally, artificial ecosystems like CWs are pivotal in restoring degraded habitats and promoting the proliferation of various species, ultimately contributing to the restoration of ecological balance in areas affected by man-made activities.

This chapter will explore a wide range of CWs, including surface flow systems where water flows over planted beds and subsurface flow systems encompassing water through soil or gravel. Moreover, it will illuminate hybrid designs that combine multiple treatment methods to effectively address complex pollutant mixtures efficiently. CWs are utilized for primary and secondary treatment to address diverse wastewater sources, such as domestic, agricultural, coal drainage, petroleum refineries, compost and landfill leachates, fishpond discharges, and industrial wastewater from pulp and paper mills, textile mills, and seafood processing sites. Adopting CWs is a cost-effective treatment option with precise design, operation, and maintenance (Davis et al. 1995). CWs utilize various ways to remove pollutants involving physical, chemical, and biological processes. Several materials may be used to improve the efficiency of CWs. These materials include marble chips, iron powder, soil, and biochar derived from rice husks, cocopeat, cupola slag, recycled bricks and stones, lightweight expanded clay aggregates, gravel, sand, sawdust, coal, and zero-valent iron (Witthayaphirom et al. 2020). The first free water surface CWs were reportedly from Hungary as early as 1968 for various types of domestic and municipal wastewater (Kadlec and Wallace 2008). In the United Kingdom, an underground flow reed-bed system involving a bed full of gravel as the bedding material and an inclined base creates a hydraulic slope in the bed. The long-lasting performance and reliable efficiency of CWs have been evident over the years. CWs have been used effectively for the treatment of different types of wastewater including agricultural wastewater (Koskiaho et al. 2003; Saeed et al. 2018), dairy wastewater (Yazdani and Golestani 2019), effluents from the textile industry (Saeed and Sun 2013), wastewater from the pulp and paper industry (Parde et al. 2021), and wastewater from acid mine drainage (Kleinmann and Girts 1987). The effectiveness and efficiency of CWs depend on many factors, such as the composition of the growth substrate, the presence of microorganisms and vegetation, and the water movement patterns within the wetland system in removing pollutants from wastewater. Wetland hydrology incorporates both surface water and subsurface flows, which greatly influence the wetland system's performance and functioning. The categorization of CWs is further enhanced by plant growth, which is based on dominant vegetation, including emergent, submerged, floating-leaved, and free-floating species. Consequently, the flow of water within subsurface wetlands is horizontal or vertical, which leads to variations in the distribution of nutrients, the availability of oxygen at different locations, and the behavior of microbes in the system (Vymazal 2014; Salvati et al. 2012).

It is feasible to integrate various types of CWs to form hybrid systems that utilize the unique advantages of each system using a viable synergistic approach. The configuration of CWs significantly determines the performance of pollutant removal mechanisms, involving different variables such as water velocity, depth, fluctuation, detention time, circulation pattern, distribution and turbulence, and wave action. The goal is to aid in distributing the wastewater more evenly and support increased interactions between the wastewater, substrate material, and vegetation for further removal. Most studies have emphasized improving the filter material through different modifications. For example, biochar addition can enhance water retention, enable anaerobic conditions, and provide electrons for denitrification (Tian et al. 2019; Peng et al. 2022). Several factors can affect the efficiency of a particular technology and its success, such as the intensity and conditions of many variables and the materials used earlier involved at all stages of its operation, both before and after its use. Optimization aims to find the best solution to the challenges that persist during treatment. This leads to the enhancement and improvement of procedures and engineered devices to boost their performance.

11.2 Types of Constructed Wetlands

CWs are categorized based on their design characteristics and processes for removing pollutants. The performance of CWs depends upon various parameters, including the types of wetlands, vegetation, applied hydraulic load, and media used in the bed. The most significant design criteria are as follows:

- Hydrology: Open water surface flow and subsurface water flow (Fig. 11.1).
- Macrophyte growth type: Emergent, submerged, floating-leaved, and free-floating.
- Subsurface wetland: Horizontal flow and vertical flow.

These are used individually and in combination with each other depending on desired efficiency removal. The combination of these is known as a hybrid system.

11.2.1 Free Water Surface (FWS) Systems

These systems can be natural and artificial. In this type of CWs, the water level is always kept above the surface of the medium (Fig. 11.2). In FWS CWs, water flows through a certain path, including inlet and outlet. The region has the following characteristics:

- Low water depth.
- Low flow velocity.

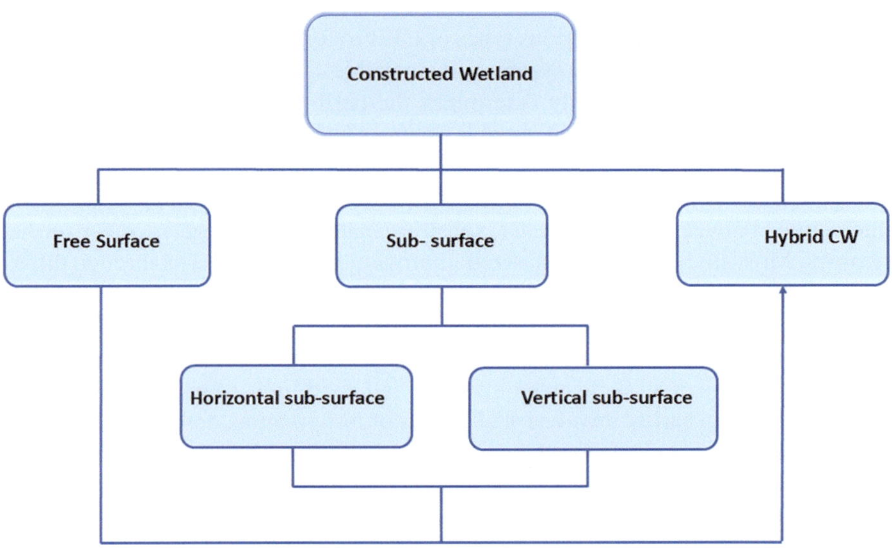

Fig. 11.1 Types of CWs

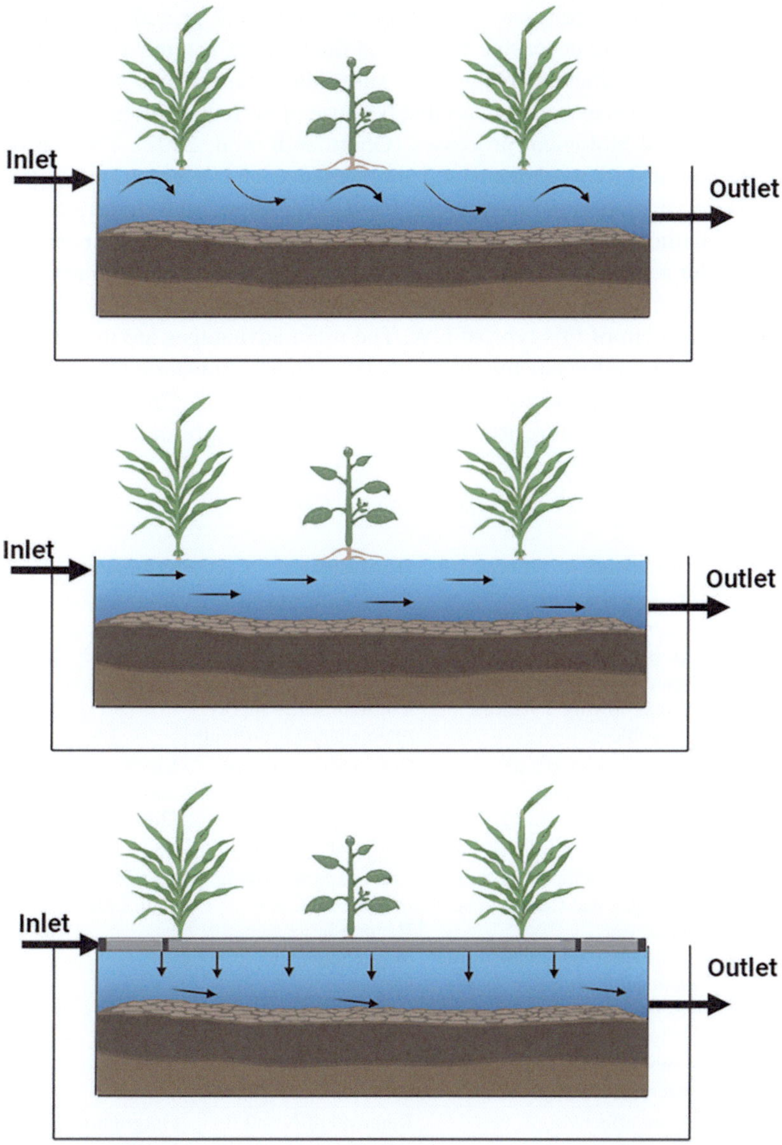

Fig. 11.2 Diagrammatic illustration of CW with FWS CW, HSF CW, VSF CW, and arrows indicating the direction of water flow

- Presence of water bodies.

While designing the FWS CWs, the primary focus is to ensure that wastewater remains in continuous contact with active microbial surfaces. It allows an efficient hydraulic retention time of wastewater in the system. Continuous contact between wastewater and microbial surfaces prevents the formation of hydraulic short circuits

(Vymazal 2008). This behaves like natural wetlands and provides aerobic conditions at the surface and anaerobic surfaces at the bottom. It removes pollutants like COD, BOD, heavy metals, pathogens, suspended particles in wastewater, etc. Removal efficiency is extremely variable because of various combinations of physical, chemical, and biological processes (Ghermandi et al. 2007). These processes include sediment microorganisms and the presence of UV radiations in the deep waters where vegetation is unavailable. FWS CWs show high performance in the removal of pathogenic microorganisms. Nitrogenous and organic substances are removed under aerobic conditions at the surface and anoxic conditions at the bottom surface. Nitrogen and phosphorous removal require a significantly large area, which is the main problem of this type of CW. The main advantages are that it has minimum mechanical equipment requirements, less energy consumption, and no need for special operator supervision, as mentioned in Table 11.2. These are also used to enhance ecological habitat restoration and recreational opportunities.

Table 11.2 Advantages and disadvantages of different CWs

Type of wetland	Advantage	Disadvantage	References
Horizontal subsurface	Longer life cycle Removal of N and P is possible, and nutrient gradients can establish Long flowing distances possible	Not very tolerant of cold climates The area required is very large Constant water supply maintenance is difficult	Greenway et al. (2007) and Gorgoglione and Torretta (2018)
Vertical subsurface flow	Higher oxygen transfer rate Small footprint	Susceptible to more clogging High capital and maintenance costs	Saeed and Sun (2012) and Bakhshoodeh et al. (2017)
Hybrid systems	Effective removal of total nitrogen and COD can be achieved Sufficient oxygen diffusion and enough carbon source Submerged plants can improve nutrient removal Submerged macrophytes reduce turbidity	More expensive than non-hybrid CWs Require large area Very complex to operate	Guo et al. (2014) and Dai et al. (2012)
Free water surface	Cost-effective Require less area Low maintenance cost Minimum equipment needs Significant removal of pollutants like COD, BOD, SS, heavy metals, etc.	Removal of N and P requires more area for CW Anoxic conditions result in poor nitrification Require long retention time in the removal of N and P The chances of mosquito production are high	Gorgoglione and Torretta (2018)

11.2.2 *Subsurface Flow Constructed Wetlands (SF CWs)*

It is considered a comparatively new technology for treating wastewater. These types of constructed wetlands can be planted or unplanted. Plants used in SF CWs can enhance removal efficiency by providing favorable conditions for microbial growth and oxygenated conditions, especially in semiarid regions (Taylor et al. 2011). This system has no direct contact between the surrounding atmosphere and the water pool. According to Zhu et al. (2011), availability of vegetation impacts much less efficiency in removing pollutants, including phosphorus, nitrogen, and total suspended solids. SF CWs are further divided into two types based on water flow.

11.2.2.1 Horizontal Subsurface Flow Constructed Wetlands (HSF CWs)

Water flow is in the horizontal direction, as the name indicates (Fig. 11.2). The wastewater flows through the inlet, then moves to the porous medium, and flows toward the outlet (Chavan and Mutnuri 2021). Under different aerobic and anaerobic conditions, the various organic compounds are degraded with the help of bacterial species. Continuous saturated conditions of the medium or bed result in anoxic conditions. Therefore, aerobic decomposition occurs at the surface, and contaminant removal occurs through anaerobic processes (Garcia et al. 2010). HSF CWs have limited efficiency for phosphorus and ammonium removal. The efficiency can be enhanced using different appropriate mediums. Effective pretreatment is preferred because direct wastewater can lead to solids that result in clogging and reduced removal efficiency. These are very prone to the cold climatic conditions. The performance of contaminant removal is reduced during cold weather conditions. Therefore, insulation is done to keep the influent warm and maximize microbial activity. Natural insulating materials such as dead plants can also be used. It is considered a good practice to prevent freezing in tanks under very-low-temperature conditions.

11.2.2.2 Vertical Subsurface Flow Constructed Wetlands (VSF CWs)

The basic difference between VSF CWs and HSF CWs is wastewater flow. The wastewater is circulated over the complete surface and then moves vertically downward into the media (Engida et al. 2021; Turcios et al. 2021). Wastewater is submerged from the top and drained out to the bottom (Fig. 11.2). Gravels covered in soil act as a bed, and macrophytes planted on the surface are the essential components of VSF. These include aerobic conditions for contaminant removal; therefore, the removal efficiency is high (BOD, COD, high nitrification, and other pollutant removal). The filling media comprises finer inert particles than HSF CWs, resulting in low water percolation. Coarse sand has appropriate hydraulic conductivity for slow vertical filtration. It provides a larger volume-to-surface area ratio than gravel to increase biomass attachment. The main advantage of VSF CWs is that they require a small land area compared to HSFCW (Witthayaphirom et al. 2020). These wetlands produce low GHGs during wastewater treatment compared to other treatment facilities. Recently, these were proven effective in removing micropollutants. Different vegetation types are involved in different CWs, as mentioned in Table 11.1.

11.2.3 Hybrid CWs

Different types of technologies are joined together to increase the pollutant removal efficiency of CWs. The design of hybrid CW mainly comprises vertical and horizontal flow beds sequentially (VSF-HSF system). Depending on the pollutant removal, these can be used individually and in combination. The VSF removes suspended solids and organic compounds to enhance the nitrification process. HSSF additionally removes suspended solids, organic compounds, and denitrification. Various technologies are combined to enhance nitrogen removal efficiency in constructed wetlands (Vymazal et al. 2007).

This technology effectively removes organic compounds and suspended solids from domestic wastewater, landfill leachate, agricultural runoff, and hospital runoff. It uses different configurations at the lab, pilot, and field scales. Because of its high flexibility and effective potential of removing pollutants, this technology is considered one of the most effective methods of treating wastewater. Different types of constructed wetlands have different advantages and disadvantages, as mentioned in Table 11.2.

Ye and Li (2009) designed a hybrid CW to enhance nitrogen removal from domestic wastewater. The combination of HSF CW and FWS CW was used. HSF CW was used in the first and third stages, while FWS CW was used in the second stage to remove various pollutants effectively. The efficiency of different CWs in removing pollutants differs based on the source of wastewater and types of CWs and characteristics of different type of constructed wetlands, as shown in Tables 11.3 and 11.4.

Table 11.1 Characteristics of CWs (horizontal and vertical wetlands) for wastewater treatment

Type of vegetation	Water level concerning the surface	Water flow direction
Submerged	Above	Horizontal
Emergent	Above Below	Horizontal Horizontal and vertical
Free-floating	Above	Horizontal
Floating-leaved	Above	Horizontal

Table 11.3 Removal efficiency of various pollutants in wastewater using different types of CWs

Type of wastewater	Type of wetland	TSS (%)	COD (%)	BOD (%)	NH4 (%)	TOTAL N (%)	TOTAL P (%)	References
Domestic wastewater	VSFCW	–	78	80	75	48	60	Masharqa et al. (2023)
Domestic wastewater	Hybrid CW	–	77	90	83	56	60	Masharqa et al. (2023)
Synthetic wastewater	VSFCW	–	65	–	89	78	–	Xu et al. (2023)
Wastewater treatment plant	Hybrid CW	81	89	98	–	–	–	Marzec et al. (2024)
Municipal wastewater	Hybrid CW	82	80		–	71	88	Haydar et al. (2020)

Table 11.4 Characteristics of different types of CWs (Malaviya and Singh 2012)

Type of wetland	Characteristics	Bed depth
Vertical SSF CWs	Wastewater is applied using different feeding mechanisms, and treated effluent (or water) is collected to maintain a vertical flow direction	2–3 m
Horizontal SSF CWs	The substrate is sustained flooded through continuous application of wastewater	Less than 0.6 m
Surface flow CWs	Thick vegetation by different species of plants	Less than 0.4 m

11.3 Applications of Constructed Wetlands

11.3.1 Removal of Toxic Elements from Industrial Wastewater Using CWs

CWs are considered the most used treatment system for treating wastewater. Industrial wastewater contains a high pollution load. Because of this, the treatment of wastewater using CWs has increased over the last few years (Korkusuz 2005). Various mechanisms are involved in removing toxic metals, including physical, chemical, and biological (Fig. 11.3).

The efficiency of heavy metal removal is highly affected by the pH concentration in the CWs. During nitrification, ammonium gets converted into nitrites, and this results in the production of protons. The coprecipitation of cationic metals (especially Cd, Cu, Ni, and Zn) is enhanced by creating an alkaline environment. A high nitrification rate results in lower efficiency of constructed wetlands (Lee and Scholz 2007).

Lesage et al. (2007) outlined four strategies for efficient metal removal from wetlands. These are listed in the following order:

1. Adsorption into fine-textured sediments and organic materials.

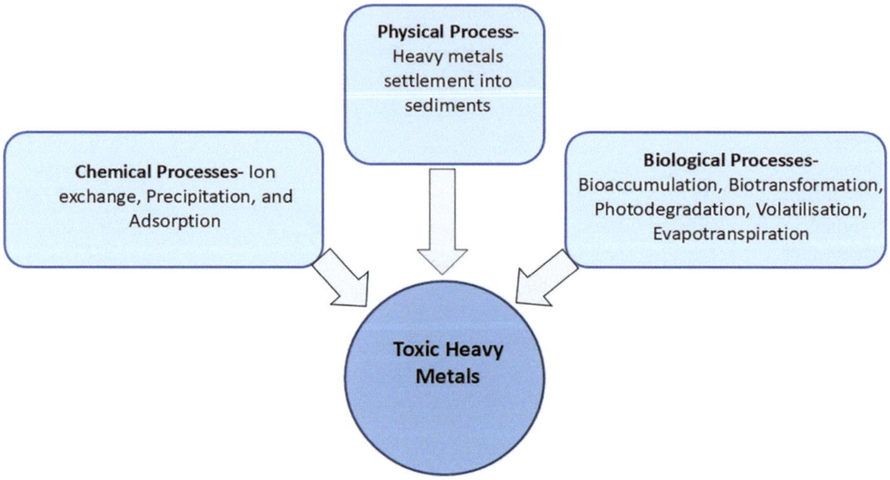

Fig. 11.3 Removal mechanisms of toxic metals

2. Precipitation as insoluble salts (mainly for the compounds of oxyhydroxides and sulfides)
3. Using various plants and microorganisms to concentrate and trigger changes in biogeochemical cycles.
4. Suspended solid deposition due to low flow rate.

11.3.2 Utilization of Plant Species in Wastewater Management

Using plants in constructed wetlands is a serious concern because plants must withstand the potentially harmful effects of wastewater and its variations. Various types of plants are used to eliminate toxic substances. Depending on the type of vegetation, plants are widely employed as emergent, floating, and submerged (Table 11.5). Different plants remove toxic metals using different mechanisms.

11.3.3 Urban Stormwater Management

Stormwater runoff in urban and suburban areas significantly contributes to poor water quality. Pollutants are the major concern in stormwater. These pollutants include polycyclic aromatic hydrocarbons, organic and particulate matter, nutrients, herbicides, and heavy metals. These pollutants are harmful to human life and also cause technical problems. Therefore, appropriate management practices are required for the mitigation of stormwater pollution.

Table 11.5 Different types of plant species in the removal of toxic elements in CWs

Type of plant	Species used	Removal mechanism	References
Emergent plants	*Typha domingensis, Phragmites karka, and Typha latifolia*	Plant species immobilize and store the rhizosphere in their roots	Juwarker et al. (1995) and Lesage et al. (2007)
Floating plants	*Salvinia herzogii, Eichhornia crassipes,* and *Pistia stratiotes*	Metals are stored in biomass and later absorbed into substrate	Vymazal et al. (1998)
Submerged plants	*Ceratophyllum demersum* and *Hydrilla verticillata Potamogeton* spp., and *Myriophyllum spicatum*	Through rhizosphere	Bunluesin et al. (2007)

The biological, chemical, and physical characteristics of wetlands may give significant attributes for the remediation of contaminants. Treating stormwater in wetland areas provides the advantage of storing water, sediment settling, nutrient cycling, breakdown of organic compounds, metal sequestration, and other benefits, including recreational facilities and functional wildlife habitat (Malaviya and Singh 2012). As a result, stormwater constructed wetlands (SCWs) are widely employed to treat stormwater runoff. These wetlands treat stormwater before dumping it into aquatic bodies, causing contamination. SCWs are natural process-based systems. Natural processes such as microbial assemblages, soils, and vegetation can help remediate urban stormwater (Vymazal 2010). The stormwater treatment uses a variety of SCWs. Other examples include floating treatment wetlands, green roofs, infiltration basins (for stormwater management downstream), ditches, and valleys (Gregoire and Clausen 2011). Various studies suggested that the performance of SCWs in stormwater treatment is the function of detention time (D_t) and inflow or hydraulic loading rate (HLR). HLR and D_t are the functions of runoff volume, length of the prior dry period, wetland size, area, storm intensity, and wetland volume (Carleton et al. 2001).

SCW performances are highly influenced by the weather conditions and hydrology of the catchment area. The physicochemical status of SCW highly affects the removal efficiency of redox-sensitive compounds such as heavy metals. The mobility of toxic heavy metals is influenced by the soil's acidity and the substrate's oxidation-reduction status. Heavy metals are partially immobilized in an anaerobic and reduced environment. During immobilization, simple and complex ligands are formed with the organic matter (He et al. 2019). Therefore, longtime oxidations result in the release of heavy metals.

In the Staten Island Bluebelt Drainage Basins (New York City), wetland systems were installed within the watershed to store stormwater temporarily. These SCWs treat over 350,000 GPD of stormwater during storm events, covering 14,000 acres, and help save more than $80 million in conventional sewer costs (Economides 2014).

The advantages of stormwater management using SCWs are as follows:

- Runoff volume reduction.
- Duration and peak flow.
- Downstream water resources safety.
- Flood reduction.
- Dropping the risks related to combined sewer overflow.
- Improvement in water quality.
- Improvement in groundwater recharging.
- Sediment stabilization.
- Wildlife restoration and improved recreational activities.

11.3.4 Habitat Restoration and Biodiversity Conservation

The primary goal of constructed wetlands is to achieve benefits for the environment. The presence of vegetation and water in constructed wetlands makes them ideal for developing natural habitats. These also help to rebuild the disrupted ecosystem by attracting wildlife (mainly green areas and birds). These CWs are also used as food and fiber sources, as well as public recreation and educational facilities (Sundaravadivel and Vigneswaran 2001).

11.4 Innovations in Constructed Wetlands for Wastewater Treatment

11.4.1 Thermal Insulation

The working of CW depends on biological and physiochemical processes. These processes are highly affected by climatic conditions. Cold conditions affect the nutrient content. Nitrogen is the most affected as compared to phosphorous and organic matter. Low temperatures result in deficient oxygen conditions, which inhibits microbial activities. To overcome these problems, the concept of thermal insulation was introduced to retain the operation and long-term performance of CW in cold climatic conditions (Meng et al. 2014).

Subsurface flow (SSF) CWs are one of the most suitable types of constructed wetlands that can be used in cold climates. These CWs insulate the air between the ice cover and the water surface level. It minimizes the energy loss (which takes place due to the convection and evaporation during the treatment process). When SSF CW is fully covered with ice, achieving satisfactory treatment performance is very difficult. To achieve desired performance, CWs are covered with insulating mulch materials. These mulching materials are used to avoid freezing and increase pollutant removal efficiency. The most common mulching materials used are pine straw, bark, waste compost, wood chips, plastics, and agricultural straw (Wallace et al. 2001).

11.4.2 Bioengineering and Plant Species

Macrophytes are the most commonly used type of vegetation. These are considered a constructed system's most significant design elements (Villa et al. 2014). Various considerations are to be taken for efficient treatment during the selection of plant species. Therefore, plant species easily adapted to grow in unnatural conditions are preferred (Stefanakis et al. 2014).

Various factors are to be taken into consideration during plant species selection:

- Climatic conditions of the site.
- Wastewater parameters which are to be treated.
- Quality of effluent.

11.4.3 ReedBox

A compact constructed wetland unit is known as ReedBox. It was developed using an aerated wetland design. These units of CW are easy to use because of their easy transportation and site installation at the desired location. This type of CW is used in an area where:

1. Less dense area with a small population of 40–75 persons producing approx. 7–15 m^3/day, respectively.
2. Public sewage treatment plants are not available.
3. Limited space.
4. The footprint requirement is minimum (the unit's area demand is 0.4 m^2/PE).

These treat domestic wastewater with a typical composition and high strength of nitrogen, organic matter, and suspended solids. The main advantage of this framework is that it is easy to install without any requirement for permanent infrastructure and earthworks. The substrate material used in this is light in weight, such as high-density polyethylene. These substrate materials are applied on the top over which wetland plants are established (Stefanakis and Prigent 2019). Artificial aeration is used to purify wastewater by integrating sludge accumulation and dewatering. The significant advantage of this unit is that it avoids the primary treatment (e.g., aseptic/sedimentation tank), which increases the efficiency.

11.4.4 Role of Different Substrates in CWs

Substrate plays an important role in CW. It acquires the largest volume during construction in CW for sewage treatment. It supports the microorganisms and plants in CW. Different substrates are used depending on their availability and removal efficiency of pollutants. Different substrates have various advantages and disadvantages (Table 11.6).

The choice of substrate is based on the hydraulic conductivity and absorption potential of a particular pollutant. Low hydraulic conductivity results in major blockage and reduces capacity and long-term extraction efficiency. Substrates are

Table 11.6 Advantages and disadvantages of different substrates

Sr No.	Types of substrates	Advantages	Disadvantages
1.	Zeolite	High adsorption The rate of removal of COD and N is also high Ion exchange capacity is high	P adsorption capacity is low
2.	Maifanite	Removal efficiency of COD and TN is high Small particle size and more surface area increase the adsorption rate	The removal efficiency of phosphorus is much less compared to iron, gravel, and shale
3.	Soil	Absorption capacity of N and P is high	The low COD removal efficiency
4.	Gravel		Low cation exchange capacity The adsorption capacity of gravel on NH_4^+-N was poor
5.	Shale	Least affected by pH and temperature High potential substrate for NH_4^+-N removal	Adsorption of NH_4^+-N affected by pH and temperature
6.	Volcanic	The highly porous structure and rich source of calcium and other minerals make it a suitable substrate The high P removal efficiency A rough surface provides adhesion to bacteria and biofilm production	Limitation of oxygen
7.	Bio-ceramic	High adsorption capacity, affinity, and stability to microbial biofilm biomass High removal efficiency of both phosphorus and nitrogen	Removal efficiency of bio-ceramic for COD is low
8.	Biochar	High removal efficiencies of TP, TN, and COD Good adsorption stability	Low anion exchange capacity Lack of effective functional groups Limited its ability to adsorb phosphate

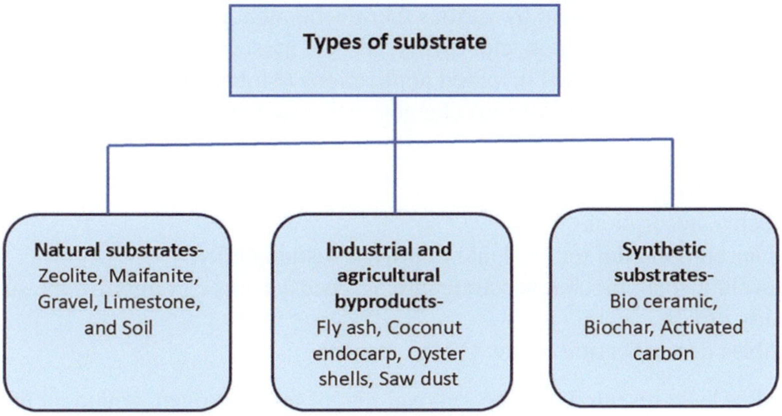

Fig. 11.4 Types of substrates used in CWs (Zhong et al. 2022)

broadly classified into three categories depending on their sources (Fig. 11.4): natural substrates, industrial and agricultural by-products, and synthetic substrates. These substrates have exceptional physiochemical properties and different removal mechanisms. The selection of substrate plays a critical role in the removal of contaminants.

11.5 Methods of Enhancing the Removal Efficiency of CWs

11.5.1 Addition of Biochar in CWs

Substrate plays a critical role in wastewater treatment. The main role of the substrate is to interrupt significant contaminants in wastewater through adsorption, sedimentation, and filtration (Dordio et al. 2007). Organic carbon and nitrogen in the wastewater are removed by nitrification and denitrification. The process of denitrification and nitrification is done by microorganisms present in the substrate (Wang et al. 2016). The most commonly used filter types include coal ash, gravel, limestone, zeolite, and industrial by-products. Various studies revealed that different substrate types have different mechanisms and efficiency for water purification (Lu et al. 2016). The main drawbacks of the abovementioned filters are as follows (Li et al. 2017):

- Not ideal for high-strength wastewater.
- Less effective in the removal of total nitrogen and phosphorous.
- Results in large production of nitrogen oxides.

To overcome the major drawbacks, finding a suitable substrate to increase wastewater treatment is a serious issue. The use of biochar to increase the efficiency of wastewater treatment is attracting close attention. Biochar is a product of agricultural waste produced by various thermochemical processes such as gasification, pyrolysis, carbonization, etc. Pyrolysis is the most common method of biochar production. Biochar is used in varied applications (Mohan et al. 2014) for environmental applications due to various characteristics such as:

- Large surface area.
- High porosity.
- Effective adsorption.
- Enhanced pollutant removal and improved sustainability.
- Recycling spent biochar substrate into agroecosystems can improve the sustainability of CW.
- Enables denitrification of low C/N wastewater.

Studies have revealed that the removal rate of total nitrogen, ammonia nitrogen (NH_4^+-N), COD, and nitrate nitrogen (NO_3—N) in CWs with biochar addition was more efficient in comparison to conventional CWs (De Rozari et al. 2015).

11.5.2 Combined CWs with Microbial Fuel Cells (MFCs)

The combination of these two technologies is compatible as both are based on the bacteria's action for removing contaminants from wastewater. In MFC, the cathode is exposed to oxygen, and the anode remains in an anoxic condition. Redox conditions are naturally developed in the CWs. The main aim of this technology is to improve wastewater treatment efficiency and simultaneously production of energy. This concept has originated in recent years, and a combination of these two is termed CW-MFC. Plants are also used in MFC based on two recognized processes:

1. Ability of microorganisms to produce electricity from organic compounds.
2. Depositions of organic compounds in the rhizosphere.

In terms of system design, PMFCs and CW-MFCs are very similar. The fundamental differences between the two systems are the source of organics and the principal function of the reactors. Consider that electrogenic bacteria in PMFCs are primarily fed by rhizodeposits and plant exudates to produce green electricity (Strik et al. 2008), whereas, in CW-MFCs, the electrogenic bacteria are fed with both rhizodeposits and wastewater, with the primary function being wastewater treatment. The most significant advantage of integrating the two systems may be improved efficiency in treating wastewater in anaerobic zones inside the wetland.

11.5.3 Combined CW and Other Emerging Technologies

Traditional CW has several disadvantages, including limited applicability and long-term stability. One of the most pressing challenges is substrate selection. Clogging the substrate when treating wastewater with a high concentration of suspended particles and organic matter is a serious issue (Ruiz et al. 2010). Furthermore, CWs can have low nitrogen removal efficiency; they may be nitrified due to insufficient oxygen transfer or denitrified due to a lack of accessible organics (Wu et al. 2014), particularly at high nitrogen loading rates. Furthermore, several recalcitrant contaminants and heavy metals found in industrial effluent can impair CW efficacy. With the deterioration of the environment and stringent discharge standards, including an emphasis on effluent reuse, CW systems operating as standalone technologies are, in some cases, unable to meet the requirements of these new guidelines (Winward et al. 2008), despite improvements in design and operational strategies and the use of intensified systems such as flow direction reciprocation, artificial aeration, tidal flow CWs, and so on.

As a result, in recent years, CWs have been integrated with other existing developing technologies such as membrane bioreactors (MBBR) (Mutamim et al. 2012) and electrochemical oxidation (Anglada et al. 2010), among others, to maximize individual reward in terms of wastewater treatment. These technologies are effective in the treatment of certain contaminants as well as in green energy recovery procedures.

11.6 Conclusions

Urbanization, industrialization, and increased population growth resulted in a considerable increase in the volume of wastewater worldwide. Conventional wastewater treatment facilities and inappropriate control of wastewater sources affect the environment and living beings. Conventional treatment techniques are not cost-effective, require a large area, and are not environmentally friendly. Therefore, such technology is needed for high performance, long-term viability, and effective wastewater treatment. CW is an innovative technology that promotes more sustainable, environmentally friendly techniques for wastewater treatment. The pollutant removal efficiency in CWs is very high and depends upon the type of wetland. The performance can be enhanced using different substrates, operational conditions such as pH, temperature, hydraulic retention time, and modification by combining various technologies such as CWs with MFC, MBBR, and electrochemical oxidation. CWs can be adopted to treat various types of wastewater with high COD, BOD, and TSS removal efficiency. This book chapter focused on wastewater treatment using CWs, their different types, and removing pollutants. It also focused on various types of substrates' roles in removing pollutants and different technology combinations to enhance pollutant removal efficiency. Therefore, it is considered a sustainable technology to treat wastewater because of its low cost, sustainability, low energy consumption, less investment, and less maintenance cost.

References

Anglada A, Ortiz D, Urtiaga AM, Ortiz I (2010) Electrochemical oxidation of landfill leachates at pilot scale: evaluation of energy needs. Water Sci Technol 61(9):2211–2217

Bakhshoodeh R, Alavi N, Majlesi M, Paydary P (2017) Compost leachate treatment by a pilot-scale subsurface horizontal flow constructed wetland. Ecol Eng 105:7–14

Bunluesin S, Kruatrachue M, Pokethitiyook P, Upatham S, Lanza GR (2007) Batch and continuous packed column studies of cadmium biosorption by Hydrilla verticillata biomass. J Biosci Bioeng 103(6):509–513

Carleton JN, Grizzard TJ, Godrej AN, Post HE (2001) Factors affecting the performance of stormwater treatment wetlands. Water Res 35(6):1552–1562

Chavan R, Mutnuri S (2021) Domestic wastewater treatment by constructed wetland and microalgal treatment system for the production of value-added products. Environ Technol 42(21):3304–3317

Chen Y, Wen Y, Zhou Q, Vymazal J (2014) Effects of plant biomass on denitrifying genes in subsurface-flow constructed wetlands. Bioresour Technol 157:341–345

Cheng G, Li Q, Su Z, Sheng S, Fu J (2018) Preparation, optimization, and application of sustainable ceramsite substrate from coal fly ash/waterworks sludge/oyster shell for phosphorus immobilization in constructed wetlands. J Clean Prod 175:572–581

Davis L, Edwards R, Garber L, Isaacs B (1995) A handbook of constructed wetlands, Volume 1: General considerations. US Environmental Protection Agency (EPA), Washington, DC

De Rozari P, Greenway M, Hanandeh A (2015) An investigation into the effectiveness of sand media amended with biochar to remove BOD5, suspended solids and coliforms using wetland mesocosms. Water Sci Technol 71(10):1536–1544

Dordio AV, Teimão J, Ramalho I, Carvalho AP, Candeias AE (2007) Selection of a support matrix for the removal of some phenoxyacetic compounds in constructed wetlands systems. Sci Total Environ 380(1–3):237–246

Economides C (2014) Green infrastructure: sustainable solutions in 11 cities across the United States. Columbia University Water Centre, New York

Engida T, Alemu T, Wu J, Xu D, Zhou Q, Wu Z (2021) Microbial community dynamics and activity in constructed wetland treating agro-industrial and domestic wastewater: a review. Appl Ecol Environ Res 19(4):2667–2687

Garcia J, Rousseau DP, Morato J, Lesage ELS, Matamoros V, Bayona JM (2010) Contaminant removal processes in subsurface-flow constructed wetlands: a review. Crit Rev Environ Sci Technol 40(7):561–661

García-Ávila F, Añazco A, Torres R, Quito A, Galabay M, Ortega H, Arévalo C (2023) Application of ornamental plants in constructed wetlands for wastewater treatment: a scientometric analysis. Case Stud Chem Environ Eng 7:100307

Ghermandi A, Bixio D, Thoeye C (2007) The role of free water surface constructed wetlands as polishing step in municipal wastewater reclamation and reuse. Sci Total Environ 380(1–3):247–258

Gorgoglione A, Torretta V (2018) Sustainable management and successful application of constructed wetlands: a critical review. Sustain 10(11):3910

Gregoire BG, Clausen JC (2011) Effect of a modular extensive green roof on stormwater runoff and water quality. Ecol Eng 37(6):963–969

Guo YM, Liu YG, Zeng GM, Hu XJ, Xu WH, Liu YQ, Huang HJ (2014) An integrated treatment of domestic wastewater using sequencing batch biofilm reactor combined with vertical flow constructed wetland and its artificial neural network simulation study. Eco Eng 64:18–26

Haydar S, Anis M, Afaq M (2020) Performance evaluation of hybrid constructed wetlands for the treatment of municipal wastewater in developing countries. Chin J Chem Eng 28(6):1717–1724

He Y, Men B, Yang X, Li Y, Xu H, Wang D (2019) Relationship between heavy metals and dissolved organic matter released from sediment by bioturbation/bioirrigation. J Environ Sci 75:216–223

Kadlec RH, Wallace S (2008) Treatment wetlands. CRC Press, Boca Raton
Kleinmann RL, Girts MA (1987) Acid mine water treatment in wetlands: an overview of an emergent technology. In: Reddy KR, Smith WH (eds) Aquatic plants for water treatment and resource recovery. Magnolia Publishing Co, Orlando, pp 255–261
Korkusuz EA (2005) Manual of practice on constructed wetland for wastewater treatment and reuse in Mediterranean countries, Report, MED-REUNET II Support Programme (EC Project No: INCO-CT-2003–502453). AGBAR Foundation
Koskiaho J, Ekholm P, Räty M, Riihimäki J, Puustinen M (2003) Retaining agricultural nutrients in constructed wetlands—experiences under boreal conditions. Ecol Eng 20(1):89–103
Lee BH, Scholz M (2007) What is the role of Phragmites australis in experimental constructed wetland filters treating urban runoff. Ecol Eng 29(1):87–95
Lesage E, Rousseau DP, Meers E, Tack FMG, De Pauw N (2007) Accumulation of metals in a horizontal subsurface flow constructed wetland treating domestic wastewater in Flanders, Belgium. Sci Total Environ 380(1–3):102–115
Li M, Wu H, Zhang J, Ngo HH, Guo W, Kong Q (2017) Nitrogen removal and nitrous oxide emission in surface flow constructed wetlands for treating sewage treatment plant effluent: effect of C/N ratios. Bioresour Technol 240:157–164
Lu S, Zhang X, Wang J, Pei L (2016) Impacts of different media on constructed wetlands for rural household sewage treatment. J Clean Prod 127:325–330
Malaviya P, Singh A (2012) Constructed wetlands for management of urban stormwater runoff. Crit Rev Environ Sci Technol 42(20):2153–2214
Marzec M, Listosz A, Malik A, Kulik M, Jóźwiakowski K (2024) Organic pollutants removal in a hybrid constructed wetland wastewater treatment plant with an aeration system. Water 16(7):947
Masharqa A, Al-Tardeh S, Mlih R, Bol R (2023) Vertical and hybrid constructed wetlands as a sustainable technique to improve domestic wastewater quality. Water 15(19):3348
Meng P, Pei H, Hu W, Shao Y, Li Z (2014) How to increase microbial degradation in constructed wetlands: influencing factors and improvement measures. Bioresour Technol 157:316–326
Mohan D, Sarswat A, Ok YS, Pittman CU Jr (2014) Organic and inorganic contaminants removal from water with biochar, a renewable, low cost and sustainable adsorbent–a critical review. Bioresour Technol 160:191–202
Mutamim NSA, Noor ZZ, Hassan MAA, Olsson G (2012) Application of membrane bioreactor technology in treating high strength industrial wastewater: a performance review. Desalination 305:1–11
Parde D, Patwa A, Shukla A, Vijay R, Killedar DJ, Kumar R (2021) A review of constructed wetland on type, treatment and technology of wastewater. Environ Technol Innov 21:101261
Peng Y, Deng S, Kong Z, Yuan Y, Long H, Fang J, Chai H (2022) Biochar and woodchip amended bioreactor extending reactive volume for enhanced denitrification in stormwater runoff. J Water Process Eng 46:102541
Rai UN, Tripathi RD, Singh NK, Upadhyay AK, Dwivedi S, Shukla MK, Nautiyal CS (2013) Constructed wetland as an ecotechnological tool for pollution treatment for conservation of Ganga river. Bioresour Technol 148:535–541
Ruiz I, Díaz MA, Crujeiras B, García J, Soto M (2010) Solids hydrolysis and accumulation in a hybrid anaerobic digester-constructed wetlands system. Ecol Eng 36(8):1007–1016
Saeed T, Afrin R, Al Muyeed A, Sun G (2012) Treatment of tannery wastewater in a pilot-scale hybrid constructed wetland system in Bangladesh. Chemo 88(9):1065–1073
Saeed T, Sun G (2013) A lab-scale study of constructed wetlands with sugarcane bagasse and sand media for the treatment of textile wastewater. Bioresour Technol 128:438–447
Saeed T, Muntaha S, Rashid M, Sun G, Hasnat A (2018) Industrial wastewater treatment in constructed wetlands packed with construction materials and agricultural by-products. J Clean Prod 189:442–453
Salvati L, Perini L, Sabbi A, Bajocco S (2012) Climate aridity and land use changes: A regional-scale analysis. Geograph Research 50(2):193–203

Stefanakis AI, Prigent S (2019) Reedbox: an innovative compact, mobile constructed wetland unit for wastewater treatment. In: Proceedings of the Oman Water & Wastewater Conference
Stefanakis A, Akratos CS, Tsihrintzis VA (2014) Vertical flow constructed wetlands: eco-engineering systems for wastewater and sludge treatment. Elsevier, Amsterdam
Strik DP, Hamelers HVM, Snel JF, Buisman CJ (2008) Green electricity production with living plants and bacteria in a fuel cell. Int J Energy Res 32(9):870–876
Sundaravadivel M, Vigneswaran S (2001) Constructed wetlands for wastewater treatment. Critical reviews in environ sci and technol 31(4):351–409
Taylor CR, Hook PB, Stein OR, Zabinski CA (2011) Seasonal effects of 19 plant species on COD removal in subsurface treatment wetland microcosms. Ecol Eng 37(5):703–710
Tian J, Jin J, Chiu PC, Cha DK, Guo M, Imhoff PT (2019) A pilot-scale, bi-layer bioretention system with biochar and zero-valent iron for enhanced nitrate removal from stormwater. Water Res 148:378–387
Turcios AE, Miglio R, Vela R, Sánchez G, Bergier T, Bergier A, Papenbrock J (2021) From natural habitats to successful application-role of halophytes in the treatment of saline wastewater in constructed wetlands with a focus on Latin America. Environ Exp Bot 190:104583
Villa JA, Mitsch WJ, Song K, Miao S (2014) Contribution of different wetland plant species to the DOC exported from a mesocosm experiment in the Florida Everglades. Ecol Eng 71:118–125
Vymazal J (2008) Constructed wetlands, surface flow. In: Encyclopedia of ecology. Elsevier, Amsterdam, pp 765–776
Vymazal J (2010) Constructed wetlands for wastewater treatment. Water 2(3):530–549
Vymazal J (2014) Constructed wetlands for treatment of industrial wastewaters: a review. Ecol Eng 73:724–751
Vymazal J, Svehla J, Kröpfelová L, Chrastný V (2007) Trace metals in Phragmites australis and Phalaris arundinacea growing in constructed and natural wetlands. Sci Total Environ 380(1–3):154–162
Vymazal J (1998) Removal of BOD in constructed wetlands with horizontal sub-surface flow: Czech experience. Water Sci and Technol 40(3):133s138
Wallace S, Parkin G, Cross C (2001) Cold climate wetlands: design and performance. Water Sci Technol 44(11–12):259–265
Wang W, Ding Y, Ullman JL, Ambrose RF, Wang Y, Song X, Zhao Z (2016) Nitrogen removal performance in planted and unplanted horizontal subsurface flow constructed wetlands treating different influent COD/N ratios. Environ Sci Pollut Res 23:9012–9018
Winward GP, Avery LM, Frazer-Williams R, Pidou M, Jeffrey P, Stephenson T, Jefferson B (2008) A study of the microbial quality of grey water and an evaluation of treatment technologies for reuse. Ecol Eng 32(2):187–197
Witthayaphirom C, Chiemchaisri C, Chiemchaisri W, Ogata Y, Ebie Y, Ishigaki T (2020) Long-term removals of organic micro-pollutants in reactive media of horizontal subsurface flow constructed wetland treating landfill leachate. Bioresour Technol 312:123611
Wu S, Kuschk P, Brix H, Vymazal J, Dong R (2014) Development of constructed wetlands in performance intensifications for wastewater treatment: a nitrogen and organic matter targeted review. Water Res 57:40–55
Xu W, Yang B, Wang H, Wang S, Jiao K, Zhang C, Wang H (2023) Improving the removal efficiency of nitrogen and organics in vertical-flow constructed wetlands: the correlation of substrate, aeration and microbial activity. Environ Sci Pollut Res 30(8):21683–21693
Yazdani V, Golestani HA (2019) Advanced treatment of dairy industrial wastewater using vertical flow constructed wetlands. Desalin Water Treat 162:149–155
Ye F, Li Y (2009) Enhancement of nitrogen removal in towery hybrid constructed wetland to treat domestic wastewater for small rural communities. Ecol Eng 35:1043–1050
Zhong H, Hu N, Wang Q, Chen Y, Huang L (2022) How to select substrate for alleviating clogging in the subsurface flow constructed wetland. Sci Total Environ 828:154529
Zhu L, Li Z, Ketola T (2011) Biomass accumulations and nutrient uptake of plants cultivated on artificial floating beds in China's rural area. Ecol Eng 37(10):1460–1466

Chapter 12
Macrophytes as a Potential Sustainable Strategy for Decentralised Wastewater Treatment

Renu Meena, Jyoti Yadav, and Smita S. Kumar

Abstract Domestic wastewater management in emerging nations is a critical issue these days. While it is challenging to avoid the generation of wastewater, efficient treatment of this wastewater can yield potable water and preserve freshwater resources. Urban areas in developing countries often rely on centralised sanitation systems which are unstable and lead to environmental problems; on the other hand, decentralised systems are more flexible as they independently collect, treat and then dispose of the treated wastewater. Decentralised wastewater systems can be classified into three types: aerobic systems, anaerobic systems and natural treatment systems. This chapter specifically focuses on nature-based solutions, such as the use of aquatic macrophytes to treat wastewater, as they offer a cost-effective approach to decentralised wastewater treatment. It is important to understand the structure and function of macrophytes to conserve biodiversity and maintain ecological balance in aquatic systems. These plants are viable for sustainable wastewater treatment solutions because they play a vital role in eliminating contaminants through various mechanisms. This chapter delves into the efficiency, mechanisms, development and implementation of treatment systems based on floating macrophytes for treating wastewater.

Keywords Macrophytes · Decentralised wastewater system · Nature-based solution · Sustainability

---

R. Meena · J. Yadav · S. S. Kumar (✉)
Department of Environmental Studies, Central University of Haryana, Mahendergarh, Haryana, India

S. S. Kumar, V. Kumar (eds.), *Nature-Based Solutions for Decentralised Liquid Waste Management*, Environmental Challenges and Solutions,
https://doi.org/10.1007/978-3-032-08362-3_12

12.1 Introduction

One of the main issues in emerging nations is the management of domestic wastewater. The generation of wastewater is unavoidable since it is a necessary component of the value chain in every area of life. Wastewater is used water that contains contaminants from domestic, industrial, commercial or agricultural sources and requires treatment before being released into the environment. In metropolitan areas, wastewater is causing major environment-related problems, particularly in developing countries (Mudasir et al. 2020). Approximately ten barrels of effluent are produced for each barrel of crude oil processed in refineries (Pendashteh et al. 2011). Potable water may be made available to everyone, and freshwater resources can also be augmented if wastewater is treated efficiently. It appears as the most obvious solution to the problem of water scarcity (Tetteh et al. 2020). Reduction in organic load and nutrition recovery are two main goals of effective wastewater treatment of both types, industrial and municipal, as an environmental management strategy to prevent pollution from tarnishing the water bodies. Municipal wastewater is commonly treated using attached or suspended growth mixed bacterial cultures in trickling filters, moving or membrane bed biofilm reactors and activated sludge systems. Phytoremediation, constructed wetlands and microbial fuel cells are some of the new technologies that are emerging to remove complicated emerging toxins (Justin et al. 2022). According to a recent Central Pollution Control Board (CPCB) assessment, India's urban sewage generation was 72,368 million litres per day (MLD) in 2020–2021. The present built capacity for sewage treatment is 31,841 MLD. However, the actual capacity is 26,869 MLD, significantly lower than the load generated. From this, only 28% of total urban sewage (20,236 MLD) was treated, which shows that without any treatment, 72% of wastewater is discharged into rivers, lakes and groundwater (NitiAayog 2022).

Treating wastewater is essential nowadays to maintain sustainability and environmental health. One novel approach for this process is using floating macrophytes such as *Pistia*, *Lemnoideae* (duckweed) and *Eichhornia crassipes* (water hyacinth) to naturally clean wastewater. These plants are viable for sustainable wastewater treatment solutions as they are vital in eliminating contaminants through various mechanisms. This chapter explores the efficiency, mechanisms, case studies and challenges of utilising floating macrophytes for the treatment of wastewater.

12.1.1 The History of Untreated Wastewater and Its Impact on Civilisation

For more than 10,000 years, people have existed alongside sewage, an inescapable result of our civilisation. Everything was renewable in ancient times when humans were hunters and gatherers. Water used to drink came from natural sources, and settlements were required to access nearby water. They returned the urine and faeces to the environment, where they became part of a natural cycle. However, as we humans began to produce and trade during the Neolithic period (about 8500 B.C.), population centres expanded. Cities consolidated trade, waste and political power. The first places for sanitation were simple pits or cesspools (around 4000 B.C.), where excreta were collected. Clay pipelines delivered waste products to these pits, while normal hydraulic systems transported water. The first latrines, which are connected to sewage systems, appeared around 3000 B.C. in Mohenjo-daro. Citizens started flushing their excreta with water, and a sewage system was formulated for the collection of wastewater, which was directed to the sumps or Indus River. This signified a huge advancement in wastewater management. Due to water scarcity, ancient Greece reused wastewater for agricultural purposes. Cities diverted their sewage to disposal dumps on the outskirts which were before piped to agricultural fields. After some time, the Romans advanced sanitary ideals by segregating wastewater with the help of sewage systems in the streets (Salgot and Folch 2018). They even started using recycled water in their thermal baths.

12.1.2 Decentralised Wastewater Systems

Rapid urbanisation and inadequate wastewater management in developing countries make centralised sanitation systems unsustainable for many locations (Chirisa et al. 2017). Centralised systems use huge pipes, equipment, access routes, construction, pumping systems and treatment facilities to collect and treat significant amounts of wastewater for towns and industrial/residential platforms. On the other hand, decentralised systems independently collect, treat and then dispose of, or on-site, and reuse the treated wastewater near or around the generating source, providing flexibility in management (Sharma et al. 2013). Treating wastewater near its source is advantageous for both the environment and the economy, as it avoids the need for costly sewage infrastructure. Decentralised wastewater treatment systems (DWWTS) are becoming increasingly popular in areas with limited sanitation services. Domestic wastewater treatment can be decentralised at several levels, ranging from individual on-site systems to semi-centralised plants that treat effluent from isolated neighbourhoods (Iribarnegaray et al. 2018). DWWTS can be classified into three types: aerobic, anaerobic and natural treatment systems.

1. Aerobic systems: Aerobic treatment involves microorganisms using oxygen to degrade organics into carbon dioxide and water. Although these systems have a smaller footprint than natural systems, they still consume much energy. Compared to natural systems, aerobic systems produce high-quality effluents that fulfil discharge regulations. Aerobic systems are ideal for wastewater treatment, particularly in low-income and developing nations, because they require semi-skilled workers for operation. Aerobic systems such as extended aeration system, submerged aerobic fixed film reactor, oxidation ditches, membrane bio-

reactor, moving bed biofilm reactor, rotating biological contractor and sequencing batch reactor have been fully deployed and documented.
2. Anaerobic systems: Anaerobic treatment systems are biological systems that require little energy to operate. Despite their limited organic and nutrient removal capacities, anaerobic treatment systems are inexpensive and easy and can be used to generate electricity. However, these systems produce low to average effluent quality and require a lengthy start-up period (about 3–4 months), whereas aerobic systems require only 2–4 weeks. The aerobic systems typically follow these systems to meet discharge requirements. Anaerobic baffled reactors, UASBs and septic tanks are examples of such systems, but most anaerobic systems are used in conjunction with other aerobic or natural processes.
3. Natural wastewater treatment systems: This is part of nature-based systems (NBS) and offers cost-effective and sustainable options for wastewater treatment. An example of a natural wastewater system is a pond system, which uses the symbiotic interaction between algae and bacteria to purify wastewater. Other typical systems rely on natural media, such as soil and aquatic macrophytes, for filtration and biochemical processes. These methods can be utilised as secondary or combined with the primary treatment. Some are also used with tertiary treatment options, including duckweed pond systems (DPS), WSP, facultative ponds and artificial wetlands (CWs).

12.1.3 Nature-Based Systems

IUCN defines nature-based systems (NBS) as acts to preserve, sustainably manage and restore natural or modified ecosystems that efficiently and adaptively solve societal concerns while simultaneously promoting human well-being and biodiversity benefits (Cohen-Shacham et al. 2016). According to the European Commission, they can also be defined as "living solutions inspired by, continuously supported by, and using nature, which are designed to address various societal challenges in a resource-efficient and adaptable manner and to simultaneously provide economic, social, and environmental benefits", including wastewater management with minimal energy and chemical inputs (Martin et al. 2020). More precisely, NBSs are methods that imitate naturally occurring processes in urban environments that remove contaminants from wastewater by using plants, soil, bacteria and other natural materials and processes; this lowers the need for chemicals and saves energy. NBSs are cost-effective, energy-efficient, low-impact, easy and environmentally sustainable to treat wastewater. Developing countries can profit from implementing these solutions to decentralised wastewater treatment in areas with inadequate public sewage access or water scarcity. Some NBSs are constructed wetlands, floating wetlands, green roofs, waste stabilisation ponds, etc.

12.1.4 Conventional Methods

Water pollution has recently become a severe hazard to the environment and human health because of many contaminants, such as waterborne bacteria, viruses, heavy metals, etc. Microbial applications, filtration, adsorption, phytoremediation and sedimentation are all techniques used to remove undesired components from wastewater, and among all of them, phytoremediation is the most effective and sustainable technique for the removal of heavy metals (Hammer 1989). Conventional techniques include a variety of biological, physical and chemical techniques. Such methods aim to remove particles, nutrients and organic matter from wastewater (Bairagi and Ali 2020).

12.1.5 New Plant-Based Technologies

The processes such as membrane biofilm reactors, activated sludge treatment, photocatalytic decomposers, trickling filters and microbial fuel cells effectively remove organic matter and safely dispose of treated wastewater; they come with high energy consumption, sludge production and operational costs (Fletcher et al. 2020). Additionally, they contribute to global warming by increasing greenhouse gas emissions (Koutsou et al. 2018). Phytoremediation refers to the use of plants to remove chemicals from contaminated soils as well as water. In the case of water, large aquatic plants that can be seen through the naked eye are used, which are known as macrophytes. This new technology uses specialised plants to process wastewater from the natural ecosystem (Bhat et al. 2018). Various types of vegetation of macrophytes are used in wastewater treatment, such as submerged, emergent, free-floating and floating leaves. The roots of the floating leaf type get stagnant in saturated media, and their complete body can be seen above the water level. Emergent types may be noticed at a considerable height above water level, and their roots are planted in soil under the water, which significantly stabilises the substrate (Mahajan et al. 2023). Local macrophytes are favoured since they can withstand the conditions of a specific area, which will maximise the outcomes.

12.2 The Role of Macrophytes in Remediation

Macrophytes play an important part in remediation as they help sequestrate and uptake pollutants and toxins in water bodies. So many studies have shown that aquatic macrophytes are a cost-effective wastewater treatment option. Aquatic macrophytes can remove pollutants from polluted water by consuming them as nutrients (Kumar 2015). These aquatic macrophytes, which include submerged, floating and emergent species, have been demonstrated to successfully remove heavy metals like cadmium and chromium from wastewater, with the use of species like *Eichhornia crassipes* and *Pistia stratiotes* displaying great effectiveness in remediation.

Furthermore, macrophytes help to maintain aquatic ecosystem functioning by converting pollutants, detoxifying contaminants and improving nutrient uptake by improving water quality, as seen in treating eutrophic lakes with submerged macrophytes. Their propensity to absorb contaminants such as nutrients, organic compounds and heavy metals makes them important components in phytoremediation, which provides sustainable and environmentally acceptable options for wastewater treatment in a variety of settings. Submerged macrophytes such as *Myriophyllum spicatum*, *Ceratophyllum demersum* and *Vallisneria natans* have been shown to have good purification efficiency in artificial eutrophic wastewater, with increasing macrophyte density favouring improved water purification performance (Ding et al. 2020).

Macrophytes help to remove and decompose organic contaminants and hydrolysable metals from constructed wetlands, as well as help in improving treatment effectiveness through a variety of mechanisms, including physical filtration, better hydrodynamics and biogeochemical reactions. Overall, macrophytes' ability to absorb contaminants and nutrients, as well as their interactions with bacterial communities, highlights their relevance in water purification processes (Li et al. 2019).

12.2.1 Overview of Macrophytes

Macrophytes are enormous aquatic plants usually found in a diverse water body. Aquatic macrophytes are a broad group of photosynthetic organisms that can be easily seen with the naked eye and belong to various divisions of the Plantae kingdom, including Chlorophyta, Bryophyta, Pteridophyta and Spermatophyta (Sayanthan et al. 2024). They play a crucial role in freshwater ecosystems by shaping their structure and function. The presence or absence of specific macrophyte species results in specific environmental conditions, as they are also crucial for rehabilitating the freshwater ecosystem. Each macrophyte's characteristics influence the success of water ecosystem rehabilitation (Soloviy and Malovanyy 2019). They are important components of aquatic ecosystems as they provide food and habitat and influence water quality (Sesin et al. 2023). They are categorised based on growth forms such as emergent, floating-leaved, submerged and free-floating plants, with varying morphologies, as shown in Fig. 12.1, and adaptations to different aquatic habitats. Research has shown that submerged macrophytes are crucial in managing contaminated aquatic habitats (Izzati et al. 2022). Macrophytes are important markers of ecosystem health, and they also help in environmental risk assessments for pollutants, particularly *Lemma* spp. being regularly used in experiments on phytotoxicity (Babko et al. 2023). Environmental factors such as temperature, light, herbivory and nutrients all influence the number and variety of these plants in lakes. Studies on communities of macrophytes in rivers have revealed structural changes caused by stormwater runoff, which impact species composition

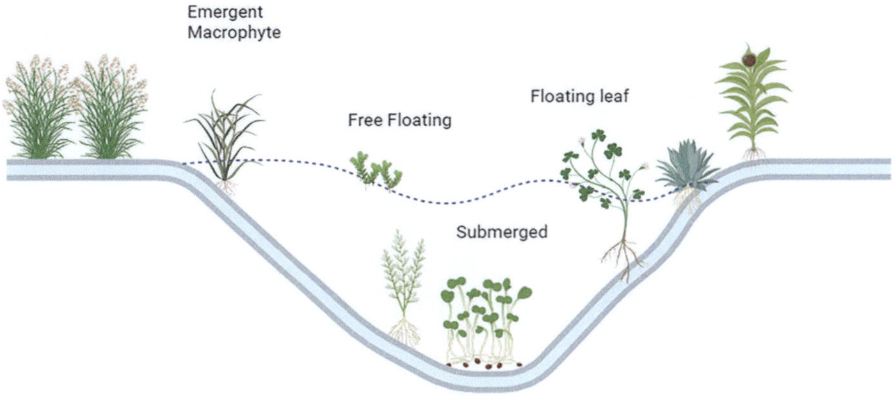

Fig. 12.1 Showing various types of vegetation of macrophytes

Table 12.1 Functions of different parts of macrophytes in wastewater

Macrophyte plant parts	Function
Plant part submerged in water	Filters out large debris Helps in better sedimentation and reduces resuspension It provides a surface area that encourages the growth of microbial biofilms
Aerial part	Reduces phytoplankton growth by reducing light Provides insulation during winter Reduces resuspension by reducing wind speed
Roots or rhizomes	Reduce erosion Uptake of nutrients Prevent clogging Increase degradation and nitrification by oxygen release

Source: Adapted from Rehman et al. (2017)

and projective covers in nearby regions (Ismail et al. 2021). Understanding these macrophyte's structure and function is critical for the conservation of biodiversity and ensuring ecological equilibrium in aquatic environments. Various studies also indicate that both submerged and free-floating macrophytes can significantly improve water quality by eliminating harmful nutrients and heavy metals (Srivastava et al. 2008), and their functions are mentioned in Table 12.1.

12.3 Macrophyte-Mediated Wastewater Treatment Mechanisms

Using macrophytes for wastewater treatment is an economical and environmentally sustainable method of removing nutrients from contaminated water sources. Macrophytes mostly employ the following mechanisms: nutrient uptake, biofiltration, oxygenation, aeration and bioaccumulation of contaminants, as shown in Fig. 12.2.

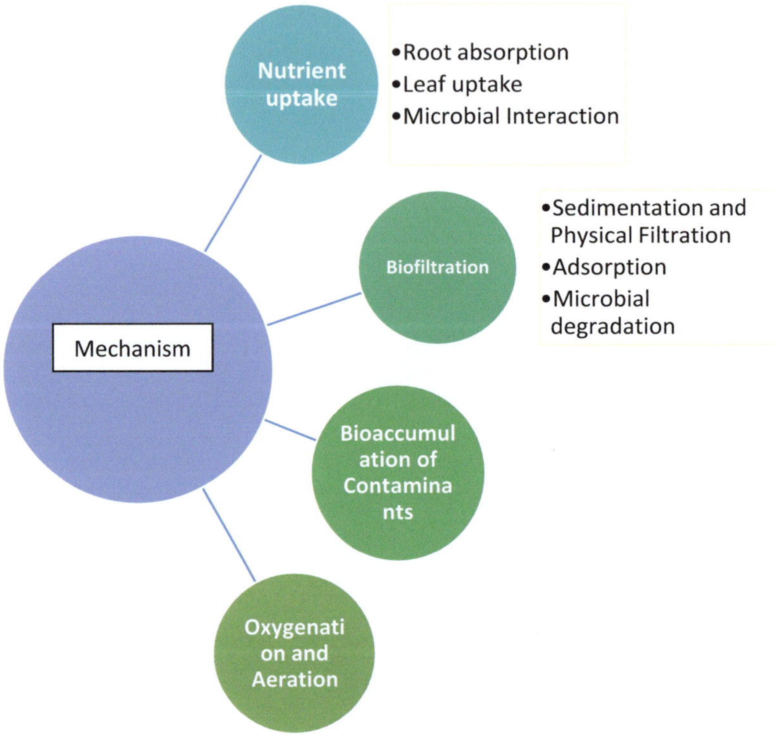

Fig. 12.2 Macrophyte-mediated wastewater treatment mechanisms

Floating macrophytes use a variety of techniques to treat wastewater successfully. One of the key methods is physical filtration, which is done by the plant canopy and helps trap and remove suspended materials from water. Furthermore, these aquatic plants absorb nutrients and heavy metals with the help of their roots, lowering the contaminants' concentration in water. Floating macrophytes are a versatile and efficient wastewater treatment alternative as they do both physical filtration and nutrient absorption.

12.3.1 Nutrient Uptake Mechanism

Macrophytes uptake nutrients with the help of roots, leaves and microbial interactions:

Root Absorption Macrophytes take nutrients like phosphorus and nitrogen through roots. When the roots come into contact with the wastewater rich in nutrients, they start spreading into the sediment. By the mechanism of ion exchange and active transport, they absorb nutrients in this process. Nitrogen is absorbed primarily in the form of nitrate (NO_3^-) and ammonium (NH_4^+). Ammonium is favoured because of its lower energy requirement than nitrate for assimilation (Vymazal 2007).

Leaf Uptake Certain macrophytes can actively take nutrients from the surrounding water through their leaves as they have specialised structures like stomata and leaf surfaces; these plants can absorb nutrients directly from water. This mechanism is primarily followed by submerged and floating macrophytes. Some macrophytes such as *Hydrilla* spp. and *Elodea* spp. absorb nutrients through the entire leaf surface.

Microbial Interaction The area around the roots of plants is known as the rhizosphere. It plays a very important role as a hotspot for microbial activity. Macrophytes release organic substances into the environment, which promotes microbial growth, and they also help accelerate the breakdown and transformation of nutrients (Tanner 1996).

12.3.2 Biofiltration

Biofiltration is a natural wastewater treatment process involving biological, physical and chemical mechanisms to remove contaminants from water. This process includes adsorption, microbial degradation, filtration and sedimentation:

Sedimentation and Physical Filtration In this process, a dense root network of macrophytes functions as physical barriers that trap particulate matter by slowing down water flow, which allows suspended materials to settle, and after settling they get trapped into the rhizosphere (Vymazal 2011).

Adsorption They can adsorb contaminants present in wastewater by the roots. Organic contaminants, heavy metals and nutrients can be adsorbed onto a large surface area in their roots and related biofilms (Hadad et al. 2006).

Microbial Degradation The rhizosphere of macrophytes is rich in microbial populations, and these microbes help in digesting all the organic contaminants present in the rhizosphere. These large aquatic plants emit organic substances (exudates), which stimulate microbial activity, which results in the enhancement of the breakdown of organic contaminants (Kadlec and Wallace 2008).

12.3.3 Oxygenation and Aeration

In artificial wetlands, oxygen is released by the roots or rhizomes of macrophytes, resulting in an oxygen-rich environment that helps in the treatment of wastewater. This process is known as macrophyte oxygenation and aeration. They have wide root systems that penetrate sediment in wetlands or artificial treatment systems. These roots form channels that allow oxygen to pass through the sediment and reach the anaerobic zones; this process is known as root zone oxygenation (Opitz et al. 2021). This encourages the proliferation of aerobic bacteria, which aids in organic matter decomposition and improves water quality. The rhizospheres serve as a route for oxygen exchange.

When oxygen-rich air enters the rhizosphere via leaves and stems, the production of methane and other gases takes place, which is called rhizosphere aeration. Aeration in roots helps with pollutant elimination, nutrient availability, pollutant elimination and overall plant health in the wastewater treatment system (Siddiqui et al. 2023). Macrophyte roots release oxygen through photosynthesis and directly from the atmosphere, which improves gas exchange chemistry, nutrient uptake and pollutant elimination. Furthermore, subsurface rhizomes of emergent macrophytes enhance oxygen transfer to apical buds, generating oxidised phyllospheres that protect against phytotoxins in waterlogged soils (Armstrong et al. 2006).

According to Reddy et al. (1990), some laboratory investigations show that aquatic macrophytes productively transport oxygen into wastewater, improving water quality by lowering biological oxygen demand and raising dissolved oxygen levels, with some plants transporting up to 90% of total oxygen into the effluent. Additionally, the oxygen regime in macrophyte-dense vegetation varies depending on ecological groups and the age of the water body, and submerged vegetation regions have the highest oxygen concentrations.

12.3.4 Bioaccumulation of Contaminants

Macrophytes can absorb and accumulate heavy metals and metalloids from water and sediment, by a process known as bioaccumulation. Numerous studies have shown that the bioaccumulation patterns of various macrophyte species vary according to environmental conditions, including *Ludwigia* sp., *Nymphaea* sp., *Nelumbo* sp., *Lemna minor, Lemna trisulca, Salvinia natans* and *Azolla filiculoides* (Regmi et al. 2022; Monferrán et al. 2023). These macrophytes have different abilities to accumulate elements such as Cd, Co, Cu, Mg, Na, Pb, Zn, As, Cr, Fe, Mn and Ni, and some other species show stronger bioaccumulation capacity for specific pollutants. The bioaccumulation process is crucial for understanding the toxicity, distribution and ecological impacts of contaminants in aquatic ecosystems; it highlights the importance of using macrophytes as bioindicators for monitoring trace element pollution and implementing effective remediation strategies (Mthombeni 2022). Macrophytes can eliminate and transform nutrients, pharmaceutical substances, heavy metals, endocrine-disrupting chemicals (EDCs), microbes and radionuclides. This can be accomplished using phytoremediation techniques such as phytostabilisation, phytoextraction, hemofiltration, phytodegradation and phytovolatilisation (Sayanthan et al. 2024).

12.4 Selection and Cultivation of Macrophytes for Waste Treatment Systems

The primary characteristics of the macrophytes that are selected and cultivated for treatment of wastewater include the following: Macrophytes, such as duckweed (*Lemna minor*) and water hyacinth (*Eichhornia crassipes*), have a high growth rate, allowing for effective pollutant removal and biomass generation. Aquatic plant features, such as growth rate, salinity tolerance, oxygen production and consumption, nutritional value and preferred food sources, significantly impact the efficacy of ecosystem management and should be carefully selected and cultivated (Izzati et al. 2022). Selected macrophytes should have high removal efficiency for nutrients (nitrogen and phosphorus), organic matter (BOD and COD) and other impurities from wastewater (Shah et al. 2014). Additionally, these macrophytes should be well adapted to the local environment; they can flourish under the treatment system's present temperature, pH and salinity conditions, indicating their tolerance towards harsh environmental conditions. The resulting macrophyte biomass can be utilised for many applications, like animal feed, bioenergy production or conversion into valuable compounds. Thus, they also help in promoting circular economy activities (Kurniawan et al. 2021). Usually, locally accessible macrophyte species are recommended to ensure a regular supply and simplicity of cultivation.

12.4.1 Equitable Species of Macrophytes for Various Environments

Macrophytes have been used to clean wastewater in constructed/man-made wetlands and natural aquatic habitats. It is crucial to consider the uptake rate and assimilation of wastewater elements into the plant's biomass to effectively use them for phytoremediation. Previous studies show that macrophytes are suitable for different types of wastewater as they are resilient to pollutants. At the time of selection of macrophytes for a man-made wetland, ensure they are locally available and tolerant of anoxic, waterlogged and hyper-eutrophic conditions because they can adapt to a wide range of environmental circumstances, which makes them appropriate for various water bodies. Therefore, selecting the appropriate macrophyte species for a given habitat is crucial for achieving optimal development and pollution elimination. Here, we discuss the importance of selecting appropriate macrophyte species for different settings. One of the most important considerations while selecting macrophytes is their capacity to grow in various environments. Water hyacinth (*Eichhornia crassipes*) thrives in warm, tropical conditions, but duckweed (*Lemna minor*) tolerates milder temperatures and may grow in a wider variety of water depths (Bubíková and Hrivnák 2018). Similarly, water lettuce (*Pistia stratiotes*) is suited to shallow environments and can survive various water temperatures and pH levels (Justin et al. 2022). Another important consideration is the macrophyte's ability to remove contaminants from the water. Pollutant removal efficiency varies from species to species, with some being more efficient at removing nutrients like nitrogen and phosphorus and others at removing organic debris. For example, *Lemna minor* is well known for its excellent nitrogen removal rates, making it an appealing option for wastewater treatment systems (Bubíková and Hrivnák 2018), and *Eichhornia crassipes* is also effective in removing nutrients, particularly carbon contaminants, from wastewater (Akowanou et al. 2023).

Choosing appropriate macrophyte species for different environmental conditions is critical for preserving ecological balance and guaranteeing effective pollutant clearance. Treatment systems can be built to successfully use macrophytes for wastewater treatment while supporting sustainable practices by considering elements such as environmental tolerance, pollutant removal efficiency, local availability and biomass utilisation.

These macrophytes shown in Table 12.2 are commonly used in a variety of wastewater treatment systems, including man-made wetlands since they can remove pollutants and toxins.

12.4.2 Factors to Consider for Growth and Maintenance of Aquatic Macrophytes

Aquatic macrophytes are widely used in wastewater treatment systems because of their capacity to remove pollutants and toxins. These plants are also quite effective at treating wastewater, particularly in polishing lagoons and small-scale sewage treatment systems. For this purpose, different factors must be considered for aquatic macrophytes to grow and sustain properly.

Various factors that should be considered for the growth and maintenance of macrophytes are:

1. *Temperature and light intensity*

The two major factors for the growth of aquatic macrophytes are temperature and light intensity. Mostly, macrophytes flourish at temperatures between 20 °C and 30 °C, with the ideal light intensities of 100–200 μmol/m^{2}s. Insufficient light might result in slower growth rates and decrease pollution removal efficiency.

2. *Depth and flow of water*

Depth and flow of water also affect the growth of macrophytes as the depth to which macrophytes grow influences their root anchoring. The species in shallow water have shorter roots, while deep-water species have longer roots for accessing nutrients and stabilising themselves. Also, deeper water reduces light penetration because of the distance from the surface. Macrophytes adapted to deem light conditions; for example, *Vallisneria* and *Elodea* thrive in deeper waters. Depth of water influences competition among macrophytes, and because of this, shallow areas may

Table 12.2 List of some common macrophytes used in wastewater treatment

S. No	Species	Common name	Growth form	References
1	*Typha latifolia*	Cattail	Emergent	Vidayanti et al. (2017) and Vymazal (2011)
2	*Mentha aquatica*	Water mint	Emergent	Justin et al. (2022)
3	*Salvinia rotundifolia*	Water fern	Free-floating	Reddy et al. (1987)
4	*Polygonum punctatum*	Dotted smartweed	Submerged	Núñez et al. (2011)
5	*Eichhornia crassipes*	Water hyacinth	Free-floating	Ajibade and Adewumi (2017)
6	*Commelina cyanea*	Scurvy weed	Emergent	Ajibade and Adewumi (2017)
7	*Phragmites australis*	Common reed	Emergent	Vymazal and Březinová (2016) and Wilikson et al. (2023)
8	*Leptodictyum riparium*	Knapwort	Submerged	Basile et al. (2012)
9	*Alocasia puber*	Elephant ear	Emergent	Thani et al. (2019)
10	*Azolla microphylla*	Mexican water fern	Free-floating	Mishra and Shukla (2016)
11	*Elodea canadensis*	American waterweed	Submerged	Basile et al. (2012)
12	*Ludwigia helminthorriza*	Water primrose	Emergent	Núñez et al. (2011)
13	*Spirodela polyrhiza*	Water duckweed	Free-floating	Justin et al. (2022)
14	*Trapa natans*	Water chestnut	Floating	Tsuchiya and Iwakuma (1993)
15	*Colocasia esculenta*	Taro	Emergent	Obi and Woke (2014)
16	*Pistia stratiotes*	Water lettuce	Free-floating	Thani et al. (2019) and Justin et al. (2022)
17	*Nymphaea tetragona*	Pygmy water lily	Floating	Kunii and Aramaki (1992)
18	*Hydrocharis dubia*	Frogbit	Floating	Tsuchiya (1989)
19	*Wolffia borealis*	Northern watermill	Free-floating	Lemon et al. (2001)
20	*Lemna minor*	Duckweed	Free-floating	Basile et al. (2012) and Lemon et al. (2001)
21	*Pontederia cordata*	Pickerelweed	Emergent	Sayanthan et al. (2024)
22	*Microsorum pteropus*	Java fern	Submerged	Sayanthan et al. (2024)
23	*Hygrophila difformis*	Water wisteria	Submerged	Shah et al. (2014)
24	*Typha domingensis*	Cattail	Emergent	Kurniawan et al. (2021)
25	*Callitriche hermaphroditica*	Water starwort	Submerged	Hadad et al. (2021)
26	*Marsilea mutica*	Water clover	Rooted floating	Shah et al. (2014)
27	*Myriophyllum spicatum*	Water milfoil	Submerged	Kurniawan et al. (2021)
28	*Alisma plantago-aquatica*	Water plantain	Emergent	Shah et al. (2014)

(continued)

Table 12.2 (continued)

S. No	Species	Common name	Growth form	References
29	*Typha domingensis*	Bulrush	Emergent	Al-thani and Yaseen (2021)
30	*Carex riparia*	Water sedge	Emergent	Kurniawan et al. (2021)
31	*Juncus effusus*	Water rush	Emergent	Sayanthan et al. (2024)
32	*Nasturtium officinale*	Watercress	Emergent	Sayanthan et al. (2024)
33	*Azolla caroliniana*	Water fern	Emergent	Kurniawan et al. (2021)
34	*Borago officinalis*	Water borage	Emergent	Kurniawan et al. (2021)
35	*Canna indica*	Indian shot	Emergent	Kurniawan et al. (2021)
36	*Carex appressa*	Tall sedge	Emergent	Sayanthan et al. (2024)
37	*Alisma plantago-aquatica*	Water plantain	Emergent	Justin et al. (2022)

become overcrowded, and ultimately this condition affects the growth of macrophytes. Moderate flow rates help by improving nutrient availability, and well-flowing water provides them with a steady source of nutrients. Flowing water replenishes oxygen around roots, which prevents anoxic conditions; however, excessive currents pose a risk of uprooting macrophytes and compromising their stability.

3. *pH*

The pH level can influence nutrient availability, enzymatic reactions and overall plant health, and most aquatic macrophytes thrive only in a pH range of 6.5–8.5. If there is a low pH in water, it will reduce nutrient availability, such as phosphorus, which affects the growth of macrophytes, while a high pH alters nutrient solubility, which can directly lead to nutrient deficiency.

12.5 Development and Implementation of Treatment Systems Based on Macrophyte

12.5.1 Constructed Wetland

Constructed wetlands (CWs) are engineered systems that treat wastewater from various sectors. These are considered a sustainable option because of their low cost, less energy consumption, long-term viability and effective land use than traditional wastewater treatment techniques. Removal efficiency depends on the type of substrate used, hydraulic retention time, water depth and various types of plant species used (Hota et al. 2023).

Macrophytes play an important role in the removal of pollutants through CW (shown in fig. 12.3). Wetlands planted with macrophytes have higher efficiency as compared to unplanted wetlands. It increases the removal efficiency of nutrients, improves microbial stimulation and increases their activities (Zhang et al. 2022).

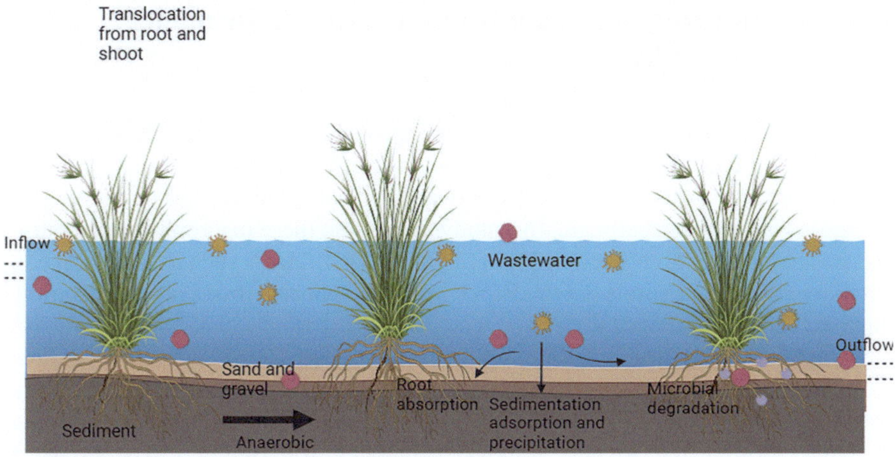

Fig. 12.3 Pollution or contaminant removal mechanisms

Various aquatic macrophytes are used to treat wastewater depending on the desired characteristics of removal efficiency.

A study was conducted in Tunisia to see the removal efficiency of CW. In this combination of macrophytes, bacterial activity and substrate were used. It efficiently removes contaminants, and the treated water meets standard criteria for irrigation. Furthermore, this treated water does not have any detrimental impact on environmental health. Macrophytes play a crucial role in contaminant removal and uptake during the treatment process as reeds, cattails and papyrus help remove nitrogen, phosphorus and organic matter from wastewater (Mahmoudi et al. 2024).

12.5.2 Floating Treatment Wetland

Floating treatment wetlands (FTWs) are an on-site treatment method for revitalising aquatic basins that often receive wastewater. In FTW, there is not much difference compared to conventional wetlands. The unique aspect of FTWs lies in the fact that the plants grow on a free-floating mat and extend their roots into polluted water. In doing so, they serve as natural biological filters, helping to purify the water. Water hyacinth (*Eichhornia crassipes*) and duckweed (*Lemna minor*) are the most common floating macrophytes used in FTW to effectively remove contaminants. Despite being an invasive plant, water hyacinth is a valuable resource for wastewater management. It spreads and expands quite quickly. The FTW is intended to maximise the development of a biofilm. Plants experience aerobic, anaerobic and anoxic environments beneath them. Plant roots improve dissolved oxygen levels, while plants and biofilm absorb nutrients and heavy metals, reducing BOD and COD.

12.6 Monitoring and Optimisation of the Performance of Macrophytes

Findings from a few research demonstrating the effectiveness of macrophytes:

A review of Justin et al. (2022) evaluated the effectiveness of locally occurring Nigerian aquatic plants, such as water hyacinth, water lettuce and duckweed, for domestic wastewater treatment. This study shows that these macrophytes can purify wastewater and are highly efficient in reducing contaminants such as electrical conductivity (EC), total solids, COD, BOD, total phosphorus, dissolved oxygen, total nitrogen and heavy metals.

In another study, three distinct aquatic macrophytes—water hyacinths, duckweed and water lettuce—were assessed for their potential to clean municipal wastewater collected from Taxila, Pakistan. When treating municipal wastewater, water hyacinth showed the highest removal efficiency for BOD5 (50.61%), COD (46.38%), nitrogen (40.34%) and phosphorus (18.76%). Significant pollutant removal was also demonstrated by duckweed and water lettuce. Duckweed achieved removal efficiencies of 33.43% for BOD5, 26.37% for COD, 17.59% for nitrogen and 15.25% for phosphorus. Water lettuce had comparable removal efficiencies. For BOD5, it was 33.43%, 26.37% for COD, 17.59% for nitrogen and 15.25% for phosphorus, which shows that in municipal wastewater treatment, water hyacinth had the highest removal efficiency for various pollutants, followed by duckweed and water lettuce (Shah et al. 2014). This study highlights the great potential of macrophyte-based systems as a workable and affordable substitute for treating municipal sewage; these systems have minimal operating and maintenance costs and perform similarly to traditional treatment plants.

Another study by Sayanthan et al. (2024) also shows that phytoremediation with aquatic macrophytes is a sustainable way of eliminating pollutants from wastewater. Aquatic macrophytes such as *Azolla*, duckweed and water hyacinth remove contaminants through bioaccumulation and biosorption. Their biomass can generate bioenergy, animal feed and fertiliser, contributing to a circular economy. Proper treatment of macrophyte biomass is vital to avoid potential environmental health risks. The study underlines the advantages of employing macrophytes for wastewater treatment, emphasising its economic viability, contaminant removal efficiency and environmental sustainability.

12.6.1 Parameters for Assessment

When evaluating macrophyte performance for wastewater treatment, numerous critical characteristics are considered. These criteria contribute to determining the effectiveness of macrophyte-based systems.

Macrophytes play an important role in removing nutrients (such as nitrogen and phosphorus) from wastewater. The efficiency of macrophytes absorbing nutrients is an important metric, and nutrient removal reduces eutrophication and improves water quality. BOD is also an important parameter. It refers to the amount of oxygen required by microorganisms during the breakdown of organic materials in water, and macrophytes help to reduce BOD levels by increasing microbial activity, which breaks down organic materials. COD indicates the total organic load in wastewater. Effective macrophyte systems reduce COD, indicating enhanced water quality. Some macrophytes can also absorb heavy metals (such as copper, zinc and cadmium) from wastewater, and assessing their capacity to absorb heavy metals is critical for pollution control. The duration of wastewater's stay in the macrophyte system is called hydraulic retention time (HRT). Longer HRT improves pollution clearance and nutrient uptake. It is also an important factor for the performance of macrophytes with all the above-mentioned factors: pH, alkalinity, temperature and climate are also considered factors.

12.6.2 Strategies for Improving Efficiency

- Macrophytes interact with the microbial populations that surround them. One strategy is to promote beneficial interactions between macrophytes and microorganisms to improve pollutant degradation. This can lead to better removal of contaminants before release into natural or recycled water bodies.
- Macrophyte roots are essential for nutrient intake and pollution elimination because this constructed wetland can be designed with an efficient root zone to improve phytoremediation, increase nutrient removal efficiency and improve pollutant degradation.
- Advances in genetic engineering have the potential to improve phytoremediation performance. More research may be done on genetic alterations to improve macrophyte performance in wastewater treatment. Apt use of this strategy can significantly improve the pollutant removal capability of potential macrophytes.
- Some invasive macrophyte species have pollutant removal capabilities and can be employed in controlled built wetland environments. Efficient elimination of pollutants can be obtained, for example, water hyacinth.

12.7 Conclusions

This chapter discusses the significant potential of macrophytes in purifying wastewater. Various literature reviews have documented macrophyte's effectiveness in organic, chemical and heavy metal contaminant reductions. Therefore, these are considered vital tools for treating carcinogenic problems from water pollution. Floating macrophytes offer an environmentally friendly green approach to eliminating pollutants from wastewater. Phytoremediation technology is more effective than conventional treatment methods since it is economically viable, nature-based, sustainable, effective in removing contaminants and environmentally acceptable. Furthermore, plant-based applications in wastewater treatment are primarily focused on contaminant removal, making it a promising area to explore the technology of using macrophytes to treat domestic wastewater at various levels (institutional, government and individual). This is because the plants are not difficult to find. Duckweed (*Lemna minor*) can be found in almost any stagnant water, while water hyacinth (*Eichhornia crassipes*) and water lettuce (*Pistia stratiotes*) are usually found along slow-moving rivers and lakes and other dominant aquatic macrophytes can be found in any lake and river basin. Various macrophytes can be utilised to clean up wastewater in the secondary and tertiary stages of treatment. Floating aquatic macrophytes such as *Azolla*, duckweed and water hyacinth play an important role in pollutant removal and can effectively remove contaminants through bioaccumulation and biosorption. However, the resulting macrophyte biomass might harm environmental health if not treated properly. In some situations, it can produce bioenergy, animal feed and fertiliser, aiding the circular economy. Using aquatic plant treatment systems involves a complicated interaction between plants and microbes. The operator must thoroughly understand standard treatment system responses, fertiliser requirements and plant requirements to identify and address potential issues early, as each plant has its own demands. In conclusion, using floating macrophytes for wastewater treatment provides a long-term, sustainable, nature-based and highly effective pollutant removal option. Researchers and practitioners can improve the use of floating macrophytes in wastewater treatment systems by studying their mechanisms, efficiency and challenges. Further study and innovation in this field are critical for leveraging the full potential of these natural plant-based systems in reducing environmental contamination and increasing water quality.

References

Ajibade FO, Adewumi JR (2017) Performance evaluation of aquatic macrophytes in a constructed wetland for municipal wastewater treatment. FUTA J Eng Eng Technol 11(1):1–11

Akowanou AVO, Deguenon HEJ, Balogoun KC, Daouda MMA, Aina MP (2023) The combined effect of three floating macrophytes in domestic wastewater treatment. Sci Afr 20:e01630

Al-Thani RF, Yasseen BT (2021) Perspectives of future water sources in Qatar by phytoremediation: biodiversity at ponds and modern approach. Int J Phytoremediation 23(8):866–889

Armstrong J, Jones RE, Armstrong W (2006) Rhizome phyllosphere oxygenation in Phragmites and other species in relation to redox potential, convective gas flow, submergence and aeration pathways. New Phytol 172(4):719–731

Babko R, Diachenko T, Zaburko J, Danko Y, Kuzmina T, Szulżyk-Cieplak J, Łagód G (2023) Macrophyte communities as bioindicator of stormwater pollution in rivers: a quantitative analysis. PeerJ 11:e15248

Bairagi S, Ali SW (2020) Conventional and advanced technologies for wastewater treatment. In: Environmental nanotechnology for water purification. Wiley, pp 33–56

Basile A, Sorbo S, Conte B, Cobianchi RC, Trinchella F, Capasso C, Carginale V (2012) Toxicity, accumulation, and removal of heavy metals by three aquatic macrophytes. Int J Phytoremediation 14(4):374–387

Bhat RA, Dervash MA, Qadri H, Mushtaq N, Dar GH (2018) Macrophytes, the natural cleaners of toxic heavy metal (THM) pollution from aquatic ecosystems. In: Environmental contamination and remediation. Cambridge Scholars Publishing, Newcastle upon Tyne, pp 189–209

Bubíková K, Hrivnák R (2018) Relationships of macrophyte species richness and environment in different water body types in the central European region. Annales de Limnologie/Int J Limnol 54:35. EDP Sciences

Chirisa I, Bandauko E, Matamanda A, Mandisvika G (2017) Decentralized domestic wastewater systems in developing countries: the case study of Harare (Zimbabwe). Appl Water Sci 7:1069–1078

Cohen-Shacham E, Walters G, Janzen C, Maginnis S (2016) Nature-based solutions to address global societal challenges, vol 97. IUCN, Gland, pp 2016–2036

Ding Y, Sun Y, Tang H, Song X (2020) Effects of macrophyte species and density on algae inhibition and water purification in submerged macrophyte ponds. Pol J Environ Stud 29(5)

Eulalia, Gómez Martín María, Máñez Costa Kathleen, Schwerdtner Máñez (2020) An operationalized classification of Nature Based Solutions for water-related hazards: From theory to practice Ecological Economics 167106460. https://doi.org/10.1016/j.ecolecon.2019.106460

Fletcher J, Willby N, Oliver DM, Quilliam RS (2020) Phytoremediation using aquatic plants. Springer, New York, pp 205–260

Hadad HR, Maine MA, Mufarrege MM (2006) Bioaccumulation of heavy metals in the water hyacinth (Eichhornia crassipes). Environ Sci Pollut Res 13(4):295–301

Hadad HR, Maine MA, Mufarrege MM, Di Luca GA, Sánchez GC, Nocetti E (2021) Macrophyte importance in contaminant treatment and biomonitoring. In: Removal of emerging contaminants through microbial processes. Springer, Singapore, pp 435–452

Hammer DA (1989) Constructed wetlands for wastewater treatment. Lewis Publishers, Chelsea, pp 702–709

Hota A, Patro SGK, Obaid AJ, Khatak S, Kumar R (2023) Constructed wetland challenges for the treatment of industrial wastewater in smart cities: a sensitive solution. Sustain Energy Technol Assess 55:102967

Iribarnegaray MA, Rodriguez-Alvarez MS, Moraña LB, Tejerina WA, Seghezzo L (2018) Management challenges for a more decentralized treatment and reuse of domestic wastewater in metropolitan areas. J Water Sanit Hyg Dev 8(1):113–122

Ismail SN, Abd Hamid M, Mansor M (2021) An overview of macrophytes in the tropical wetland ecosystem. Indones J Limnol 2(1):25–34

Izzati M, Soeprobowati TR, Prasetyo S (2022) Characterization of three selected macrophytes–an ecological engineering approach for effective rehabilitation of Rawapening Lake. J Ecol Eng 23(9):277–287

Justin LD, Olukanni DO, Babaremu KO (2022) Performance assessment of local aquatic macrophytes for domestic wastewater treatment in Nigerian communities: a review. Heliyon 8(8):e10093

Kadlec RH, Wallace S (2008) Treatment wetlands. CRC Press, Boca Raton

Koutsou OP, Gatidou G, Stasinakis AS (2018) Domestic wastewater management in Greece: greenhouse gas emissions estimation at country scale. J Clean Prod 188:851–859

Kumar PA (2015) Waste water management through aquatic macrophytes. Int Res J Environ Sci 4(3):41–46

Kunii H, Aramaki M (1992) Annual net production and life span of floating leaves in Nymphaea tetragona Georgi: a comparison with other floating-leaved macrophytes. Hydrobiologia 242:185–193

Kurniawan SB, Ahmad A, Said NSM, Imron MF, Abdullah SRS, Othman AR et al (2021) Macrophytes as wastewater treatment agents: nutrient uptake and potential of produced biomass utilization toward circular economy initiatives. Sci Total Environ 790:148219

Lemon GD, Posluszny U, Husband BC (2001) Potential and realized rates of vegetative reproduction in Spirodela polyrhiza, Lemna minor, and Wolffia borealis. Aquat Bot 70(1):79–87

Li L, Yue CL, Zhang H, Li HP, Yang L, Wang J (2019) Correlation between water purification capacity and bacterial community composition of different submerged macrophytes. Huan Jing ke Xue Huanjing Kexue 40(11):4962–4970

Mahajan B, Shastri S, Londhe S (2023) Waste removal efficiencies of floating macrophytes for restoration of polluted stream: an experimental analysis. Urban Sci 7(1):27

Mahmoudi A, Hannachi C, Mhiri F, Hamrouni B (2024) Performances of constructed wetland system to treat whey and dairy wastewater during a macrophytes life cycle. Desalin Water Treat 318:100364

Mishra VK, Shukla R (2016) Aquatic macrophytes for the removal of heavy metals from coal mining effluent. In: Phytoremediation: management of environmental contaminants, vol 3. Springer, Cham, pp 143–156

Monferrán MV, da C. Souza I, Arrivabene HP, Amé MV (2023) Metal (loid) s in Macrophytes from the Americas. In: The ecotoxicology of aquatic macrophytes. Springer, Cham, pp 53–74

Mthombeni TS (2022) The evaluation of the macrophyte species in the accumulation of selected elements from the Varkenslaagte drainage line in the west Wits, Johannesburg South Africa. In: Vegetation dynamics, changing ecosystems and human responsibility. IntechOpen

NitiAayog (2022). https://www.niti.gov.in/sites/default/files/2022-09/Waste-WaterA4_20092022.pdf

Núñez SR, Negrete JM, Rios JA, Hadad HR, Maine MA (2011) Hg, Cu, Pb, Cd, and Zn accumulation in macrophytes growing in tropical wetlands. Water Air Soil Pollut 216:361–373

Obi C, Woke J (2014) The removal of phenol from aqueous solution by Colocasia esculenta araesia Linn Schott. Sky J Soil Sci Environ Manag 3(6):59–66

Opitz J, Alte M, Bauer M, Peiffer S (2021) The role of macrophytes in constructed surface-flow wetlands for mine water treatment: a review. Mine Water Environ 40(3):587–605

Pendashteh AR, Fakhru'l-Razi A, Madaeni SS, Abdullah LC, Abidin ZZ, Biak DRA (2011) Membrane foulants characterization in a membrane bioreactor (MBR) treating hypersaline oily wastewater. Chem Eng J 168(1):140–150

Reddy MR, Tucker MR, Dunn SJ (1987) Effect of manganese on concentrations of Zn, Fe, Cu and B in different soybean genotypes. Plant Soil 97:57–62

Reddy KR, D'angelo EM, DeBusk TA (1990) Oxygen transport through aquatic macrophytes: the role in wastewater treatment. 19(2):261–267. American Society of Agronomy, Crop Science Society of America, and Soil Science Society of America

Regmi D, Bhatta R, Gurung S, Tuladhar S, Dahal BM, Raut N et al (2022) Bioaccumulation of heavy metals by macrophytes in Ghodaghodi Lake of western Nepal. Nepal J Environ Sci 12(1):7–17

Rehman F, Pervez A, Khattak BN, Ahmad R (2017) Constructed wetlands: perspectives of the oxygen released in the rhizosphere of macrophytes. CLEAN Soil Air Water 45(1)

Salgot M, Folch M (2018) Wastewater treatment and water reuse. Curr Opin Environ Sci Health 2:64–74

Sayanthan S, Hasan HA, Abdullah SRS (2024) Floating aquatic macrophytes in wastewater treatment: toward a circular economy. Water 16(6):870

Sesin V, Arnolds JL, Arts GH (2023) Global perspective for the use of aquatic macrophytes in regulatory risk assessment for contaminants. In: The ecotoxicology of aquatic macrophytes. Springer, Cham, pp 75–108

Shah M, Hashmi HN, Ali A, Ghumman AR (2014) Performance assessment of aquatic macrophytes for treatment of municipal wastewater. J Environ Health Sci Eng 12:1–12

Sharma AK, Tjandraatmadja G, Cook S, Gardner T (2013) Decentralised systems definition and drivers in the current context. Water Sci Technol 67(9):2091–2101

Siddiqui AJ, Jahan S, Adnan M, Ashraf SA, Singh R (2023) Macrophytes and their role in wetland ecosystems. In: Aquatic macrophytes: ecology, functions and services. Springer, Singapore, pp 119–138

Soloviy K, Malovanyy M (2019) Freshwater ecosystem macrophytes and microphytes: development, environmental problems, usage as raw material. Rev Environ Probl 3(4):115–124

Srivastava J, Gupta A, Chandra H (2008) Managing water quality with aquatic macrophytes. Rev Environ Sci Biotechnol 7:255–266

Syeed, Mudasir Shah, Ishfaq Ruqia, Bhat Gousia, Jeelani Shabeena, Farooq Shah, Fouziya Baba, Uqab Rouf Ahmad, Bhat Khalid Rehman, Hakeem (2020) Bioremediation and Biotechnology Vol 4 Techniques for Noxious Substances Remediation Application of Macrophytes for Remediation of Wastewater in Constructed Wetlands Springer International Publishing Cham 225–238

Tanner CC (1996) Plants for constructed wetland treatment systems—a comparison of the growth and nutrient uptake of eight emergent species. Ecol Eng 7(1):59–83

Tetteh EK, Obotey Ezugbe E, Rathilal S, Asante-Sackey D (2020) Removal of COD and SO42− from oil refinery wastewater using a photo-catalytic system—comparing tio2 and zeolite efficiencies. Water 12(1):214

Thani NSM, Ghazi RM, Amin MFM, Hamzah Z (2019) Phytoremediation of heavy metals from wastewater by constructed wetland microcosm planted with alocasia puber. Jurnal Teknologi 81(5)

Tsuchiya T (1989) Growth and biomass turnover of Hydrocharis dubia L. cultured under different nutrient conditions. Ecol Res 4:157–166

Tsuchiya T, Iwakuma T (1993) Growth and leaf life-span of a floating-leaved plant, Trapa natans L., as influenced by nitrogen flux. Aquat Bot 46(3–4):317–324

Vidayanti V, Choesin DN, Iriawati I (2017) Phytoremediation of chromium: distribution and speciation of chromium in Typha angustifolia. Int J Plant Biol 8(1):6870

Vymazal J (2007) Removal of nutrients in various types of constructed wetlands. Sci Total Environ 380(1–3):48–65

Vymazal J (2011) Plants used in constructed wetlands with horizontal subsurface flow: a review. Hydrobiologia 674(1):133–156

Vymazal J, Březinová T (2016) Accumulation of heavy metals in aboveground biomass of Phragmites australis in horizontal flow constructed wetlands for wastewater treatment: a review. Chem Eng J 290:232–242

Zhang Q, Huang J, Dzakpasu M, Gao Z, Zhou W, Zhu R, Xiong J (2022) Assessment of plants radial oxygen loss for nutrients and organic matter removal in full-scale constructed wetlands treating municipal effluents. Bioresour Technol 360:127545

Chapter 13
Circularity and Nature-Based Solutions: A New Paradigm for Sustainability

Hema Diwan and Binilkumar Amarayil Sreeraman

Abstract Water is a resource that has been exploited extensively to meet the growing human needs and because of the competing needs of the community. Water footprints have increased considerably, and the need for their assessment has grown multifold for efficient management of water resources (Aldaya MM, Chapagain AK, Hoekstra AY, Mekonnen MM. The water footprint assessment manual: setting the global standard. Routledge, 2012). The chapter integrates the domain of water management from the perspective of circular economy and the role of nature-based solutions in achieving water conservation. The concept of life cycle assessment is also explored to understand the performance of water management-related interventions considering circularity principles like reuse, recycling, recovery, and reduction. The circular economy can be used as a model for the management of water and reducing the water footprint.

Keywords Water · Nature-based solutions · Life cycle assessment · Circularity

13.1 Introduction

Water is critical for human survival and well-being and plays an important role in many sectors of the economy. Water, because of the competing users like agriculture, industry, commercial, and even ecosystem services, is one of the most vulnerable resources. Water availability, scarcity, and management are one of the high-rated risks globally, according to the World Economic Forum. Water footprint assessment entailing water apportionment as blue, green, and grey water is required to assess the quantum of risk attributed to resource scarcity (Jamshidi et al. 2024; Rodríguez-Chueca et al. 2024; EU 2020; Guerra-Rodríguez et al. 2020). Water scarcity is

H. Diwan (✉) · B. A. Sreeraman
Indian Institute of Management Mumbai, Mumbai, India
e-mail: diwanhema@iimmumbai.ac.in

S. S. Kumar, V. Kumar (eds.), *Nature-Based Solutions for Decentralised Liquid Waste Management*, Environmental Challenges and Solutions,
https://doi.org/10.1007/978-3-032-08362-3_13

typically measured per capita renewable water supply per year (Falkenmark 1989; Brown and Matlock 2011). The global statistics claim that about 40% of shortage would be there if we operate in the business-as-usual scenario and no path-breaking interventions are taken (World Economic Forum 2015; EU 2020).

13.2 Water Footprint

Water footprint is a concept introduced by UNESCO, which is an index of quantum of freshwater needed to produce goods or provide services to be used by individuals, business, or community. Water footprints can be categorised as green, blue, and grey. Green water footprint denotes rainwater consumed; blue water footprint refers to the surface and groundwater consumed and it is an indicator of water that evaporates, water that is incorporated in a product, etc.; and grey water footprint quantifies the volume of freshwater required to assimilate pollutant load as per the existing ambient water quality standards (Mekonnen and Hoekstra 2016; Jenicek et al. 2012).

13.2.1 Blue Water Footprint

Blue water footprint represents water available in fresh surface water sources and catchments like rivers or groundwater resources like aquifers (Lam et al. 2020). The blue water footprint is thus the volume of surface water consumed for human activities, and consumptive use refers to the volume of freshwater consumed and evaporated or incorporated in a product.

$$\mathrm{WF}_{\mathrm{proc,blue}} = \mathrm{BlueWaterEvaporation} + \mathrm{BlueWaterIncorporation} + \mathrm{LostReturnflow}\left[\frac{\mathrm{volume}}{\mathrm{time}}\right]$$

13.2.2 Green Water Footprint

The green water footprint is an index of human use of green water that is sourced from precipitation and stored in soil, etc. Green water footprint amounts to the precipitation water that gets stored in soil or vegetation avoiding the runoff or groundwater recharge. Green water footprint can be said as water that is precipitated in the root zone or is taken up by plants (Chapagain 2017; Bohra et al. 2022).

$$WF_{proc,blue} = BlueWaterEvaporation + BlueWaterIncorporation + LostReturnflow\left[\frac{volume}{time}\right]$$

13.2.3 Grey Water Footprint

The grey water footprint represents the polluted water produced during the production processes or contaminated water because of leaching from agricultural operations. It can be identified as the volume of freshwater used to assimilate the pollutant load to meet the ambient water quality standards.

$$WF_{proc,grey} = \frac{L}{C_{max} - C_{nat}}\left[\frac{Volume}{time}\right]$$

where:

C_{nat} is the natural concentration of the pollutant in the water
C_{max} is the maximum acceptable concentration in the water
L is pollutant load

13.3 Nature-Based Solutions and Ecosystem Services

Nature-based solutions (NBS) are increasingly considered sustainable and pragmatic solutions to many pressing anthropogenic environmental problems. Nature-based solutions are defined as solutions that are inspired and supported by nature and are cost-effective but provide environmental, social, and economic benefits and help build resilience. Nature-based solutions can be defined as "Nature-based solutions are actions to protect, sustainably manage and restore natural and modified ecosystems in ways that address societal challenges effectively and adaptively, to provide both human well-being and biodiversity benefits (IUCN 2020). Such solutions bring more, and more diverse, nature and natural features and processes into cities, landscapes, and seascapes, through locally adapted, resource-efficient, and systemic interventions."

Nature-based solutions as an appropriate approach to address pressing societal problems have gained momentum in the recent past. NBS are also considered an efficient approach to achieving sustainable development goals (SDG) by 2030 (Bianciardi and Cascini 2023). NBS has been widely adopted all around the globe (Liu et al. 2022). NBS is also considered an effective climate change adaptation and mitigation method, particularly in urban areas (Kabisch et al. 2016; UNEP 2021).

NBS was proposed as a useful tool in ecosystem management by IUCN in 2009 but was later adopted as a common term to represent several conservation methods that are inspired by nature or work in tandem with nature (WWAP 2018). The NBS is evolving and approaches environmental issues and climate change resilience with more socioecological reliance (Dumitru and Wendling 2021). But still ambiguity exists in the definitions and application fields of NBS, which has resulted in applying the same in different disciplines (Sowińska-Świerkosz and García 2022).

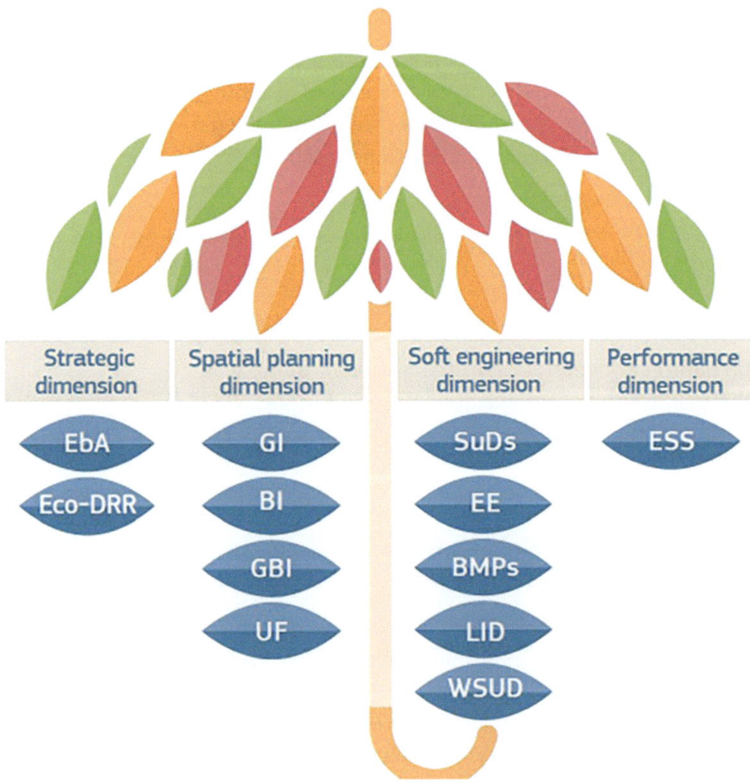

NBS depicted as an umbrella concept encompassing the existing concepts. (Source: EC 2021) *EbA* ecosystem-based adaptation, *Eco-DRR* ecosystem-based Disaster Risk Reduction, *GI* green infrastructure, *BI* blue infrastructure, *GBI* green-blue infrastructure, *UF* urban forestry, *SuDS* sustainable urban drainage systems, *EE* ecological engineering, *BMPs* best management practices, *LID* low-impact design, *WSUD* water-sensitive urban design, *ESS* ecosystem services

13.4 NBS and Water Ecosystem Services

NBS helps in addressing water-related risks such as floods and drought. It has been estimated that around 30% of the global population resides in areas and regions that are vulnerable to such extreme climatic events like floods and droughts. It has been found that ecosystem degradation is the major driver of such extreme risks. Hence, adopting ecosystem-based solutions will address such issues in tandem with nature. NBS can form a key element of policy initiatives at different levels, depending on the requirements of the regions. The COVID-19 pandemic has allowed for an understanding of the value of nature and has made public demand for greater attention to biodiversity as climate threats continue to grow at the local and global scales (EC 2021).

Therefore, NBS may realise water security by ensuring water availability and quality while reducing water-related adverse events. Further, it will also ensure more social, economic, and environmental benefits. The water-related ecosystem services are providing different essential services to people in supporting their livelihood. However, there are different challenges that affect the water-related ecosystem services, that is, water security/availability, water quality, and water-related risks. NBS is also considered a useful tool for better conservation of water as well as wastewater treatment, particularly in urban centres. In the case of water-related ecosystem services and the conservation programme targeting these resources, the adoption of NBS is still in the initial stages (WWAP 2018). However, the relationship between ecological systems and hydrological processes in the ecosystem, significantly influences the origin, movement, and quality of water in a system. NBS becomes an imperative choice, which needs to be prioritized over the artificial cost-intensive mechanisms for water management. NBS can address the issue of water supply through managing the hydrological processes like precipitation, humidity, water storage, infiltration, transmission, etc.

One of the significant features of the NBS in water conservation and management is cost-effectiveness. Water management in urban areas is a major challenge. But creating more water bodies in the urban peripheries is an excellent example of addressing the water shortage through NBS. The cost of artificial methods of water management is significantly larger compared to NBS-based methods. It will lower costs for the treatment of the water for domestic use. NBS has been increasingly used as the solution for wastewater management. It helps in cleansing wastewater by capturing and retaining pollutants and recycling nutrients most cost-effectively. NBS largely depend on ecosystem services to find practical solutions to water management issues. However, the universality of the NBS cannot be warranted. Hence, a site-specific approach needs to be adopted while implementing NBS in the field (WWAP 2018). A considerable difference exists in the impacts of various ecosystems on water or hydrological systems, depending on their location, climate, and management scenarios.

13.5 NBS and Wastewater Management: Decreasing Grey Water Footprints

NBS can be considered as alternative natural infrastructure for wastewater treatment. An ecosystem-based cleansing system can be used to treat different wastewater types, including urban wastewater, agricultural wastewater, and industrial wastewater. Developing NBS in wastewater treatment reduces the overall cost of the wastewater treatment by replacing costly wastewater treatment infrastructure (IWA 2021). Historically, NBS have been supporting wastewater treatment in different societies across the globe. Ancient Egyptian and Chinese cultures were known to use wetlands for wastewater disposal (Brix 1995). The increase in the population size and expansion of urban areas increased pollution levels in the urban water systems as well as the volume of wastewater generated from the urban settlements. It necessitated the technologies to be developed to treat the pollution levels in water or wastewater itself. Hence, many conventional wastewater treatment plants developed, which followed a combination of chemical and biological processes and operations to remove solids, organic matter, and, when needed, nutrients from wastewater (IWA 2021). But since the latter half of the previous century, we started using the natural-based approaches to treat the pollution levels in water. Treatment wetlands are the major source of such NBS approaches. They could reduce the effluent levels to the desired levels in the wastewater in a relatively cheaper and sustainable manner (Vymazal 2010; IWA 2021). Treatment wetlands are complemented with waste stabilisation ponds, which have also been created widely, particularly in relatively low-income countries and developing economies. Treatment wetlands and waste stabilisation ponds are part of popular NBS for wastewater treatment (Brix 1995).

NBS largely mimics the biological process of nature by utilising plants, soil, porous media, bacteria, and other natural elements and processes to remove pollutants in wastewater, including suspended solids, organics, nitrogen, phosphorus, and pathogens (Kadlec and Bevis 2009). Wetlands are known for their cleansing ability, for transforming contaminants into organic matter and nutrients, in wetlands being often called as "the kidneys of the landscape" (Mitsch and Gosselink 2000). This cleansing ability of wetlands can be exploited for developing NBS for wastewater management and water quality improvement (Brix 1995).

The ability of wetlands to treat the wastewater depends on the loading rate (wastewater per area per time) and on the wetland's specific hydrological and ecological characteristics (Knight 1992; Knight et al. 2003). Hence, the ability to remove the nutrients at the desired levels can be optimised by altering the loading rate to the most appropriate levels by installing a regime of fluctuating water levels and by the addition of absorbing materials to the wetland sediment. It is always advocated that constructed wetlands, particularly free-water surface constructed wetlands, are more appropriate NBS for wastewater treatment due to their higher ability in the optimum removal of nutrients (Vymazal 2010).

13.6 Circularity in Wastewater

A circular water economy works on improving the functionality of wastewater systems by introducing the concepts of reuse, recycling, treatment, recovery, and reduction (Brears 2015). Circular wastewater management optimises the use of water resources by circularity interventions and acting on the flow replenishment by providing management interventions (Smol et al. 2020). Integrated water resource management by offering strategies of closing the supply and the demand loop or production and the consumption loop has been approached as one of the best solutions to water conservation and preservation. Reuse, recycling, reduction, and recovery of water as a resource are the need of the hour. The concept of circular economy can be applied to wastewater treatment systems by rethinking the design and the processes to imbibe reuse, recycling, and recovery of water resource. To transition from a linear model to a circular model of production and consumption, water is an important element to be addressed. Various research studies have been done in this context, ranging from water footprint assessment to life cycle analysis. Hoekstra et al. (2011) used a water footprint assessment for water management. Jia et al. (2022) explored regional water scarcity assessment by water footprint and life cycle costing. Rufí-Salís et al. (2020) studied the resource circularity in wastewater treatment systems and implications of technological upgrades.

Grey water footprint is an unavoidable outcome in urban infrastructure, and circularity can be embraced in reducing this footprint. Today equal impetus is on how wastewater can be treated along with what can be recovered from the wastewater. Circular economy can increase the functionality of wastewater by recovery resources like nitrogen or phosphorus or energy (Brears 2018). Introducing circularity in wastewater treatment through reuse, recovery, and reclaiming from the resources can help in resource conservation and preservation. The Circular Economy Action Plan by the European Union has identified nutrient recovery and wastewater among the important intervention in achieving circularity (Smol et al. 2020). CE intervention can be seen in wastewater treatment plants for resource recovery in terms of nutrients, water, and energy (Voulvoulis 2018).

Conventional wastewater treatment systems help in regenerating polluted waters; however, they do have impacts on the environment at a systems level. Wastewater systems can be studied to understand and estimate the water footprints, viz. freshwater footprints vis-a-vis polluted load treatment through treatment technologies. Life cycle assessment in this regard has been used for performance measurement of these systems and further assessment. The water footprint studies in this context can provide crucial insights to evaluate the impact of a wastewater treatment plant regarding the use of freshwater.

13.7 Circularity and Life Cycle Assessment (LCA)

The water footprint assessments can be integrated with circularity, along with life cycle assessment tool principles to lessen the stress on water resources across the value chain. The assessment of the life cycle of a product/process that can identify the water footprints across the value chain of the product or a process can help in streamlining water management interventions. The circular economy can be embedded in the value chain that can help to reuse/recycle/regenerate/recover the resources within the loop. Closed-loop or open-loop circularity can be applied on various water footprints like grey/blue or green water. For example, rainwater harvesting can encourage recovery of runoff water, which can be used for charging aquifers. Life cycle assessment can help in predicting the impact of water circularity interventions across the value chain. For instance, water footprint (WF) assessment can provide complementary information to evaluate the impact of a wastewater treatment technology regarding the use of freshwater (Fig. 13.1).

13.8 Life Cycle Assessment (LCA)

Life cycle assessment is a globally recognised methodology, a standard under ISO 14000 series—ISO 14040. It can be subdivided into four categories.

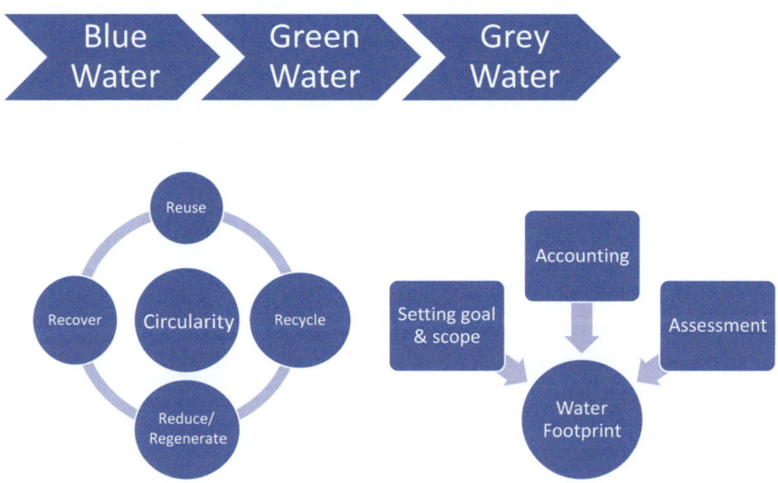

Fig. 13.1 Water footprint

13.8.1 Defining Goals and Objectives

It involves setting the objective and scope of the unit under study.

13.8.2 Life Cycle Inventory (LCI) Analysis

The life cycle inventory involves collecting data on flows, creating databases with all inflows and outflows for the system. This step involves flow data with respect to water consumption, pollution load, etc.

The assessment of a water footprint is defined in the ISO 14046 standard, defining water scarcity footprint and water availability footprint. A "water scarcity footprint" and a "water availability footprint" assess impacts associated with water use only, whereas a water footprint assesses all relevant impacts related to water.

13.8.3 Impact Assessment

This consists of classification and characterisation, normalisation, and weighing (optional). During classification, the collected data are categorised, and characterisation involves the determination of individual emissions. The normalisation and weighing steps are optional steps under the ISO standard.

Water footprint impacts include impacts along the value chain ranging from water use to impacts on water availability for various services. These impacts can be measured by using LCA impact categories (e.g. freshwater eutrophication, freshwater acidification, human toxicity, and eco-toxicity).

13.8.4 Interpretation and Improvement

The results are analysed to identify the hotspots of improvement and further decision-making.

A host of LCA studies have been done related to water as a unit of study. Life cycle assessment can be applied to study the impact of wastewater treatment technologies, benchmarking performance concerning resource (water) use, water quality, water quantity, and water discharge. It can also be used to conduct the water footprint assessments of blue water and grey water (Chen et al. 2020; Nguyen et al. 2022). LCA studies have been conducted for urban water infrastructure involving wastewater treatment plants. LCAs have been conducted to understand the environmental and health risks from wastewater, understand the performance of waste treatment technologies, identify hotspots of improvement, and make informed decisions. Smol et al. (2020) conducted a scenario analysis of wastewater treatment plants under no-treatment, secondary treatment, and tertiary treatment scenarios. Similar work was done by Chapagain (2017) in the sustainability assessment of water footprint. Shao and Chen (2013) dealt with the evaluation of the blue water footprint of wastewater. Morera et al. (2016) studied the water footprint in their work, considering that the pollution reduction is taken as a separate function of the wastewater focused on both the blue and grey to assess the water resource consumption of wastewater. Jia et al. (2022) explored regional water scarcity assessment by water footprint and life cycle costing. Studies by Morera et al. (2016) showed that treating wastewater significantly reduced the grey water footprint; however, it has implications for freshwater use (Fig. 13.2).

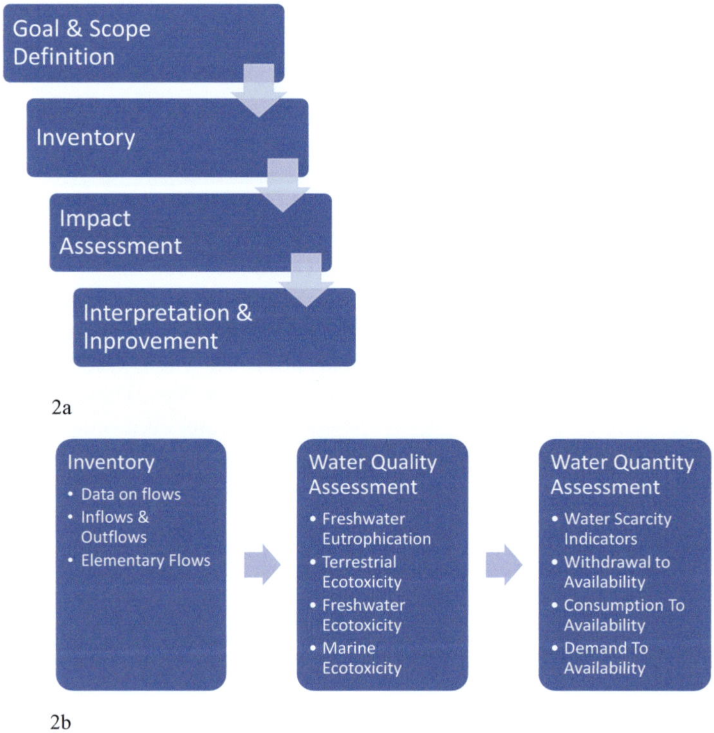

Fig. 13.2 (**a** and **b**) LCA methodology

References

Aldaya MM, Chapagain AK, Hoekstra AY, Mekonnen MM (2012) The water footprint assessment manual: setting the global standard. Routledge

Bianciardi A, Cascini G (2023) How would nature design and implement nature-based solutions? Nat Based Sol 3:100047

Bohra V, Ahamad KU, Kela A, Vaghela G, Sharma A, Deka BJ (2022) Energy and resources recovery from wastewater treatment systems. In: Clean energy and resource recovery. Elsevier, pp 17–36

Brears R (2015) The circular economy and the water-energy-food nexus, NFG policy paper series, no. 07, February 2015, NFG research group "Asian perceptions of the EU". Freie Universität, Berlin

Brears RC (2018) Blue and green cities: the role of blue-green infrastructure in managing urban water resources. Springer

Brix H (1995) Treatment wetlands: an overview. In: Toczylowska I, Mierzejewski M (eds) Proceedings conference on constructed wetlands for wastewater treatment. Technical University of Gdansk, Poland, pp 167–176

Brown A, Matlock M (2011) A review of water scarcity indices and methodologies. Sustain Consortium White Paper 106

Chapagain A (2017) Water footprint: state of the art: what, why, and how? https://doi.org/10.1016/B978-0-12-409548-9.10164-2

Chen KH, Wang HC, Han JL, Liu WZ, Cheng HY, Liang B, Wang AJ (2020) The application of footprints for assessing the sustainability of wastewater treatment plants: a review. J Clean Prod 277:124053

European Commission (2021) Evaluating the impact of nature-based solutions: a handbook for practitioners. Luxembourg, European Union

Falkenmark M (1989) The massive water scarcity now threatening Africa: why isn't it being addressed? Ambio 18:112–118

Guerra-Rodríguez S, Oulego P, Rodríguez E, Singh DN, Rodríguez-Chueca J (2020) Towards the implementation of circular economy in the wastewater sector: challenges and opportunities. Water 12(5):1431

Hoekstra AY, Chapagain AK, Aldaya MM, Mekonnen MM (2011) The water footprint assessment manual: setting the global standard. Earthscan, London

IUCN (2020) Guidance for using the IUCN global standard for nature-based solutions. A user-friendly framework for the verification, design and scaling up of nature-based solutions, 1st edn. IUCN, Gland

Jamshidi S, Farsimadan M, Mohammadi H (2024) A holistic approach for performance evaluation of wastewater treatment plants: integrating grey water footprint and life cycle impact assessment. Water Sci Tech 89(7):1741–1756

Jenicek P, Bartacek J, Kutil J, Zabranska J, Dohanyos M (2012) Potentials and limits of anaerobic digestion of sewage sludge: energy self-sufficient municipal wastewater treatment plant? Water Sci Technol 66(6):1277–1281

Jia X, Shahzad K, Klemeš JJ, Jia X (2022) Changes in water use and wastewater generation influenced by the COVID-19 pandemic: a case study of China. J Environ Manag 314:115024

Kabisch N, Frantzeskaki N, Pauleit S, Naumann S, Davis M, Artmann M, Bonn A (2016) Nature-based solutions to climate change mitigation and adaptation in urban areas: perspectives on indicators, knowledge gaps, barriers, and opportunities for action. Ecol Soc 21(2)

Kadlec RH, Bevis FB (2009) Wastewater treatment at the Houghton Lake wetland: vegetation response. Ecol Eng 35(9):1312–1332

Knight RL (1992) Ancillary benefits and potential problems with the use of wetlands for nonpoint source pollution control. Ecol Eng 1(1–2):97–113

Knight RL, Gu B, Clarke RA, Newman JM (2003) Long-term phosphorus removal in Florida aquatic systems dominated by submerged aquatic vegetation. Ecol Eng 20(1):45–63

Lam TWL, Ho HT, Ma AT, Fok L (2020) Microplastic contamination of surface water-sourced tap water in Hong Kong—a preliminary study. Appl Sci 10(10):3463
Liu Y, Gao Y, Liu L, Song C, Ai D (2022) Nature-based solutions for urban expansion: integrating ecosystem services into the delineation of growth boundaries. Habitat Int 124:102575
Mekonnen MM, Hoekstra AY (2016) Four billion people facing severe water scarcity. Sci Adv 2(2):e1500323. https://doi.org/10.1126/sciadv.1500323. PMID: 26933676; PMCID: PMC4758739
Mitsch WJ, Gosselink JG (2000) Wetlands, 3rd edn. Wiley, New York
Morera S, Corominas L, Poch M, Aldaya M, Comas J (2016) Water footprint assessment in wastewater treatment plants. J Clean Prod 112:4741–4748
Nguyen MD, Thomas M, Surapaneni A, Moon EM, Milne NA (2022) Beneficial reuse of water treatment sludge in the context of circular economy. Environ Tech Innov 28:102651
Regulation EU (2020) 741 of the European Parliament and of the council of 25 may 2020 on minimum requirements for water reuse. Off J Eur Union Brussels, Belgium 177:32–55
Rodríguez-Chueca J, Bañuelos AC, Rodríguez JP (2024) Life cycle assessment to evaluate the integral water cycle in industrial supply: a real case study. Sci Total Environ 931:172891
Rufí-Salís M, Petit-Boix A, Villalba G, Sanjuan-Delmás D, Parada F, Ercilla-Montserrat M, Arcas-Pilz V, Muñoz-Liesa J, Rieradevall J, Gabarrell X (2020) Recirculating water and nutrients in urban agriculture: an opportunity towards environmental sustainability and water use efficiency? J Clean Prod 261:121213
Shao L, Chen GQ (2013) Water footprint assessment for wastewater treatment: method, indicator, and application. Environ Sci Technol 47:7787–7779
Smol M, Adam C, Preisner M (2020) Circular economy model framework in the European water and wastewater sector. J Mater Cycles Waste Manage 22:682–697
Sowińska-Świerkosz B, García J (2022) What are nature-based solutions (NBS)? Setting core ideas for concept clarification. Nat Based Sol 2:100009
United Nations Environment Programme (2021) Adaptation gap report 2020. UNEP, Nairobi
United Nations World Water Assessment Programme, WWAP (2018) The United Nations world water development report 2018: nature-based solutions for water. UNESCO, Paris
Voulvoulis N (2018) Water reuse from a circular economy perspective and potential risks from an unregulated approach. Curr Opin Environ Sci Health 2. https://doi.org/10.1016/j.coesh.2018.01.005
Vymazal J (2010) Constructed wetlands for wastewater treatment. Water 2(3):530–549
Wendling DAL (2021) Evaluating the impact of nature-based solutions: a handbook for practitioners. Luxembourg, European Commission
World Economic Forum (2015) Global risks 2015, 10th edn. World economic Forum, Geneva

Part III
Case Studies

Chapter 14
A Low-Cost Appropriate Sanitation Technology for Rural India: A Case Study of Vadgaon Maval, India

Chandana N and **Bakul Rao**

Abstract A sustainable sanitation chain is essential for clean, healthy, and liveable conditions in developing countries. Under the Swachh Bharat Mission Gramin, the Indian government constructed 180 million toilets with onsite sanitation technologies (OSTs), but these often fall short in managing faecal sludge (FS). This study systematically characterizes FS to assess its resource potential and develop a treatment design in a rural area of Maharashtra, India. The collected FS contains nutrients like nitrogen, phosphorus, and potassium, qualifying it as an organic fertilizer per Indian specifications. However, high pathogenic content requires pathogen elimination and liquid–solid separation. A 1000 m^3 drying bed and 20 m^3 pathogen elimination chamber are needed for FS treatment in Vadgaon Maval, aiding in establishing FSM regulations for India.

Keywords Faecal sludge · Faecal sludge treatment · Characterization · Unplanted drying beds · Rural sanitation

14.1 Introduction

The sustainable development of human society depends on resource recovery and environmental care (Atun et al. 2025; Adlakha et al. 2022). Onsite sanitation technologies generate a large amount of faecal sludge (FS), which requires proper treatment and disposal due to the high concentration of pathogens present in it (Chandana

Chandana N (✉)
Center for Emerging Technologies for Sustainable Development, Indian Institute of Technology Jodhpur, Jodhpur, Rajasthan, India
e-mail: chandana@iitj.ac.in

B. Rao
Centre for Technology Alternatives for Rural Areas, Indian Institute of Technology Bombay, Mumbai, India
e-mail: bakulrao@iitb.ac.in

S. S. Kumar, V. Kumar (eds.), *Nature-Based Solutions for Decentralised Liquid Waste Management*, Environmental Challenges and Solutions,
https://doi.org/10.1007/978-3-032-08362-3_14

and Rao 2021a, b). Faecal sludge management (FSM) is critical because billions worldwide depend on onsite sanitation technologies (Chaudhuri and Mimi 2016). The practice of utilizing untreated FS as a fertilizer in agricultural lands has been a source of increased disease transmission in developing countries (Dasgupta et al. 2021). Proper management and beneficial use of FS can significantly enhance sanitation and support the aquaculture and agriculture sectors. The generation of revenue from FS products can incentivize the adoption of FSM in rural areas, ultimately leading to better and more sustainable sanitation management (Garn et al. 2017; Chandana and Rao 2021c). FS can serve multiple purposes, including (i) soil conditioner, (ii) energy source, (iii) construction material, and (iv) protein source (Gautam et al. 2021).

Although FS has been historically used by humans, its high pathogen content necessitates adequate treatment before being released into the environment. The current treatment technologies available for FS aim for four main objectives: (i) solid–liquid separation, (ii) sludge separation, (iii) nutrient enhancement, and (iv) pathogen reduction (Chandana and Rao 2021c). However, it should be noted that all the processes mentioned above should happen in onsite sanitation technology if they have been built efficiently. The main reason for existing FS treatment with this process is mainly due to the transfer of wastewater technologies directly to FS treatment without much technical investigation and research on FS characteristics (Halcrow et al. 2014; Strande et al. 2014; Tilmans et al. 2015).

India has undergone an interesting journey of sanitation coverage, from 1% rural sanitation coverage in 1980 to maximum toilet construction by 2019 (Chandana and Rao 2021d). Predominantly, the SBM Gramin constructed more than 180 million toilets in India from 2014 to 2019 (Manga et al. 2021; Patel et al. 2019; Routray et al. 2015). However, these toilets lack the required management systems for collecting, emptying, transporting, and safely disposing of the FS (O'reilly et al. 2017; Wolf et al. 2014). A lack of chemical–physical–biological–mineralogical characterization of FS, which influences the design of treatment technologies for resource recovery and sustainable implementation, has hindered the process. In India, where rural areas are dependent on onsite sanitation technologies, low-cost FS treatment is critically needed. The current study outlines the design of FS treatment facility based on the detailed characterization of FS from a small town in Maharashtra, India.

14.2 Methodology

14.2.1 Study Area

Between 2018 and 2019, Maharashtra's SDG score remained stagnant according to NITI Aayog's assessment. Additionally, the state consistently stands out as the top 3 concerning waste generation. Hence, Maharashtra has been chosen as a representative case to underscore the critical need for effective faecal sludge management

(FSM). As per the Centre for Environmental Planning and Technology, India, municipal corporations and councils require FSM based on population criteria. According to the criteria, about 259 cities with a 30.2 million population require FSM in Maharashtra. It is estimated that around 7,036,600 m^3/year of FS is generated in municipal corporations and councils of Maharashtra. This FS, which has 900×10^2–3900×10^2 *E. coli* CFU/mL and 112×10^4–210×10^5 SESC CFU/mL, has been disposed into open lands, agricultural lands, water bodies, and dumping yards, which indicates the magnitude of the problem because of a lack of FSM. Maharashtra, with a population of 112 million, is one of the most urbanizing states. It should be noted from Fig. 14.1 that the census town, Nagar panchayats, and containment boards are increasing rapidly in numbers and population every decade. The census town, Nagar panchayats, containment boards, and rural areas that have not been equipped with underground sewers (due to economic and technical reasons) are growing at higher growth rates indicating the urbanization phenomena and thus requiring a large and additional number of OSTs.

The guiding idea was to select the fieldwork area in a non–/peri-urban area of India, i.e. a region that does not come under an urban centre (i.e. population less than 99,999) according to the Reserve Bank of India. The present research investigated the influence of urbanization, population density, community-level understanding of sanitation processes, existing sanitation infrastructure, and the end-use pathway of FS generated on sanitation in Vadgaon Maval, Maharashtra, India, with 3800 HHs. A comparative analysis of the underlying socioeconomic and

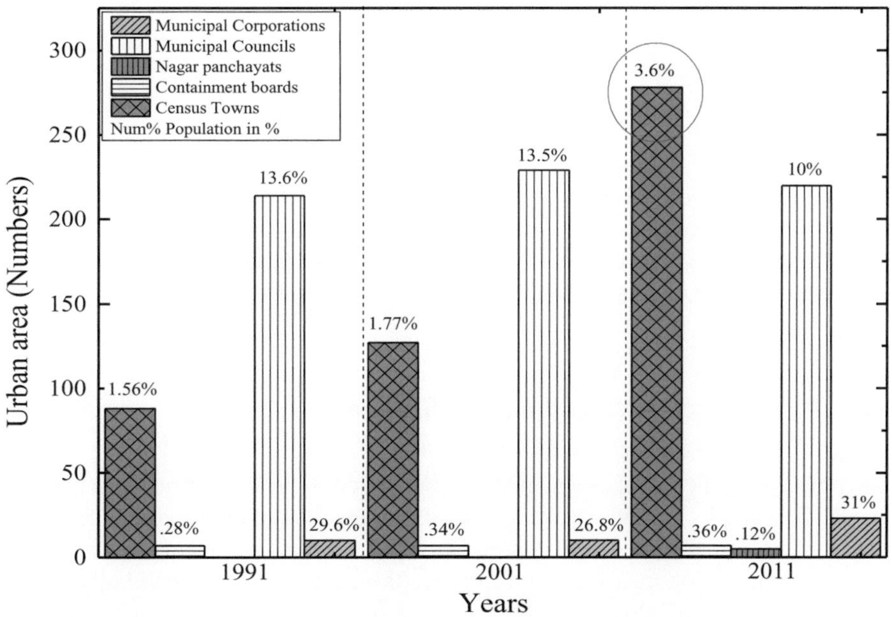

Fig. 14.1 The growth of urban bodies in Maharashtra, India, from 1991 to 2011

Table 14.1 Demographic profile and socioeconomic features of the study area

Name	Vadgaon Maval
Population	14,595
Migration rate	High
Number of HHs	3800
% pucca HHs[a]	94
% HHs with own house	53
Major livelihood	Private employees
% social class General SC/ST[a] OBC[a]	55 20 15
% residing in place >15 years	33
% BPL[a]	39
Water supply/week	7 days
Water body	Indrayani River
% private taps/borewells	81

[a]Pucca HH (masonry HH), *ST* scheduled cast, *SC* scheduled tribes, *OBC* other backward castes, *BPL* below poverty line

demographic disparities across the three study areas is presented in Table 14.1. Vadgaon Maval, a NAC in Pune District (18.74°N, 73.64°E), has experienced 67% household in-migration over 15 years as shown in Table 14.1, driven by proximity to Pune, indicating rapid urban expansion and escalating pressure on local sanitation infrastructure.

14.2.2 Origin of Faecal Sludge

FS samples incorporated in the current study were retrieved from varied domestic onsite sanitation facilities (i.e. septic tanks, lined pits, unlined pits) from a town in Maharashtra, India. The collection, transportation, and laboratory analysis of each FS sample are detailed in the paper by Chandana and Rao (2021a).

14.2.3 Characteristics of Faecal Sludge

The physicochemical properties, nutrient content, concentration of intrinsic bacteria in each FS sample, and other relevant details are detailed in the paper by Chandana and Rao (2021b). The electrical conductivity, pH, total solids, and organic matter of the FS samples were measured by following the 2520 F code of Standard Method APHA-AWWA-WEF 2012 and ASTM D-2974-14 standards. Carbon, hydrogen, nitrogen, and sulfur elements were evaluated through a CHNS analyser (Thermo Finnigan, FLASH EA 1112 series, Milan, Italy) in IIT Bombay.

14.3 Results

14.3.1 Characterization of Faecal Sludge in Vadgaon Maval, India

The various physicochemical properties of faecal sludge (FS) are presented in Fig. 14.2. Among these, COD (g/L) and pathogens (g/L) were found to be the highest in FS. The analysed FS exhibited nutrient concentrations such as nitrogen (3.0% ± 0.4%), phosphorus (29.07 ± 4.48 mg/L), potassium (43.57 ± 0.74 mg/L), and organic carbon (33% ± 3.2%) and favourable physicochemical features, which are in accordance with the permissible limits set by Indian guidelines for natural nutrient-rich soil amendments. However, the pathogen load of 6 log CFU/g and 6–13 viable helminth ova/g exceeds permissible thresholds under US land application standards. The physicochemical properties, nutrient content, concentration of intrinsic bacteria in each FS sample, and other relevant details are detailed in the paper by Chandana and Rao (2021a).

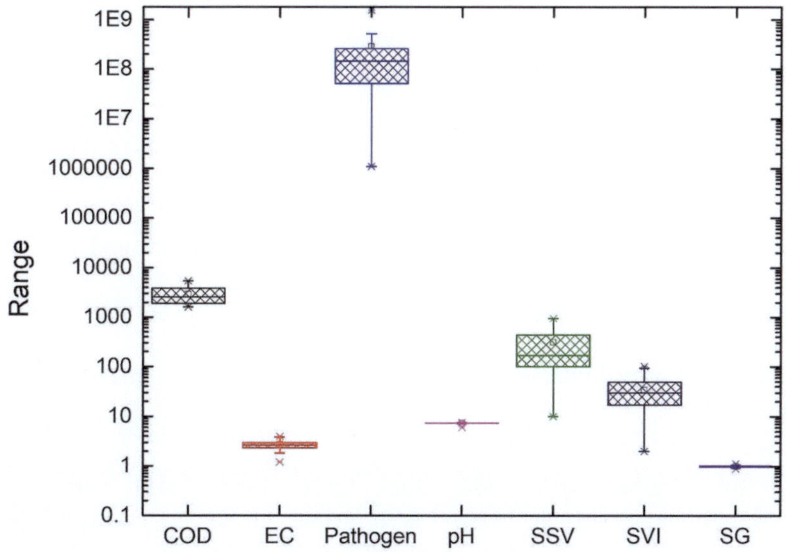

Fig. 14.2 Characteristics of faecal sludge from Vadgaon Maval

14.3.2 Design Period for Vadgaon Maval Faecal Sludge Treatment Plant

The planned operational life of a faecal sludge treatment plant (FSTP) is generally considered to be 20 years, corresponding to the anticipated durability of its main structural elements. To optimize capital costs while ensuring the FSTP design remains compatible with the characteristics of FS, a design horizon of 10 years has been adopted, with the facility expected to operate at full capacity for 20 years following that period. The total population of Vadgaon Maval was 14,595 in 2019. Taking 2022 as the base year for implementation, and using the population increase method, considered more accurate than both arithmetic and geometric approaches, the projected populations for 2032 and 2042 are estimated to be 26,930 and 37,601, respectively. The population to be considered in 2022, therefore, will be 32,265.

14.3.3 Determination of Plant Capacity for Vadgaon Maval Faecal Sludge Treatment Plant

The sizing of the FSTP is based on projected FS volumes anticipated over the design horizon. The initial load estimations are derived from the population baseline of 2022, with an annual increment rate of 1.1% applied to project future capacity requirements. Additionally, implementing a desludging frequency ranging between 2 and 2.5 years has been shown to significantly mitigate reliance on manual sludge removal practices in Vadgaon Maval. As presented in Table 14.2, the FS accumulation rate in the area is estimated at 233 L/person/year.

1. The per capita annual FS generation is estimated at 233 L/year.
2. The proportion of FS collectable from OST that is desludgable is in the range of 60% to 80%.
3. Projected yearly FS produced: $0.6 \times (32265) \times 0.233 = 4511\ m^3$.
4. Including an estimated 10% contribution from commercial and institutional sources: 451 m^3.
5. Assuming a 6-day work week over 280 working days, the daily volume is 17.70 m^3.
6. Based on an annual increase of 1.1%, the projected daily volume of sludge handled shall be:

2022	2023	2024	2025	2026	2027	2028	2029	2030	2031	2032
17.70	19.50	21.40	23.60	25.90	28.50	31.40	34.50	37.90	41.70	45.90

Thus, a daily design load of 46 m^3 at the end of the 10-year phase is considered.

Table 14.2 Faecal sludge accumulation in onsite sanitation technologies of Vadgaon Maval in litre/person/year

OST no.	Age (year)	Dimensions (m*m*m)	Volume of OST (m^3)	Height of FS (m)	Volume of FS present (m^3)	Residence in the HH	FS accumulation (L/person/year)
1	<1	3.1*1.5*2.1	9.8	NA		5	
2	<1	2.1*3.1*1.5	9.8	0.8	5.2	5	1040
3	2	1*1.5*1.5	2.3	1.3	1.9	5	950
4	3	2.1*1.2*1.5	3.8	1.4	3.5	5	233
5	3	1.5*1.5*2.1	4.7	1.2	2.7	10	90
6	5	0.3*3.7*2.1	2.3	1.1	1.2	5	48
7	5	1.5*0.6*1.5	1.4	1.1	0.9	5	46
8	10	2.1*3.1*2.1	13.7	1.25	8.5	10	85
9	>10	2.1*3.1*2.1	13.7	0.4	5.5	5	73
10	NA	2.4*1.5*1.5	5.4	0.8			

14.3.4 Treatment Combination for Vadgaon Maval Faecal Sludge Treatment Plant

Based on the FS characterization, the following treatment combinations have been suggested for the FS of Vadgaon Maval: screen + pathogen elimination chamber + unplanted drying beds.

14.3.4.1 Design of Inlet Channel with Screens

To enable direct FS discharge from trucks, a receiving pit connected to an inlet channel is essential. This setup facilitates the transfer of FS to the pathogen reduction unit for subsequent treatment. The inlet channel must be hydraulically optimized to ensure smooth flow of the slurry without sedimentation, while a bar screen is required to intercept and remove floating solids. For operational efficiency, dual channels should be constructed to allow alternating use for cleaning and loading cycles. Based on field assessments and literature insights, a RCC header pit of 1 m × 1 m with a longitudinal slope of 1% is recommended to guide FS flow towards the bar screening section of the pathogen elimination chamber.

14.3.4.2 Bar Screen

A coarse bar screen constructed from SS304 grade stainless steel, featuring 20 mm spacing and measuring 1.5 m × 1.5 m in plan, will be installed facing the sludge channel.

Considering an inlet discharge = 5 lps = 0.005 m3 /s.

The width of the channel, denoted as b, is taken as 300 mm.

Verifying the assumed dimensions for scoring velocity, using Manning's equation

$$Q = A * \frac{1.49}{n} * R^{\frac{2}{3}} * \sqrt{S}$$

Q = flow rate.
n = Manning's roughness coefficient.
A = flow area.
R = hydraulic radius.
S = slope of energy gradient.

Adopt Manning's 'n' = 0.014 (concrete with surface punning)

$$AR^{\frac{2}{3}} = n \times \frac{Q}{\sqrt{s}}$$

$$AR^{\frac{2}{3}} = 0.014 \times \frac{0.005}{\sqrt{0.01}} = 0.007$$

Assuming a rectangular cross section, the hydraulic radius (R) = (by / b + 2y)

$$AR^{\frac{2}{3}} = (by) \times \frac{by}{b+2y}^{\frac{2}{3}}$$

$$AR^{\frac{2}{3}} = \frac{by^{\frac{5}{3}}}{(b+2y)^{\frac{2}{3}}}$$

$$0.007 = \frac{by^{\frac{5}{3}}}{(b+2y)^{\frac{2}{3}}}$$

The width of the channel, denoted as b, is taken as 300 mm.

$$0.007 = \frac{0.3y^{\frac{5}{3}}}{(0.3+2y)^{\frac{2}{3}}}$$

Solving the equation for y, ⇒ y = 0.15 m or 150 mm = depth of flow (OK)

Wetted area

$$A = by$$

$$A = 0.045\,\text{m}^2$$

Wetted perimeter

$$A = \frac{by}{b + 2y}$$

$$A = 0.075\,\text{m}$$

V = velocity of flow in discharge channel

$$V = \frac{1}{0.014} * \frac{0.3 * 0.15}{0.3 + 2(0.15)}^{\frac{2}{3}} * \sqrt{0.01}$$

The designed velocity is greater than 0.8 m/s (scouring velocity); thus, the assumed dimensions are OK.

14.3.4.3 Design of Pathogen Elimination Chamber

The addition of 10–20 g/L lime to FS increased its pH above 10 and temperature to more than 30 °C, whereas 200–400 g/L bauxite residue could achieve the same result. The main objective of the pathogen elimination chamber is pathogen reduction, not solid–liquid separation. Based on the interviews with the desludging agency in Vadgaon Maval, the peak FS collection occurs around 10 a.m. and mainly in the summer season. Considering FSTP works for 6 days, 8 hours, assumed daily peak coefficient as 1.6 and assumed an inlet velocity as 0.5 m/h based on the experience obtained from Bhubaneswar and Puri FSTPs.

Peak flow

$$Q_p = Q \times \frac{C_p}{\text{time}}$$

Q_p = peak flow
C_p = daily peak flow
Q = Daily design flow = 46 m^3

$$Q_p = 46 \times \frac{1.6}{8}$$

Surface area required

$$S = \frac{Q_p}{V_c}$$

$$S = \frac{9.2}{0.5}$$

$$S = 18.4\,\text{m}^2$$

Two tanks of 20 m^2 each are to be constructed, with a depth of 1 m. A pump capable of discharging 25 m^3 of sludge onto the drying bed will be required.

14.3.4.4 Design of Unplanted Drying Filter Bed

Unplanted drying beds are widely utilized for separating solids from liquids, efficiently reducing moisture content in faecal sludge (FS) through filtration and drainage. The performance of unplanted drying beds is influenced by multiple parameters, including (i) sludge loading rate (SLR), (ii) total exposed filtration surface of the bed, (iii) deposited sludge depth, (iv) the features of influent FS, (v) the nature of the filtration media matrix, and (vi) ambient climatic variables such as temperature and humidity. All these parameters must be considered during the design of an unplanted drying bed to ensure its effectiveness. A sand layer is generally positioned above the gravel to enhance drainage and reduce clogging by preventing sludge from entering the gravel's pore spaces. During the dewatering process, biosolids and liquid fractions are produced.

The sand–gravel filter matrix facilitates surface-level solid retention and promotes percolation of the liquid fraction towards the base of the drying system. FS is applied to these beds and remains until it reaches a target moisture level. Once dried, the sludge is manually or mechanically removed for either disposal or resource recovery. The drying process operates on two main mechanisms: initially, leachate drains through the filter layers via percolation—a rapid process especially effective for FS with high free water content, occurring over hours to a few days. The second mechanism is evaporation, which removes moisture bound within the sludge matrix, and typically requires several days to weeks.

Heinss (1994) reported that drainage accounts for about 50–80% of total volume reduction, while evaporation contributes to an additional 20–50%. The drainage layer is structured with graded layers of gravel and sand. The selection and configuration of these materials are influenced by local availability. For instance, Cofie et al. (2006) utilized a base layer of 15 cm gravel with a diameter of 19 mm, topped by 10 cm of 10 mm gravel. An intermediate layer of smaller gravel is sometimes incorporated to prevent finer sand particles from migrating into the coarser gravel. It is also important to recognize that the sand layer will gradually deplete, as some sand adheres to the sludge and is removed during desludging, necessitating periodic replacement. Many trials of beds were made based on previous literature. However, none were suitable for FS collected from Vadgaon Maval. Thus, a new five-layered unplanted drying bed was created, as shown in Fig. 14.3 made of 16 cm of gravel (6-19 mm diameter) and 14 cm sand (0.42–3.35 mm diameter).

For a 1 m depth of the bed, 0.53 m of gravel (19–6 mm diameter) and 0.47 m sand (0.42–3.35 mm diameter) are required for an unplanted drying bed. The loading rate of FS ranged from 9.4 g TS/m^2/h to 20.5 g TS/m^2/h (82–180 kg TS/m^2/year). The efficiency of the unplanted drying bed is provided in Fig. 14.4.

Proper removal of FS requires it to be sufficiently dry for shovelling. However, field data on FS characteristics, SLR, and FS thickness applied on unplanted drying beds are often limited, with written records rarely available. Consequently, this study focuses on designing a laboratory-scale unplanted drying bed, testing FS collected from rural areas to determine suitable SLR and sludge thickness, and exploring various loading rates under Indian conditions.

FS loading rate of 82–180 kg TS/m^2/year was obtained in a laboratory-designed unplanted drying bed. A SLR ranging from 100 kg TS/m^2/year to 200 kg TS/m^2/year is considered suitable for tropical climatic conditions. In Vadgaon Maval, where the

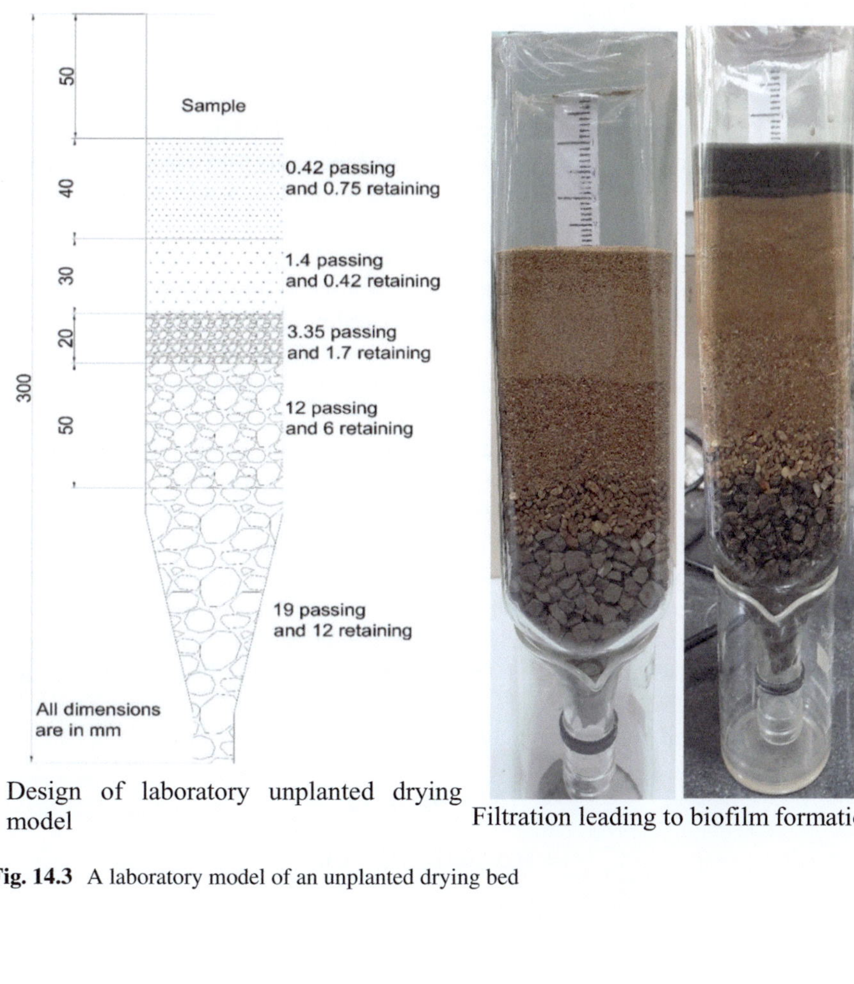

Design of laboratory unplanted drying model

Filtration leading to biofilm formation

Fig. 14.3 A laboratory model of an unplanted drying bed

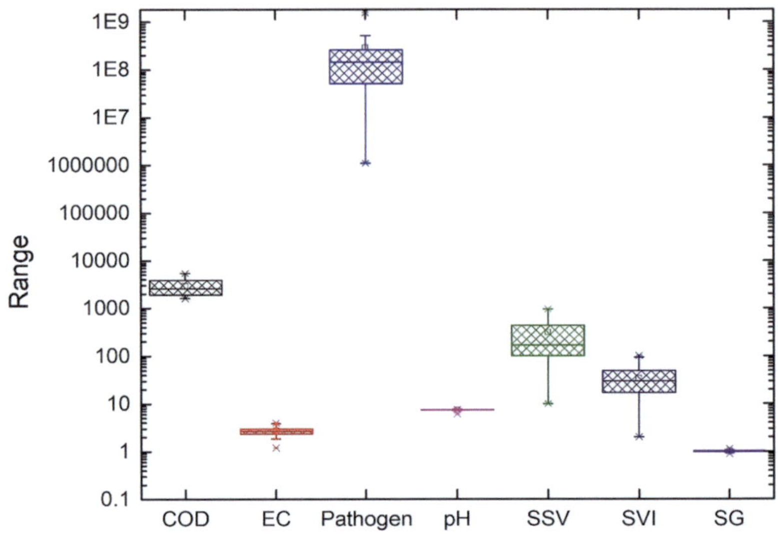

Fig. 14.4 Efficiency of unplanted drying bed of faecal sludge collected from Vadgaon Maval, Maharashtra, India

number of sunny days exceeds 200 annually, a higher loading rate can be implemented. Considering the values obtained in the laboratory and based on the literature, a loading rate of 150 kg TS/m^2/year has been adopted. A thickness of 20 cm is adopted. The drying bed sidewalls must be adequately elevated to accommodate varying FS loading volumes. For instance, a 20 cm sludge layer with 90% moisture content results in a significantly higher initial volume before drainage. When FS is directly discharged from trucks rather than settling tanks, wall heights should exceed the typical 20–30 cm design to prevent overflow. Given operations occur 6 days (excluding Sunday) per week across 52 weeks, structural capacity must align with projected hydraulic loadings.

Annual sludge loading (kg TS/year) is computed as the product of average total solid concentration in incoming FS (g TS/L), the volumetric inflow per discharge event (m^3/day), and the annual frequency of deliveries. Since the pathogen elimination chamber is before unplanted drying beds and if red mud or lime has been used for pathogen elimination, then two calculations occur: (i) for lime as pathogen elimination agent (where the flow and TS of the FS arriving at unplanted drying bed remains the same from the vacuum trucks) and (ii) for bauxite residue as a pathogen elimination agent (where flow and TS will be double as the FS from the vacuum trucks) (Fig. 14.5).

Lime as pathogen elimination agent

$$= 32.3^{*}\, 46^{*}\, 6^{*}\, 52$$

$$= 463{,}569\,\text{kg TS / year}$$

Fig. 14.5 Effluent and dried FS from unplanted drying bed of faecal sludge collected from Vadgaon Maval, Maharashtra, India

Taking the yearly FS load into account, a drying bed with a surface area of

$$A = 3090\,\text{m}^2$$

Bauxite residue as a pathogen elimination agent

$$= 32.3^{*}\ 92^{*}\ 6^{*}\ 52$$

$$= 927{,}140\,\text{kg TS / year}$$

Taking the yearly FS load into account, a drying bed with a surface area of

$$A = 6180\,\text{m}^2$$

Assuming a sludge depth of 0.1 m and a daily loading volume of 46–92 m^3, the required surface area ranges from 460 m^2/day to 920 m^2/day. With each bed accommodating 1000 m^2/day, a minimum of three to six beds is needed to manage a cumulative treatment area of 3090–6180 m^2. A depth of 1.5 m was proposed based on, the depth of the filter bed will be 1 m with 0.53 m of gravel (19–6 mm diameter) and 0.47 m sand (0.42 mm and 3.35 mm diameter), as shown in Fig. 14.3. The periodic sand replacement becomes necessary when organic accumulation leads to clogging and impaired drainage efficiency. A typical drying bed facility includes inlet–outlet structures, a leachate collection and drainage network, and an adjoining area allocated for further drying and temporary storage of dewatered faecal sludge. After pathogen elimination, FS is being pumped into the drying beds. Alternatively, faecal sludge (FS) may be discharged directly from vacuum trucks onto drying beds, provided pathogen reduction has been achieved at the source. Infrastructure configurations vary; some designs incorporate dual-bed systems fed via a single inlet and a flow divider, while others integrate inlet ramps for easier FS transfer. The application of a splash plate is essential to ensure uniform sludge distribution and to safeguard the integrity of the sand filter layer, which is otherwise vulnerable to erosion during initial loading. Drying beds are generally rectangular, in-soil excavations with impervious linings and a sloped base to facilitate leachate flow towards a drainage point or further treatment. For effective removal, the sludge must reach a sufficiently desiccated state to allow manual or mechanical collection. Access ramps are required to enable transport equipment, such as wheelbarrows, to enter the bed. Rainfall prior to complete dewatering may cause rehydration of the sludge, thereby delaying drying. However, once cracks in the FS reveal the underlying sand layer, rainwater can percolate without significantly impeding the process. To mitigate rewetting, the CPHEEO recommends the use of fibre-reinforced plastic (FRP) canopies over the beds.

14.4 Conclusion

The sustainable development of human society depends on resource recovery and environmental care. Onsite sanitation systems produce significant volumes of faecal sludge (FS), often laden with high pathogen loads, necessitating safe treatment and disposal. Effective FS management remains essential, given that a substantial global population relies on these decentralized sanitation solutions. The practice of the utilization of untreated FS as a fertilizer in agricultural lands has been a source of increased disease transmission in developing countries. Appropriate management and resource-oriented reuse of faecal sludge can enhance sanitation outcomes and support nutrient recovery in aquaculture and agriculture. The focus of this study was to understand the effectiveness of toilets built, onsite sanitation technologies, its faecal sludge management, and what are the improvements required for non−/peri-urban parts of India, especially in Maharashtra. This study analysed the physico-chemical characteristics, nutrients, and pathogenic content of an FS and provide an appropriate treatment design by understanding the nature of FS.

This study explores the application of unplanted drying beds for the secondary treatment of pre-processed FS with additives like bauxite residue or lime, with the objective of enabling its safe reuse as an organic fertilizer. These drying beds promote solid–liquid separation by allowing dewatering and drying through gravitational flow and filtration mechanisms. Thus, a new five-layered unplanted drying bed with 0.53 m of gravel (19–6 mm diameter) and 0.47 m of sand (0.42–3.35 mm diameter) was created for solid and liquid separation. The FS loading rate varied between 9.4 and 20.5 g total solids (TS)/m^2·h, corresponding to an annual rate of 82–180 kg TS/m^2·year. FSM system for Vadgaon Maval was designed considering the FS quantification and qualities obtained from analysis of FS samples collected from field. The FSTP design tailored to FS characteristics with a design period of 10 years to serve a population of 32,265. An analysis of inter- and intra-variation in FS from OSTs in Vadgaon Maval indicates that implementing a scheduled desludging interval of 2–2.5 years can eliminate the need for manual emptying, with an estimated FS accumulation rate of 233 L/person/year. The FSTP design of Vadgaon Maval consists of a bar screen of 1.5 m width and 1.5 m depth, with a two-number pathogen elimination chamber of 20 m^3 and a three- to six-layered unplanted drying bed of 1000 m^3.

References

Adlakha A, Pathak P, Kumar A, Pandey P (2022) Antecedents and consequences of adopting CLTS among tribal communities to become open defecation free: case study on Indian Swachh Bharat Abhiyan. Environ Sci Pollut Res Int

Atun, Roy Choudhury Neha, Singh Sumanth, Chinthala Jitesh, Lalwani Sri Kalyana Rama, J. Chandana, N. Sankar Ganesh, Palani Mohammad, Mehdizadeh V. Vinayaka, Ram Azam, Akhbari (2025) Environmental health hazards attributed to deteriorated indoor air quality caused by inferior construction practices Environmental Science: Atmospheres 5(9) 941–961 https://doi.org/10.1039/D5EA00027K

Chandana N, Rao B (2021a) Assessing inter and intra-variation in the characteristics of faecal sludge from Vadgaon Maval, Maharashtra: for better faecal sludge Management in India. J Environ Manag

Chandana N, Rao B (2021b) A critical review on sludge management from onsite sanitation systems: a knowledge to be revised in the current situation. Environ Res 203:111812

Chandana N, Rao B (2021c) Evaluating the physicochemical, nutrient, and pathogenic characteristics of Fecal sludge for fertilizer application: case from Vadgaon Maval, Maharashtra, India. J Environ Eng 147(3):04021003

Chandana N, Rao B (2021d) Status of sustainable sanitation chain in rural, semi-urban, and urban regions: a case study of Maharashtra, India. J Water Sanit Hyg Dev 11(1):112–125

Chaudhuri S, Mimi R (2016) Overview of rural water supply sector in West Bengal, India: challenges and concerns. Int J Innov Res Sci Eng Technol 5(6)

Cofie O, ASM Strauss, H Esseku, A Montangero E Awuah and KD (2006) "Solid-liquid separation of faecal sludge using drying beds in Ghana: implications for nutrient recycling in urban agriculture." Water Research.1, 75–82. https://doi.org/10.1016/j.watres.2005.10.023

Dasgupta S, Agarwal N, Mukherjee A (2021) Moving up the onsite sanitation ladder in urban India through better systems and standards. J Environ Manag 280:111656

Garn JV, Sclar GD, Freeman MC, Penakalapati G, Alexander KT, Brooks P, Rehfuess EA, Boisson S, Medlicott KO, Clasen TF (2017) The impact of sanitation interventions on latrine coverage and latrine use: a systematic review and meta-analysis. Int J Hyg Environ Health 220(2):329–340

Gautam M, Wankhade K, Sarangan G, Sudhakar S (2021) Framework for addressing occupational safety of desludging operators: a study in two Indian cities. J Environ Manag 289:112243

Halcrow G, Yetsho T, Nguyen N, Tshering G (2014) Developing behaviour change communication for improving faecal sludge management in Bhutan. J Water Sanit Hyg Dev 4(1):164–170

Heinss U, SA Larmie and M Strauss (1994) "Characteristics of Faecal Sludges and their Solids-Liquid Separation." Management of Sludges from On-Site Sanitation, S.A. Larmie, EAWAG/ SANDEC, Duebendorf, Switzerland

Manga M, Camargo-Valero MA, Anthonj C, Evans BE (2021) Fate of faecal pathogen indicators during faecal sludge composting with different bulking agents in tropical climate. Int J Hyg Environ Health 232:113670

Manga M, Kolsky P, Rosenboom JW, Ramalingam S, Sriramajayam L, Bartram J, Stewart J (2022) Public health performance of sanitation technologies in Tamil Nadu, India: initial perspectives based on E. Coli release. Int J Hyg Environ Health 243:113987

O'reilly K, Dhanju R, Goel A (2017) Exploring "the remote" and "the rural": open defecation and latrine use in Uttarakhand, India. World Dev 93:193–205

Patel R, Gupta A, Chauhan S, Bansod DW (2019) Effects of sanitation practices on adverse pregnancy outcomes in India: a conducive finding from recent Indian demographic health survey. BMC Pregnancy Childbirth 19(1):378

Routray P, Schmidt WP, Boisson S, Clasen T, Jenkins MW (2015) Socio-cultural and behavioural factors constraining latrine adoption in rural coastal Odisha: an exploratory qualitative study. BMC Public Health 15:880

Strande L, Mbéguéré M, Regulinski J, Stupin L, Kang J, Dione H, Niwagaba C, Nakato T, Muspratt AM (2014) Fuel potential of faecal sludge: calorific value results from Uganda, Ghana and Senegal. J Water Sanit Hyg Dev 4(2):223–230

Tilmans S, Russel K, Sklar R, Page L, Kramer S, Davis J (2015) Container-based sanitation: assessing costs and effectiveness of excreta management in Cap Haitien, Haiti. Environ Urban 27(1):89–104

Wolf J, Pruss-Ustun A, Cumming O, Bartram J, Bonjour S, Cairncross S, Clasen T, Colford JM Jr, Curtis V, De France J, Fewtrell L, Freeman MC, Gordon B, Hunter PR, Jeandron A, Johnston RB, Mausezahl D, Mathers C, Neira M, Higgins JP (2014) Assessing the impact of drinking water and sanitation on diarrhoeal disease in low- and middle-income settings: systematic review and meta-regression. Trop Med Int Health 19(8):928–942

Chapter 15
Application of Phytoremediation Technology as a Cost-Effective Solution for Treatment of Municipal and Industrial Wastewater: A Few Case Studies

Madhurina Majumder, Debashis Chatterjee, and Amitava Gangopadhyay

Abstract Municipal wastewater contains organic contaminants that comprise carbon and nutrients, whereas industrial effluent usually contains metals and other toxic elements besides organic contaminants, depending on the type of industry. The present chapter deals with removal of organic components by an aquatic plant *Pistia stratiotes* through different inherent processes. Wastewaters, both synthetic and real-life, were chosen for laboratory and field scale phytoremediation study. In the present study, Municipal Wastewater generated from a municipal area of South Kolkata and industrial wastewaters generated from Rice Mill and Pharmaceutical industrie. The Industrial wastes are often low with regard to toxic elements, were subjected to phytoremediation, facilitating disposal and utilization of harvested exhausted crops as fodder in environment-friendly manner. During experiments (both lab and field) plant sustains and grows in different concentrations and advocates sustainable use patterns. The study reveals that the utilization perspective of *Pistia stratiotes* as fodder is potential both in terms of protein and fibres.

Keywords Phytoremediation · Municipal Wastewater · Industrial Wastewater (Rice Mill & Pharmaceutical) · Fodder

M. Majumder
Public Health Engineering Department, Government of West Bengal, Kolkata, India

D. Chatterjee (✉)
Department of Chemistry, University of Kalyani, Nadia, India

A. Gangopadhyay (✉)
Department of Civil Engineering, Jadavpur University, Kolkata, India

S. S. Kumar, V. Kumar (eds.), *Nature-Based Solutions for Decentralised Liquid Waste Management*, Environmental Challenges and Solutions,
https://doi.org/10.1007/978-3-032-08362-3_15

15.1 Introduction

Wastewaters, both domestic and industrial, require treatment to conform water quality prescribed by regulatory authorities prior to disposal to the environment. Scientists have been conducting in-depth researches on cost-effective stabilization pond treatment system coupled with phytoremediation. Aquatic plants are used to clean up contaminated water by degradation of organic pollutants or by stabilization of metal contaminants.

Phytoremediation has been viewed as a promising green technology using plants to degrade, assimilate, metabolize, or detoxify metals, hydrocarbons, pesticides, and chlorinated solvents (Sridhar et al. 2002; Majumder 2016). Phytoaccumulation, phytoextraction, phytostabilization, phytotransformation, phytovolatilization, and rhizodegradation were identified as the salient mechanisms involved in the process of phytoremediation with definite role of enzymes in transforming organic contaminants in plants. The phytoremediation prescriptions are usually wastewater-specific viewing its potential for providing the most cost-effective and resource-conservative approach for remediating wastewaters contaminated with a variety of hazardous chemicals.

Mechanisms related to phytoremediation of organic contaminants and metals are furnished below (Table 15.1):

Remediation of most of the wastewater streams is accomplished by converting the organic components to simple end-products through biodegradation. These processes are commonly known as phytodegradation and rhizodegradation. Some organic compounds, either in the modified and/or unaltered form, are lost to the environment along with water molecules present in the stem and leaves of the plant through evapotranspiration. This process is known as phytovolatilization. The processes through which the contaminants are extracted from the water and remain assimilated in the harvested biomass are termed as phytoextraction and rhizofiltration. In brief, phytoremediation, the green technology, encompasses uptake, accumulation, and transformation of the organic pollutants and other toxic components by plant either by rhizospheric microbial degradation or by extraction and filtration. Phytoremediation process involves the introduction of plants into ecosystem thereby assimilating the contaminants in their roots and associated parts.

The relative advantages and disadvantages of phytoremediation are narrated below:

Advantages It is more economically viable.

- Disposal sites are not needed.
- It is more aesthetically pleasing.

Table 15.1 Mechanisms involved in phytoremediation of organic contaminants and metals

Type of contaminants	Mechanisms involved		
Organic contaminants	Phytodegradation	Rhizodegradation	Phytovolatilization
Metals	Phytoextraction	Rhizofiltration	Phytostabilization

- It avoids risk of spreading the contamination.
- It has the potential to treat various types of pollutants in polluted wastewaters.

Disadvantages It is dependent on climate, geology, altitude, and congenial temperature conditions required by the plant.

- Large-scale operations are required.
- Success is dependent on the tolerance of the plant to the pollutant.
- Considerable time taken to remediate polluted wastewaters.
- **Removal Mechanism** - There are six major biological reactions involved in the plant system of removal of pollutants from wastewaters. They may be mentioned as the following:
  - Photosynthesis.
  - Respiration.
  - Fermentation.
  - Nitrification.
  - Denitrification.
  - Microbial phosphorus removal.

Various researches are found to have unanimously highlighted the efficacies of aquatic plants and suggested phytoremediation as an effective means to remove carbon as well as nutrients from various types of wastewaters. The plants selected during the earlier phases of experiments are *Pistia stratiotes* L. (Wibowo et al. 2023; Akowanou et al. 2023; Nahar et al. 2021; Qin et al. 2010; Zimmels et al. 2006; Sooknah et al. 2004), *Canna indica* L. (Akansha et al. 2020; Cui et al. 2010), *Lemna* L. (Akowanou et al. 2023; Ceschin et al. 2019), *Eichhornia crassipes* (Wibowo et al. 2023; Akowanou et al. 2023; Cui et al. 2010; Sooknah et al. 2004), and *Hydrocotyle umbellata* (Sooknah et al. 2004). The suggested plant species are remarkably potential and effective to clean the polluted wastewaters to a great extent. Many of the researchers have highlighted the significant performance of *Pistia stratiotes* and advocated the removal effectiveness for utilization in the phytoremediation process.

15.2 Present Experimental Studies

Based on the findings of the researches so far undertaken, it has become necessary to concentrate on the following points:

1. Selection of a suitable substrate-specific hyperaccumulating plant, viz., *Pistia stratiotes*, for the present study.
2. Experiments designs to grow and acclimatize the plant species in the laboratory hydroponic culture medium in room temperature (upto 30 °C). This process of acclimatization has been practiced during the present phytoremediation study for municipal as well as industrial wastewaters.

3. To ascertain potential of the selected aquatic macrophytes in removing carbon and nutrients (NPK) from the wastewater, subjected as phytoremediation. The potential of *Pistia stratiotes* in removing carbon, nitrogen, and phosphorus has been well established through numerous laboratory-scale studies.
4. Experiments have been performed judiciously to study the degree of withstand of wastewater toxicity during phytoremediation. During the experiment, the growth of *Pistia stratiotes* has not been hindered within the wastewaters under study. The study proves that the characteristics of the wastewater is non-toxic.
5. The strength of Industrialwastewater is diluted with domestec wastewater to achieve desired concentration levels where macrophytes can remove contaminants effectively. The municipal wastewater does not require nutrient supplementation, but the rice mill and pharmaceutical wastewaters deserve mixing of municipal wastewater to provide requisite quantity of nutrients essential for healthy growth of the plants.

15.2.1 Work Schedule

The entire work of phytoremediation with *Pistia stratiotes* (water lettuce) was divided into four different phases:

- In the first phase of the study, the phytoremediation potential of the plant species was functionalized with synthetic solution of carbon (glucose-glutamic acid) along with nutrients provided by Hoagland nutrient solution.
- The second phase of the study was focused on the treatment of municipal wastewater by phytoremediation.
- The third phase of experimental study was intended to improve the quality of primary or secondarily treated Rice mill wastewater.
- The fourth phase of experimental study was designed to improve the quality of secondarily treated pharmaceutical wastewater.

15.2.1.1 Phytoremediation Study with Synthetic Solutions Using Plant, *Pistia* sp. Phase I)

Experiments were conducted with aquatic plant *Pistia* sp. exposed to synthetically prepared sample (glucose and glutamic acid solution, 150 mg/L each) containing nutrient solution (Hoagland). The initial characteristics of the test sample were analyzed as the following: COD, 252 mg/L; nitrogen, 95.2 mg/L as N; phosphorus, 39.26 mg/L as P; and potassium, 126.75 mg/L as K. Experimental observations were carried out for seven consecutive days, and at the end of the seventh day, the removal efficiency of COD, N, and P by the aquatic plant species were found to be about 82%, 66%, and 60%, respectively. The potassium (K) uptake (3.72%) was observed to be relatively insignificant indicating that *Pistia* hardly takes significant amount of potassium as macronutrients unlike other terrestrial plants.

15.2.1.2 Phytoremediation Study with Municipal Wastewater Using Plant, *Pistia* sp. Phase II

The initial characteristics of the test sample were analyzed as the following: COD, 114 mg/L; Nitrogen, 13.91 mg/L as NH_4^+; Nitrogen, 3.2 mg/L as NO_3^-; and Phosphorus, 2.93 mg/L as PO_4^{3-}. The municipal wastewater sample was subjected to phytoremediation with *Pistia* sp., and after 15 days, the percent removals are found to be 81, 95, 56, and 90 for COD, NH_4^+-N, NO_3^--N and PO_4^{3-}-P, respectively.

15.2.1.3 Phytoremediation Study with Rice Mill Wastewater Phase III

Rice mill wastewater samples were collected from two sources time to time and analyzed for carbon, nutrient, and metal contents. The average values of some essential parameters are furnished below for both the rice mills where Rice mill 1 is located in the central part of he state (Bardhaman) and other one Rice mill 2 is situated near river Ganga (Budge Budge) (Table 15.2).

That both the wastewaters were suitable for phytoremediation from pH, carbon, nutrient, and metal concentration points of view. Moreover the toxicity levels are well below to maintain potential plant growth. The preliminary studies were conducted through visual observation on plant sustenance in the wastewater environment with different dilution mode. It was found that there is some higher limit of organic content and lower limit of nutrients to allow the plants to grow in a befitting manner.

15.2.1.4 Phytoremediation of Raw Wastewater of Rice Mill 1 by Aquatic Plant *Pistia* sp.

Similar observations were made with respect to raw wastewater of Rice mill 1 when subjected to phytoremediation with aquatic plant *Pistia* sp. The initial characteristics of the raw test sample collected from Rice mill 1 were obtained as follows: temp., 30 °C; pH, 7.8; COD, 685 mg/L; NH_4^+-N, 19.1 mg/L; and NO_3^-N, 3.52 mg/L. This sample was subjected to phytoremediation with *Pistia* sp., and after 3 and 6 days, the concentrations of COD, NH_4^+-N, and NO_3^-N were measured. The percent removals of COD were observed as 31 and 68 after 3 and 6 days, respectively, whereas for NH_4^+-N, the corresponding values were 82 and 90, and for NO_3^-N, the values were 31 and 46.

Table 15.2 Characteristics of rice mill wastewater

Parameters	Rice mill 1 (Bardhaman)	Rice mill 2 (Budge Budge)
pH	6.17	6.84
BOD	649	465
COD	949	539
TKN as N	5.8	25.3
NH_4^+ -N as NH_4^+	<0.1	<0.1
NO_3^--N as NO_3	4.3	0.47
NO_2^- -N as NO_2^-	<0.02	<0.02
Phosphorus as PO_4^{3-}	4.25	7.37
Cd	<0.01	<0.01
Cr	0.06	<0.05
Cu	<0.05	0.054
Fe	0.57	0.55
Mn	<0.05	<0.05
Ni	<0.05	<0.05
Pb	<0.05	<0.05
Zn	<0.05	0.084

All values are in mg/L except pH

15.2.1.5 Phytoremediation of Raw Wastewater of Rice Mill 2 by Aquatic Plant *Pistia* sp.

The initial concentrations of the raw test sample from rice mill 2 are analyzed as the following: temp., 25 °C; pH, 6.3; COD, 589 mg/L; TKN, 15.57 mg/L; and PO_4^{3-}-P, 4.7 mg/L. This sample has been subjected to phytoremediation with *Pistia* sp., and after 3, 8, and 12 days, the concentrations of COD, TKN, and PO_4^{3-}-P were measured. The percent removals of COD were observed as 46, 66, and 77 after 3, 8, and 12 days, respectively, whereas for TKN, the corresponding values are 20, 65, and 91, and for PO_4^{3-}-P, the values are 58, 83, and 99, respectively.

15.2.1.6 Phytoremediation of Wastewater of Rice Mill 2 Mixed with Domestic Wastewater by Aquatic Plant *Pistia* sp.

The initial concentrations of the raw test sample collected from rice mill 2 mixed with domestic wastewater were analyzed as the following: temp., 19 °C; pH, 8.0; COD, 490 mg/L; NH_4^+-N, 12.7 mg/L; NO_3^-N, 8.15 mg/L; and PO_4^{3-}-P, 22.84 mg/L. This sample was subjected to phytoremediation with *Pistia* sp., and after 3, 8, and 14 days, the concentration of COD, NH_4^+-N, NO_3^-N, and PO_4^{3-}-P were measured. The percent removals of COD were observed as 54, 69, and 89 after 3, 8, and 14 days, respectively, whereas for NH_4^+-N, the corresponding values were 93, 98, and 99; for NO_3^-N, the values were 25, 70, and 92; and for PO_4^{3-}-P, the values were 48, 66, and 86, respectively.

15.2.1.7 Phase IV of the Study: Phytoremediation Study with Pharmaceutical Wastewater

Characterization of Raw Pharmaceutical and Municipal Wastewater

Pharmaceutical wastewater samples were collected from a pharmaceutical industry situated around Kolkata and analyzed for carbon, nutrient, and metal contents. Simultaneously, municipal wastewater from a township was also collected and analyzed before mixing with pharmaceutical wastewater for phytoremediation study.

The results of characterization for pharmaceutical and municipal wastewater are depicted in Tables 15.3 and 15.4, respectively.

15.2.1.8 Pretreatment of Pharmaceutical Wastewater

Chemical Treatment

The sample of pharmaceutical wastewater when treated with CaO showed marked decrease in COD value up to the dosage of 2.0 g per 100 mL of wastewater, and after that, further increase in CaO showed virtually no decrease in COD value. Thus, the dosage of CaO in future courses of experiments was restricted to 2.0 g per 100 mL of wastewater. Percent reduction of COD during chemical treatment with 2.0 g CaO per 100 mL of pharmaceutical wastewater was obtained as 19.35. This chemical treatment using CaO increased the pH value of the wastewater making it suitable for biological treatment.

Biological Treatment

The initial COD value of the chemically treated wastewater prior to receiving anaerobic biological treatment was 63,296 mg/L, which was mixed with sludge taken from another lab-scale anaerobic treatment unit and requisite nutrient solution. Alkaline condition was maintained to facilitate the growth of anaerobic bacteria. Initially, during acclimatization of bacteria, COD reduction was observed low; later COD was reduced up to 12,800 mg/L after 29 days.

Phytoremediation of Pharmaceutical Wastewater (Pretreated Chemically and Biologically Under Anaerobic Condition) Mixed with Municipal Wastewater by Aquatic Plant *Pistia* sp.

Pretreated pharmaceutical wastewater was mixed with municipal wastewater in the ratio 1:39, i.e., pharmaceutical wastewater was diluted 40 times with municipal wastewater.

Table 15.3 Characterization of pharmaceutical wastewater

Parameters	Value
pH	3.5
BOD	39,593
COD	91,264
Total solids	364.54
Volatile solids	179.50
Fixed solids	185.04
Cd	0.034
Cr	0.081
Cu	0.356
Fe	0.057
Mn	0.328
Ni	0.021
Pb	0.883
Zn	0.468

All values are in mg/L except pH

Table 15.4 Characterization of municipal wastewater

Parameters	Value
pH	6.9
BOD	167
COD	249
TDS	2224
TSS	213
TKN as N	30
NH_4^+ -N as NH_4^+	20
NO_3^--N as NO_3	3.4
NO_2^- -N as NO_2^-	0.5
Phosphorus as PO_4^{3-}	6.0
Conductivity	3825

All values are in mg/L except pH and conductivity. Unit of conductivity is microsiemen/cm

Initial values of different parameters and the percent reduction of few salient parameters after phytoremediation of the test sample at 35 °C for a contact period of 15 days were obtained as furnished in Table 15.5.

15.3 Discussion

The percent removals of different constituents for all type of wastewaters by *Pistia stratiotes* L. (water lettuce) found in the present study were quite high and comparable with those observed by other researchers. Wibowo et al. (2023) found that the

Table 15.5 Results of phytoremediation of pharmaceutical wastewater (pretreated chemically and biologically under anaerobic condition) mixed with municipal wastewater

Parameters	Value	Percent removal
pH	8.8	–
COD	598	78
NH_4^+ –N as NH_4^+	14.6	96
NO_3^--N as NO_3^-)	4.43	61
Phosphorus as PO_4^{3-}	6.75	66

All values are in mg/L except pH

same plant effectively removed organic pollutants and the percent removals after 5 weeks are as follows: BOD, ~98%; COD, ~99%; phosphate, ~73%; ammonia, ~70%; nitrite, ~48%; and nitrate, ~91%.

Rice mill and pharmaceutical wastewaters considered in the present study being similar in nature with that of dairy industry wastewater, undertaken for treatment effectively through combined electrocoagulation and phytoremediation Phytoremediation process by Akansha et al. (2020), adoption of the same combined treatment is viewed as good possibility of treatment of wastewaters considered in the present study.

In the present study, the removal of nutrients from the wastewaters by *Pistia stratiotes* L. (water lettuce) was found identical with that of observation by Akowanou et al. (2023), who conducted an experiment in a multiple floating macrophyte treatment system with *Pistia stratiotes* L. (water lettuce) being present as an effective one for treatment of domestic wastewater. In their study, the overall percent removals of TSS, COD, BOD_5, TKN, and PO_4^{3-} were 92, 98, 98, 87, and 80, respectively. Their study results suggest that treatment of rice mill as well as pharmaceutical wastewaters through multiple floating macrophyte treatment system will be more effective than single macrophyte treatment system adopted in the present study. Regarding Rice mill wastewater, both aquatic plants (*Pistia stratiotes* and *Lemna sp.*) have been employed to treat Rce mill wastewater (both 1 & 2) and found that the removal efficacy is varying as well as plant survival. *Pistia stratiotes* was behaving better to remove COD and aAmmoniacal Nitrogen. Where as *Lemna sp.* was effectively removing Ammoniacal Nitrogen and nutrients. The investigation indicates that Phytoremediation is effective and could be manipulated by selecting the species to achieve the quality of treated water as per standards (Bidisha et al. 2015; Majumder et al. 2014).

The parameters analyzed in the present study were quite different from those presented in the study by Nahar et al. (2021), who used the aquatic macrophyte *Pistia stratiotes* L. (water lettuce) to improve the quality of eutrophic lake water at a lab scale. In their study, the parameters like pH, DO, EC, TDS, turbidity, and NaCl were tested on a periodic basis, and substantial improvement was observed after 1 week of treatment by plant. The decrease in turbidity, TDS, EC, and NaCl with an increase in pH and DO levels showed a significant potential by the aquatic macrophyte for improving water quality parameters by removing pollutants from eutrophic lake waterbodies. Moreover, the aquatic plant survival rate was 100%, i.e., there was no toxic effect of the pollutants on the biomass. The present study results were no different from their observation.

The present study results showed similar trend as was observed by Ceschin et al. (2019). The researchers used *Lemna minor* L. (duckweeds) and concluded that this plant too could tolerate and capture diverse pollutants and large quantities of nutrients showing their suitability for bioremediation of wastewaters. The parameters like pH, temperature, oxygen, nitrates, phosphates, sulfates, and pathogenic bacteria (*Escherichia coli*) were measured once per season over a year, and removals of nutrients and other constituents were found. The nutrients were accumulated in the intracellular spaces of the plant, and year-round vegetative growth reached very high rates producing new biomass. *Lemna minor* L. (duckweeds) may be viewed as an alternative macrophyte for future experiments of phytoremediation of rice mill and pharmaceutical wastewaters.

The removal efficiencies of nitrogen (N) and phosphorus (P) from wastewaters found in the present study can be compared with those observed by Cui et al. (2010). The researchers used *Canna indica* L. (canna flower or canna lily) instead of *Pistia stratiotes* L. (water lettuce). They used simulated vertical-flow constructed wetlands. Their conclusions were that lower hydraulic loading rate or longer hydraulic retention time (HRT) resulted in a higher removal of TP, NH_4^+-N, and TN because of more contacts and interactions among nutrients, substrates, and roots under the longer HRT. Similar findings were noted regarding the duration of the test with regard to different wastewater samples in the present study also.

Similar to the present study, Qin et al. (2010) used *Pistia stratiotes* L. (water lettuce) in a constructed water detention system to remove nitrogen and phosphorus from storm water and assessed the nutrient removal efficacy and potential of the plant. The parameters like pH, electrical conductivity (EC), suspended solids, turbidity, total N and P, NO_3–N, NH_4 –N, and soluble P were tested. They found that the values of turbidity, suspended solids, and nutrient content decreased through the treatment with the said plant. They also tested with many other aquatic plants to remove nutrients, but owing to the rapid growth and high biomass yield potential of water lettuce, it proved superior to most other plants in nutrient removal efficiency. Similar conclusions were drawn in the present study, where *Pistia stratiotes* L. (water lettuce) was used to remove nutrients from different types of wastewater samples, both municipal and industrial.

The present study results were again supported by the observations of the study conducted by Zimmels et al. (2006), who used both *Pistia stratiotes* L. (water lettuce) and *Eichhornia crassipes* (water hyacinth) on laboratory- and pilot-scale experimental basis. Very low levels of BOD, COD, TSS, and turbidity were maintained by the plants all throughout the experiment. COD value decreased by 78% in 4 days of treatment with an arrangement of circulation of the effluent in the pilot plant study, whereas the same percent removal was observed in 7 days where no recirculation was adopted.

Similar to the present study with pharmaceutical wastewater, where pre-chemical and anaerobic biological treatments were adopted with subsequent dilution of the treated wastewater before being subjected to phytoremediation, Sooknah et al.

(2004) tried with anaerobically digested flushed dairy manure wastewater for remediation. The researchers used three floating aquatic macrophytes, *Pistia stratiotes* L. (water lettuce), *Eichhornia crassipes* (water hyacinth), *Hydrocotyle umbellata* (pennywort), to improve the water quality of the wastewater. All the three plants failed to grow in the undiluted wastewater with chemical oxygen demand (COD) value of 2010 mg/L. All the three plants grew successfully in 1:1 diluted wastewater. In the case of water hyacinth, percent reductions in total kjeldahl nitrogen (TKN), ammonia nitrogen, total phosphorus, and soluble phosphorus were 91.7, 99.6, 98.5, and 96.5, respectively, in a 31-day batch culture. The next best performance was exhibited by a polyculture of the three plants in 1:1 diluted abovementioned wastewater. Polyculture of three floating aquatic macrophytes may be adopted in future experiments to render better results in phytoremediation of rice mill as well as pharmaceutical wastewaters.

15.4 Conclusions

Phytoremediation is a living system of wastewater rejuvenation. The conditions likely to develop in the living system (plant) due to its interaction with the surroundings (the constituents present in the wastewater and the microbes both associated and non-associated) are hardly predicted beforehand. The experiments highlight that the removal efficiency of organics and nutrients are varying and depending on both plant uptake capacity and wastewater charactics.

Positive results obtained through the studies on the phytoremediation of synthetic solutions have given way to investigate the possibility of the same treatment system to be adopted in case of municipal wastewater as well as industrial wastewaters.

COD value of Municipal wastewater is somewhat less and it contains nutrients of its own. Small primary treatments like screening and plain sedimentation are generally provided to remove larger particles before subjecting the wastewater directly to phytoremediation. In most of the cases, nutrient supplementation is not required because it is present in the wastewater.

The entire process of paddy treatment is batch and not a continuous one. The discharge of effluents occurs mostly twice a day. In most of the cases, the timing of sampling does not match with the time when fresh discharge is made. The wide variation in the analysis results obtained in different trials. Higher BOD to COD ratios of the Rice mill wastewaters show the presence of a good amount of biodegradable organic matters, and the wastewaters contain substantial amount of nutrients, confirming that the wastewaters are amenable to both biodegradation and phytoremediation. All the heavy metal contents of the wastewaters show low value, and plants grown in significantly well in the wastewaters. Hence, the plants grown in the Rice mill effluent can be safely used as fodder crops without any adverse effects. Based on nitrogen values, plant takes up nitrogen more easily from ammonia than from nitrate.

The pharmaceutical wastewater being highly rich in organics has been undergone a series of treatments like chemical and anaerobic biological, before being subjected to phytoremediation as tertiary treatment. As this wastewater does not contains nutrients, fortification of nutrients to the wastewater is essential to make the same suitable for phytoremediation. Dilution of chemically and biologically treated pharmaceutical wastewater with municipal wastewater has been served as dual purposes. Firstly, nutrients has been added through municipal wastewater, which contains nitrogen and phosphorus, and secondly, dilution has been deminished the organic matter concentration to a great extent.

It is finally concluded that phytoremediation is suitable for all sorts of biodegradable wastewaters with no preliminary treatment or after some chemical and biological treatment depending on the pH value and extent of organic contents. The wastewaters should preferably be free from toxic elements; otherwise, it would be difficult for disposal of plant biomass contaminated with toxicity of those elements.

References

Akansha J et al (2020) Treatment of dairy industry wastewater by combined aerated electrocoagulation and phytoremediation process. Chemosphere 253:126652

Akowanou AVO et al (2023) The combined effect of three floating macrophytes in domestic wastewater treatment. Sci Afr 20:e01630

Ceschin S et al (2019) Phytoremediation performance of *Lemna* communities in a constructed wetland system for wastewater treatment. Environ Exp Bot 162:67–71

Cui L et al (2010) Removal of nutrients from wastewater with *Canna indica* L. under different vertical-flow constructed wetland conditions. Ecol Eng 36(8):1083–1988

Majumder M (2016) Ph.D Thesis, Effective Use of Phytoremediation as a Viable Alternaive in Municipal and Industrial Wastewater Treatment, Jadavpur University.: Kolkata 700032., India.

Majumder M et al (2014) Study on the phytoremediation potential of pharmaceutical wastewater spiked with nutrients through municipal wastewater - A case study in indian context. Int res j Environ Sci

Mukherjee B et al (2015) Phytoremediation of perboiled rice mill wastewater using water lettuce. INT J Phytoremediation

Nahar K et al (2021) Phytoremediation to improve eutrophic ecosystem by the floating aquatic macrophyte, water lettuce (*Pistia stratiotes* L.) at lab scale. Egypt J Aquat Res 47(2):231–237

Qin L et al (2010) Phytoremediation to remove nutrients and improve eutrophic storm waters using water lettuce (*Pistia stratiotes* L). Environ Sci Pollut Res 17:84–96. https://doi.org/10.1007/s11356-008-0094-0R

Sooknah RD et al (2004) Nutrient removal by floating aquatic macrophytes cultured in anaerobically digested flushed dairy manure wastewater. Ecol Eng 22:27–42

Sridhar S et al (2002) Phytoremediation: an ecological solution to organic chemical contamination. Ecol Eng 18:647–658

Wibowo YG et al (2023) Phytoremediation of several wastewater sources using *Pistia stratiotes* and *Eichhornia crassipes* in Indonesia. Nanotechnol Monit Manag 20:100781

Zimmels Y et al (2006) Application of *Eichhornia crassipes* and *Pistia stratiotes* for treatment of urban sewage in Israel. J Environ Manag 81:420–428

Chapter 16
Environmentally Sound Wastewater Treatment Options in Hebbal-Nagavara Valley, Bengaluru

T. V. Ramachandra, K. S. Asulabha, and V. Sincy

Abstract Lake ecosystems provide valuable services to society, including the provision of fish, fodder, groundwater recharge, recreation, moderating the local climate, etc. However, the sustained inflow of untreated or partially treated wastewater into lakes has altered the chemical integrity, affecting biological productivity. The current study investigates the water quality of lakes in the Hebbal-Nagavara valley catchment of Bengaluru, Karnataka State, India, and these lakes were chosen based on the existing wastewater treatment facilities. Water quality assessment through the water quality index (WQI) reveals that lakes such as Jakkur, Devasandra, and Thirumenahalli have good water quality, highlighting that the biological processes (using microalgae, macrophytes, and fish) for wastewater treatment represent a promising resource recovery and clean water production approach. The wastewater treatment with the recovery of value-added resources would contribute to the circular economy while minimizing contamination, paving the way for sustainable practices, and leading to a clean environment with socioeconomic prosperity.

Keywords Lakes · Water pollution · Phytochemical parameters · WQI · Wastewater treatment · Constructed wetlands · Circular economy

T. V. Ramachandra (✉)
Energy & Wetlands Research Group [CES TE15], Centre for Ecological Sciences, Indian Institute of Science, Bangalore, India

Centre for Sustainable Technologies (Astra), Indian Institute of Science, Bangalore, India

Centre for Infrastructure, Sustainable Transportation and Urban Planning (CiSTUP), Indian Institute of Science, Bangalore, India
e-mail: tvr@iisc.ac.in; energy.ces@iisc.ac.in

K. S. Asulabha · V. Sincy
Energy & Wetlands Research Group [CES TE15], Centre for Ecological Sciences, Indian Institute of Science, Bangalore, India

S. S. Kumar, V. Kumar (eds.), *Nature-Based Solutions for Decentralised Liquid Waste Management*, Environmental Challenges and Solutions,
https://doi.org/10.1007/978-3-032-08362-3_16

Article Highlights

- Assessed water quality of select lakes in the Hebbal-Nagavara valley of Bengaluru, Karnataka.
- Optimal wastewater treatment options.
- Constructed wetlands with algal ponds would help in efficient nutrient removal.
- Wastewater treatment at decentralized levels with resource recovery contributes to the circular economy and mitigates pollution of local waterbodies.

16.1 Introduction

Water is vital for sustaining life on Earth, and clean water is essential for ensuring a healthy population. Wetlands are productive ecosystems that provide a variety of ecosystem services, such as the provision of food and water, microclimate moderation, groundwater recharge, water purification, nutrient cycling, flood reduction, erosion regulation, carbon sequestration, air quality regulation, provision of habitat for flora and fauna, disease regulation, opportunities for education, aesthetics, and recreation (de Groot et al. 2012; Ramachandra et al. 2022a). The undulating terrain of the Bengaluru topography has facilitated numerous human-made lakes to meet societal demands. However, rapid urbanization, enhanced infrastructure, and industrial development have contributed to the contamination of lakes due to the discharge of untreated or partially treated wastewater (Ramachandra et al. 2020a). Lakes receive both point sources (pollution discharges from an identifiable source to the environment, like wastewater plants) and non-point sources (releases from extensive unidentifiable sources, like surface runoff) of pollution. According to the United Nations, about 80% of global wastewater is discharged into the environment without adequate treatment, causing severe environmental damage and increasing human waterborne diseases. Industries are responsible for dumping heavy metals, solvents, toxic sludge, and other wastes into aquatic ecosystems (WWAP 2017).

In Indian cities, domestic sewage and industrial waste are the leading causes of the deterioration of the water quality of lakes and groundwater aquifers (Ramachandra et al. 2020b; Herojeet et al. 2017; Jacintha et al. 2017). Domestic wastewater is generally of two types, namely, black water (wastes from toilets or urinals) and gray water (wastewater from sinks, showers, baths, dishwashers, and washing machines) (Ghaitidak and Yadav 2013). Major water pollutants from agriculture are nutrients (nitrogen and phosphorus), pesticides (herbicides, insecticides, fungicides, and bactericides), salts (sodium, chloride, potassium, magnesium, sulfate, calcium, and bicarbonate), sediments, organic carbon, pathogens, metals (selenium, lead, copper, mercury, arsenic, manganese, etc.), and drug residues (Mateo-Sagasta et al. 2017). Effluents from industries such as distilleries, including fermentation, fertilizer plants, pesticide manufacturing, iron and steel industries, automobile manufacturing, petrochemicals manufacturing, textiles, oil refineries, pulp and paper mills, tanneries, pharmaceuticals, dyes and dye intermediates, cement, lead acid battery manufacturing, chemical manufacturing, mining industry, power plants, sugar industries, and food processing industries are all major sources of water pollution (Kjellstrom et al. 2006; Ahmed et al. 2021). Hence, untreated wastewaters containing organic contaminants (pesticides, oil, etc.), inorganic pollutants (heavy metals and ions), and harmful microorganisms need to be treated before discharging into the natural environment to reduce several hazardous impacts on the environment and human health (Oteng-Peprah et al. 2018).

The requisite for wastewater treatment is increasing as enormous quantities of wastewater are generated in urbanizing landscapes—cities and towns. Wastewater is discharged into the environment without adequate treatment, impairing ecosystem health. Untreated wastewater causes detrimental effects on aquatic flora and fauna (loss of biodiversity and bioaccumulation of heavy metals in aquatic organisms) and impairs human health (Sharma and Bhattacharya 2017). Thus, there is an urgent need to protect freshwater resources and increase the number of wastewater treatment facilities to minimize the contamination of waterbodies. Considering the looming threat of water scarcity with changes in the climate coupled with the burgeoning demand, it is essential to treat wastewater and implement sustainable measures to increase clean water availability through wastewater treatment, i.e., the reuse and recycling of treated water (Kumar 2009).

The main objectives of the current study are to (a) assess the water quality of lakes in the Hebbal-Nagavara valley catchment of Bengaluru, Karnataka State, India; (b) review the existing wastewater treatment facilities in the region; and (c) recommend options to mitigate contamination of water. The outcome of this study would aid in evolving strategies toward a circular economy with the proper management of water resources.

16.2 Materials and Methods

16.2.1 Study Area

Greater Bangalore (Bengaluru), the capital of Karnataka State, lies between latitude 12°39′00″ to 13°13′00″ N and longitude 77°22′00″ to 77°52′00″ E, covering an area of 741 square kilometers. It is located at an altitude of 920 meters above mean sea level. The Bengaluru landscape, which has undulating terrain, forms three major watersheds: Koramangala and Challaghatta valley, Hebbal-Nagavara valley, and Vrishabhavathi valley. The temperature ranges from 18 °C to 38 °C in the summer and 12 °C to 25 °C in the winter. The study involves 38 lakes in the Hebbal-Nagavara valley of Greater Bangalore (Fig. 16.1). Annual rainfall in the catchment varies between 770 mm and over 1200 mm. The population in the catchment has increased from 0.7 million (1991) to 1.23 million (2001), 1.7 million (2011), and 2.6 million (2021), with a population density of 83.6 persons per hectare (Ramachandra et al. 2018a).

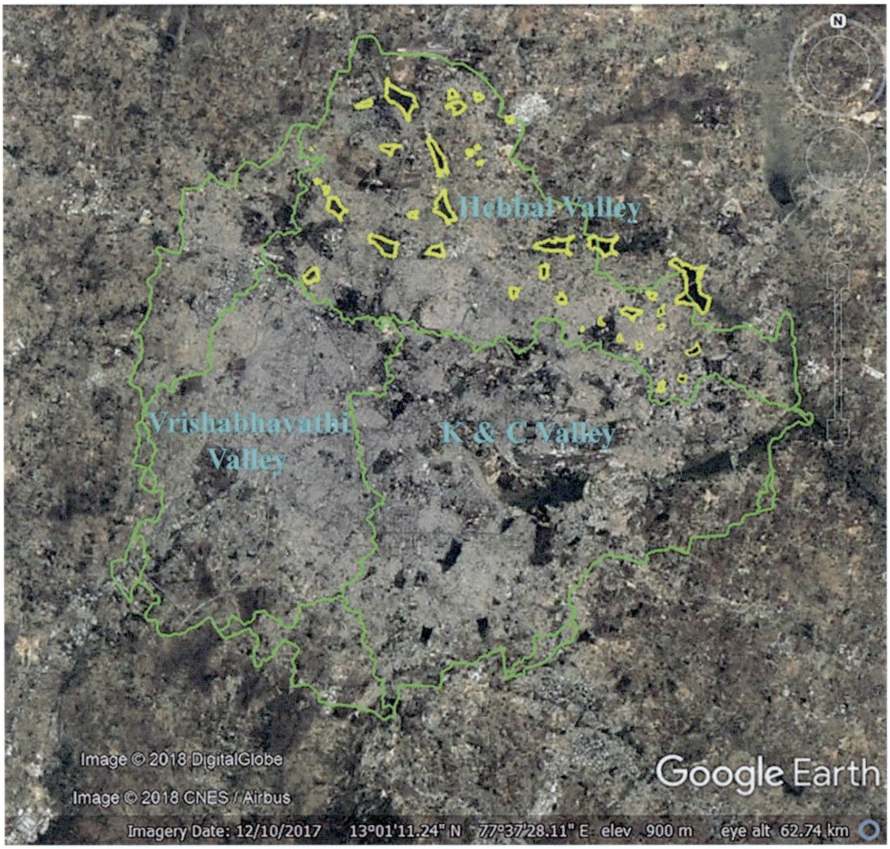

Fig. 16.1 Lakes in Hebbal-Nagavara valley of Bengaluru, Karnataka

16.2.2 Water Sampling and Physicochemical Analysis

The water samples from 32 lakes in Hebbal-Nagavara valley, Bengaluru, were collected in clean, sterile polypropylene bottles from 2017 to 2019. The physicochemical parameters like water temperature (WT), pH, electrical conductivity (EC), total dissolved solids (TDS), and dissolved oxygen (DO) were measured onsite using probes during sample collection, while other chemical and nutrient parameters were estimated in the laboratory. The remaining physicochemical parameters, such as total alkalinity (TA), calcium (Ca), magnesium (Mg), total hardness (TH), chlorides (Cl), sodium (Na), potassium (K), nitrate (N), orthophosphate (OP), chemical oxygen demand (COD), and biochemical oxygen demand (BOD), were analyzed using a standard protocol (APHA 2005) at the aquatic ecology laboratory at CES TE 15, Indian Institute of Science (IISc).

Table 16.1 Water quality rating based on WQI value

WQI	Water quality status	Possible usages
0–25	Excellent water quality	Drinking, irrigation, and industrial
26–50	Good water quality	Domestic, irrigation, and industrial
51–75	Poor water quality	Irrigation
76–100	Very poor water quality	Restricted use for irrigation
>100	Unsuitable for drinking	Proper treatment is required before use

16.2.3 Water Quality Index (WQI)

A weighted arithmetic water quality index (WQI) was computed based on the water quality parameters of 32 monitored lakes. Ten prominent water quality parameters like pH, electrical conductivity, total dissolved solids, dissolved oxygen, total alkalinity, calcium, magnesium, total hardness, chloride, and nitrate were used to compute WQI (Ramachandra et al. 2022b; Rameshbabu and Selvanayagam 2015; Devi et al. 2015). Then, the results are evaluated for a water quality status based on WQI criteria (Table 16.1).

16.3 Results

16.3.1 Water Quality Assessment of Lakes in Hebbal-Nagavara Valley

Lakes serve as a water source for domestic purposes, industrial activities, and irrigation. The lakes in Bengaluru also provide recreation and livelihood to a dependent population (fish, fodder, water, etc.). Thus, it is necessary to maintain the good water quality of lakes. Assessing lake water quality would help protect the lakes from further pollution and help support habitat quality and biodiversity.

The water temperature influences most of the water quality parameters and accelerates the metabolic activities of aquatic organisms, varying from 23.1 °C to 29.9 °C. pH ranged from 7.2 to 9.62 in lakes, indicating alkaline conditions. The total alkalinity in water varied from 100.67 to 548 mg/L, contributed by carbonic acid, carbonates, and bicarbonates, indicating the buffering capacity of water. Total dissolved solids (the amount of total solids dissolved in water) and electrical conductivity (the ability of water to conduct electricity) varied from 178.5 to 788 mg/L and 364 to 1221 μS/cm, respectively.

Dissolved oxygen indicates the amount of oxygen present in water, which is essential for the survival of aquatic organisms. The oxygen demand in lakes is indicated by the amount of dissolved oxygen (DO), biochemical oxygen demand

(BOD), and chemical oxygen demand (COD), which range from 0 to 16.53 mg/L, 4.07 to 111.79 mg/L, and 8 to 200 mg/L, respectively. Here, COD and BOD levels indicate contamination of lakes due to the sustained inflow of domestic sewage and industrial effluents. Total hardness, calcium, and magnesium varied between 90.34 and 432 mg/L, 25.1 and 138.01 mg/L, and 2.75 and 58.63 mg/L, respectively. Higher hardness, calcium, and magnesium are consequences of the discharge of domestic sewage, industrial effluents, and detergents into lakes. Higher amounts of chlorides indicate pollution due to organic matter of allochthonous and autochthonous origin (Altaf and Saltanat 2014), which varied between 49.94 and 389.55 mg/L in the present study. Sodium and potassium in lake water samples ranged from 44 to 488 mg/L and 9.25 to 119.6 mg/L, respectively. The concentration of nutrients such as nitrate and orthophosphate ranged from 0.117 to 1.364 mg/L and 0.011 to 2.785 mg/L, respectively, increasing lake productivity.

16.3.2 Water Quality Index (WQI)

The water quality index (WQI) value differed among the monitored lakes. Lakes such as Devasandra 1, Jakkur, and Thirumenahalli 1 have good water quality (Fig. 16.2).

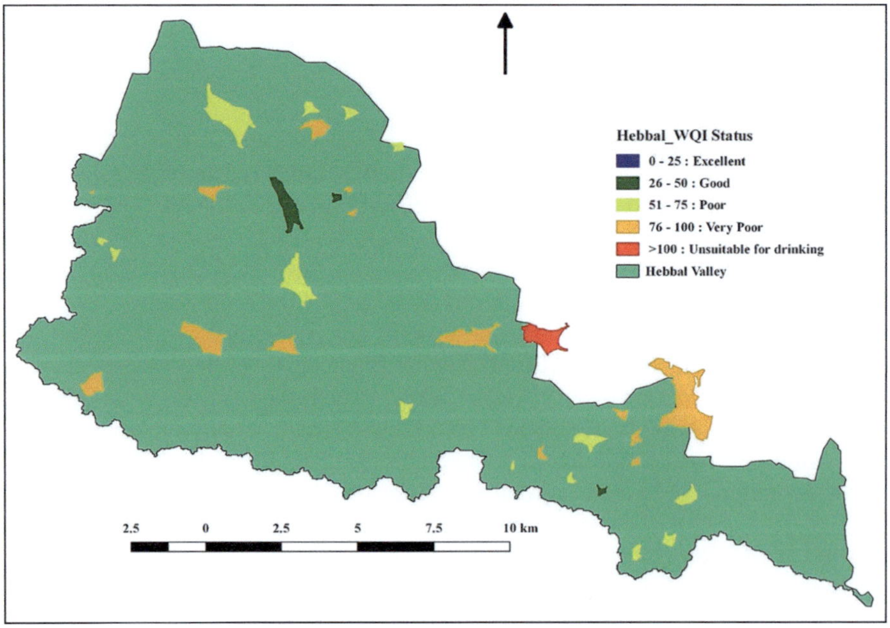

Fig. 16.2 WQI status of lakes in Hebbal-Nagavara valley, Bengaluru

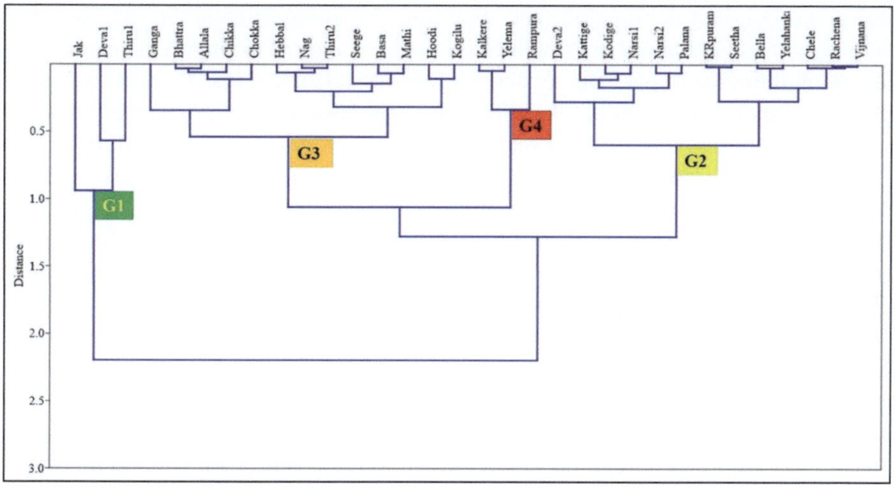

Fig. 16.3 Cluster analysis of lakes of Hebbal-Nagavara valley, Bengaluru

The cluster analysis yielded a dendrogram showing the similarities among various sampling locations, as similar locations were gathered under the same cluster. Four groups were evident in cluster analysis based on WQI (Figs. 16.2 and 16.3). Group 1 includes Devasandra 1, Jakkur, and Thirumenahalli 1 lakes with good water quality. Group 2 includes Bellahalli, Chelekere, Devasandra 2, Kattigenahalli, Kodigehalli, KR Puram, Narsipura 1, Narsipura 2, Palanahalli, Rachenahalli, Seetharamapalya, Yelahanka, and Vijnanapura lakes with poor water quality. Group 3 includes Allalasandra, Basavanapura, Bhattrahalli, Chikkabettahalli, Chokkanahalli, Gangadharkere, Hebbal, Hoodi, Kogilu, Mathikere, Nagavara, Seegehalli, and Thirumenahalli 2 lakes with very poor water quality. Group 4 includes Kalkere, Rampura, and Yelemallappa Chetty lakes, which are unsuitable for drinking.

16.4 Discussion

16.4.1 Water Quality Status of Lakes in Hebbal-Nagavara Valley, Bengaluru

WQI results revealed that only three lakes, Jakkur, Devasandra 1, and Thirumenahalli 1, have good water quality. Jakkur lake water had lower pollutants as it received treated water from an integrated wetland ecosystem consisting of a secondary treatment plant and constructed wetlands. The rest of the lakes had poor water quality with high ionic, nutrient, and organic pollutants. The common threats lakes face in the Hebbal-Nagavara valley are the entry of untreated sewage, the dumping of solid waste and building debris along the shoreline, the encroachments in the lake area, and the consequent massive growth of microalgae and macrophytes. The sustained inflow of untreated or partially treated sewage into lakes will lead to nutrient enrichment and foam formation in lakes. This nutrient enrichment has caused the widespread growth of microalgae (mainly Cyanophyceae) in lakes such as Allalasandra, Chelekere, Gangadhar Kere, Hoodi, Kodigehalli, and Kogilu. Lakes such as Amrutahalli, Puttanahalli, Doddabommasandra, Kammagondahalli, Horamavu Kere, and Horamavu Agara were fully covered with macrophytes due to nutrient enrichment and eutrophication. *Eichhornia crassipes*, *Alternanthera* sp., *Ipomoea aquatica*, and *Typha angustifolia* are the dominant macrophytes present in these lakes. Foam formation was observed in lakes such as Rampura and Yelemallappa Chetty. Both the external (through detergents from untreated sewage) and internal loading of phosphates result in foam formation and stimulate eutrophication in lakes with the prolific growth of microalgae and macrophytes. Surface-active agents present in lake water reduce the surface tension and create foam containing higher concentrations of chemical contaminants and nutrients than in lake water (Ramachandra et al. 2021). Nutrient pollution is a major factor responsible for water quality deterioration in lakes, leading to a decline in ecosystem services (Janssen et al. 2017; Liu et al. 2010).

Water pollution in lakes has increased due to rapid urbanization, consequent industrialization, and population increase, which has resulted in changes in land use and land cover (LULC), loss of interconnectivity among lakes, encroachments, and pollution. Different land use patterns and anthropogenic disturbances are critical in determining spatial water quality patterns in lakes. Temporal analyses of land use in the Hebbal valley revealed an increase in built-up area from 0.3% (1977) to 35.3% (2018), owing primarily to the extensive conversion of agriculture, grassland, and other natural resources to urban areas. Thus, agriculture and grassland declined from 86.7% (1977) to 33.7% (2018). A reduction in waterbodies from 6% (1977) to 2.4% (2018) occurred due to unplanned urbanization and city development (Ramachandra et al. 2018a). It was noticed that the buffer zone around the lakes is not maintained, and encroachments and construction activities alter its integrity. Lakes need to be restored to control pollution and improve water quality and aquatic habitats (Ramachandra et al. 2020b).

Discharging untreated waste into water resources can harm aquatic ecosystems, especially by altering natural habitats and causing biodiversity loss and human health. Consumption of contaminated water leads to waterborne diseases such as cholera, typhoid, amoebiasis, dysentery, giardiasis, hepatitis, guinea worm infections, and poliomyelitis (Praveen et al. 2016; Basavaraddi et al. 2014). Thus, the present study highlights the prerequisites for effective management of water resources through the implementation of wastewater treatment facilities that help safeguard vital wetland ecosystem services and aquatic health.

16.4.2 Treatment Facilities in Bengaluru

Wastewater treatment contributes to the circular economy through water reuse, energy generation, nutrients, and material recycling (Zarei 2020). Generally, wastewater treatment consists of the following steps: (i) preliminary treatment (screening and grit removal), which aids in the removal of floating and suspended coarse solids; (ii) primary treatment, which involves the removal of suspended solids and organic matter by gravity; (iii) secondary treatment, which involves biological treatment of wastewater to remove organic matter (e.g., trickling filters, waste stabilization ponds, etc.); and (iv) tertiary treatment, which aids in the removal of nutrients and other micropollutants (Attri et al. 2022). Wastewater treatment, water resource management, and pollution prevention require a sustainable, cost-effective approach. In India, sewage treatment plants (STPs) installed for decentralized wastewater treatment are based on the activated sludge process (ASP), extended aeration (EA), fluidized aerobic bed reactor (FAB), moving bed bioreactor (MBBR), oxidation pond (OP), sequential batch reactor (SBR), upflow anaerobic sludge blanket (UASB), waste stabilization ponds (WSP), trickling filter (TF), etc. STPs based on sequential batch reactor (SBR) and activated sludge process (ASP) technologies are predominant in India. An estimated 72,368 MLD (million liters per day) of sewage is generated in India, but the installed capacity of STPs is 31,841 MLD (43.9%). Of the total operationalized capacity (26,869 MLD), about 20,235 MLD is the actual utilized capacity. Thus, out of 72,368 MLD of sewage generated daily, only 20,235 MLD is treated in India (CPCB 2021).

Karnataka State generated 4458 MLD, whereas the installed capacity of STPs (140 installed STPs) was 2712 MLD (60.83%). Thus, the treatment capacity has a gap of 1746 MLD (39.17%). The treatment plants in Bengaluru are presented in Table 16.2. SBR and ASP technologies are predominant in Bengaluru and Karnataka State (CPCB 2021).

Sewage generated in the Hebbal catchment, Bengaluru, was 212 MLD (in 2011) and 279 MLD (in 2018) and is now estimated to be 313 MLD (Ramachandra et al. 2018a). The wastewater treatment can improve the quality of the wastewater, and the resultant treated water will be harmless and reusable, ensuring the safety of the public and the environment. If appropriately treated, recycled gray water may not threaten human health (Shaikh and Ahammed 2020). According to the national framework on the safe reuse of treated water (SRTW), various non-potable end users for SRTW are industries; municipal uses (for parks, landscaping, toilet flushing, and firefighting); agriculture and aquaculture; discharge of treated water to surface waterbodies (to maintain wetlands and environmental flow); aquifer recharge; and construction and onsite use within STPs (landscaping and cleaning activities). Water availability with sewage treatment would also help address the water scarcity in the rapidly urbanizing Bengaluru district.

Table 16.2 Sewage treatment plants (STPs) available in Bengaluru (CPCB 2021)

Sl. no.	STPs in Bengaluru urban	Year of installation	Installed capacity (in MLD)	Actual utilization of installed capacity (in MLD)	Treatment technology
STPs operating					
1	Dasarahalli	2018	20	8.218	SBR
2	Dasarahalli	2019	20	7.778	EAT
3	Dasarahalli	2020	5	3.145	SBR
4	Rajacanal	2004	40	36	ASP
5	Rajacanal	2018	40	40	EA
6	Jakkur	2018	15	15	MBBR
7	K & C valley	2005	218	183	ASP
8	K & C valley	2005	30	30	EA
9	Horamavu	2018	20	20	SBR
10	Sarakki	2019	5	4	SBR
11	Bellandur Amanikere	2018	90	90	ASP
12	Lalbagh	2006	1.5	–	ASP and UV method
13	Cubbon Park	2018	4	–	MBR
14	Hebbal	1999	60	35	ASP
15	Yelahanka	2003	10	8	ASP
16	Yelemallappa Chetty	2018	15	13	SBR
17	Chikkabanavara	2018	5	3.5	SBR
18	Mailasandra	2005	75	75	ASP
19	V Valley	2009	180	70	Two-stage TF
20	V Valley (TIP)	2004	10	10	High rate clarifier DensaDeg
21	Kempambudhi	2002	1	1	ASP
22	Mallathahalli	2011	5	5	SBR
23	Kadugodi	2018	6	4	SBR
24	Halasuru	2018	2	1.8	SBR
25	Nagasandra	2018	20	10.814	SBR
26	Nagasandra	2005	20	8.484	EA
27	K R Puram	2005	20	20	UASB
28	Doddabele	2019	40	30	SBR
29	Kadubeesanahalli	2005	50	42	ASP
30	K & C valley	2018	60	60	ASP
31	Kengeri	2018	60	40	ASP
32	Doddabele	2018	20	20	SBR
33	Hulimavu	2020	10	4.5	SBR
34	Agaram	2020	35	15	SBR
STPs under construction					
35	K & C valley	–	150	–	SBR

(continued)

Table 16.2 (continued)

Sl. no.	STPs in Bengaluru urban	Year of installation	Installed capacity (in MLD)	Actual utilization of installed capacity (in MLD)	Treatment technology
36	Chikkabegur	–	5	–	SBR
37	K.R. Puram	–	20	–	SBR
38	Hebbal	–	100	–	SBR
39	V. Valley	–	150	–	SBR

16.4.3 Efficacy of Constructed Wetlands

Most of the installed wastewater treatment plants in Bengaluru are not functioning at their maximum capacity. The treated water emanating does not comply with the standards prescribed before discharge into waterbodies, which has resulted in the contamination of land, surface water, and groundwater in Bengaluru, necessitating the adoption of economic and integrated approaches based on the natural principle of bioremediation for nutrient removal. For example, integrating a sewage treatment plant (of 10 MLD capacity) with wetland systems in Jakkur lake in Bengaluru helped sustain water to acceptable levels in accordance with CPCB (Central Pollution Control Board) norms. The wetland in Jakkur, consisting of reed beds and an algal pond, helped in the removal of COD, BOD, nitrate, and orthophosphate (Ramachandra et al. 2018b). Thus, capitalizing on bioremediation (phytoremediation or phyco-remediation) processes through constructed wetlands integrated with secondary wastewater treatment helps in curtailing pollutants (Fig. 16.4).

Phytoremediation involves wastewater remediation by using aquatic plants, as they can uptake and absorb contaminants such as inorganic pollutants, organic pollutants, and heavy metals. This process mainly depends on aspects such as environmental factors, characteristics of the polluted water, duration of exposure, and characteristics of aquatic plants (Mustafa and Hayder 2021). Macrophytes help to mitigate increasing nutrient loads (Table 16.3) due to the uptake of nutrients or the prevention of nutrient resuspension as their roots stabilize the lake sediments (Xu et al. 2014; Zhao et al. 2012). Phyco-remediation involves the use of microalgae for wastewater remediation, and the most commonly used microalgae include *Scenedesmus* sp., *Arthrospira* sp., *Botryococcus* sp., *Chlorella* sp., *Chlamydomonas* sp., etc. (Koul et al. 2022; Singh et al. 2017; Sydney et al. 2011).

Constructed wetlands are engineered systems for wastewater treatment that utilize natural processes and interact with wetland vegetation (macrophytes), substrates, and microbial activities (Vymazal 2013; Saeed et al. 2018). In recent decades, constructed wetlands (CWs) have been successfully used for wastewater treatment. They are highly efficient in pollutant removal, easy to operate and maintain, cost-effective, environmentally friendly, and sustainable (Table 16.4). The contaminant removal effectiveness of free water surface (FWS) CWs, subsurface flow (SSF) CWs, and hybrid wetland systems varied greatly (Wang et al. 2017).

Fig. 16.4 Treatment of wastewater through constructed wetlands

Phragmites australis and *Cyperus alternifolius* are freshwater wetland plants widely used in CWs (Huong et al. 2020; Ebrahimi et al. 2013; Rezania et al. 2019).

16.4.3.1 Legal Framework and Policies for Water and Waste Management

There is a need to strictly adhere to existing policies and formulate new national policies that involve pollution prevention and encourage wastewater treatment and management. The Water (Prevention and Control of Pollution) Act 1974, prohibits the disposal of any toxic or polluting matter into waterbodies. The Water (Prevention and Control of Pollution) Cess Act, 1977, proposes the levy and collection of cess on water consumed by industries and local authorities, which is used for executing the Water (Prevention and Control of Pollution) Act. The Environment (Protection) Act of 1986 ensures the protection and improvement of environmental quality and aims to prevent, control, and lessen environmental pollution. The act lays down the standards for the quality of the environment (air, water, or soil) for various areas and purposes, provides the criteria for the emission and discharge of environmental pollutants from various sources, and lays down procedures and safeguards for handling hazardous substances. The Hazardous Waste (Management and Handling) Rules 1989, aim to manage and handle hazardous substances. The National Environment

Table 16.3 Efficiency of macrophyte in pollutant removal from untreated wastewater

Macrophyte	Pollutant removal efficiency	References
Typha angustifolia	BOD (68%), COD (65%), TDS (45%)	Chandanshive et al. (2017)
Lemna minor	BOD (92%), COD (92%), EC (68%), pH (8, 9), TDS (68%)	Amare et al. (2018)
Lemna minor	TP (104.99%), TN (102.55%), BOD (85.58%), COD (90.57%), pH (97.65%)	Basiglini et al. (2018)
Vetiveria zizanioides	BOD (76%), COD (71.78%), pH (7.8)	Seroja et al. (2018)
Ipomoea aquatica	COD (80%), TP (99%), nitrate (99%), ammonia (82.7%)	Sa'at and Zaman (2017)
Salvinia cucullata	TP (81.25%), BOD (43.02%), COD (31.04%), nitrate (20%), ammonia (5.26%)	Alam and Hoque (2017)
Eichhornia crassipes	Nitrate (92%), ammonia (81%), phosphorous (67%), COD (49%)	Kutty et al. (2009)
Hydrilla verticillata	Orthophosphate (52.58%), TP (44.69%), nitrate (50.35%), nitrite (47.59%), ammonia (39.45%), calcium (24.39%), magnesium (37.16%), COD (36.14%), EC (15.66%)	Patel and Kanungo (2013)
Alternanthera sp.	COD (78.9–83.9%), BOD (87%), TN (45%), TP (36%)	Abbasi et al. (2019)
Salvinia natans	BOD (96.9%), COD (95%), TN (85.2%), ammonia (79%), nitrite (40%), orthophosphate (37%)	Ayache and Boudehane (2019)

Note: *TP* total phosphorous, *TN* total nitrogen, *TDS* total dissolved solids, *EC* electrical conductivity, *BOD* biochemical oxygen demand, *COD* chemical oxygen demand

Appellate Authority Act 1997 was created to hear appeals concerning restrictions on areas of industries, operations, processes, etc., which are carried out and subject to certain safeguards under the Environment (Protection) Act of 1986.

The Municipal Solid Wastes (Management and Handling) Rules of 2000 (and the Solid Waste Rule of 2016) stipulate measures for the collection, segregation, storage, transportation, processing, and disposal of solid wastes. The National Urban Sanitation Policy of 2008 stipulates achieving a clean India through implementing open defecation-free policies in all cities and towns of India. All human and liquid waste needs to be collected and safely treated to ensure the operation and maintenance of sanitation facilities for a healthy life. The National Water Policy (2012) suggests that recycling and reuse of water after treatment to specified standards should also be incentivized through an adequately planned tariff system. In India, programs such as the Atal Mission for Rejuvenation and Urban Transformation (AMRUT) (2015) were launched to improve citizens' quality of life in 500 selected cities by providing water supply (tap connections) and sewer connections, developing greenery and well-maintained open spaces (parks), and reducing pollution by switching to public transportation or providing facilities for non-motorized transport (e.g., walking and cycling). In 2016, the Ministry of Environment, Forest and Climate Change (MoEFCC) categorized industrial sectors into Red, Orange, Green, and White based on the pollution index, considering the levels of emissions (air

Table 16.4 Efficiency of constructed wetlands

Wetland	Nutrient removal efficiencies	References
Jakkur lake, Bengaluru, India	~77% COD, ~90% BOD, ~33% NO3-N, and ~ 75% $PO4^{3-}$-P	Ramachandra et al. (2018b)
Qingyuan River at the Minhang campus of Shanghai Jiaotong university	19.37–65.27% total phosphorus (TP) and 21.56–62.94% total nitrogen (TN)	Sheng-Bing et al. (2007)
Wetland in Czech Republic	74.9% COD, 85.8% BOD, 84.3% suspended solids, 65.0% TP, and 38.9% TN	Vymazal (2002)
Cyperus alternifolius and *vetiver* in treating the pig farm wastewater	BOD (68%), COD (64%), ammonium (20%), and TP (18%)	Liao et al. (2005)
Constructed wetland vegetated with *Cyperus alternifolius*	COD, NO_3^-N, NH_4^+-N, and $PO4^{3-}$P were reduced to 72%, 88%, 32%, and 0.8%, respectively	Ebrahimi et al. (2013)
Constructed wetland, Kothapally village, Telangana, India	COD (61.54%), phosphate (54.29%), and ammoniacal nitrogen (58.54%)	Datta et al. (2021)
Constructed wetland system Gložan, Vojvodina Province (Serbia)	Nitrogen (47.3–78.3%) and phosphorus (29.1%)	Josimov-Dunđerski et al. (2011)

pollutants), discharge of effluents (water pollutants), generation of hazardous wastes, and resource consumption. The main point is that none of the industries in the "Red" category shall be permitted in an ecologically fragile or protected area. The Plastic Waste Management Rules of 2016 mandate that plastic waste generators adopt appropriate steps to minimize plastic waste generation, prevent littering of plastic waste, ensure segregated waste storage at the source, and hand over segregated waste to approved waste collection agencies. The primary mission of Swachh Bharat Mission-Urban 2.0 (SBM-U 2.0) is to make all cities "garbage free," which involves segregation of wastes (wet waste and dry waste); 100% door-to-door collection of segregated wastes and 100% processing of municipal solid wastes; remediation of all legacy dumpsites and their conversion into green zones; reduction in air pollution from solid wastes; reduction in the use of single-use plastic; treatment of used water (including sewerage, gray water, and black water) before discharge into water resources to avoid pollution due to untreated sewage; and reuse of treated used water (Swachh Bharat Mission-Urban 2.0, 2021). This helps in achieving the Sustainable Development Goals (SDGs). The Swachh Survekshan is the largest urban sanitation survey in the world. The Ministry of Housing and Urban Affairs has launched Swachh Survekshan (SS) 2023 to achieve circularity through effectively implementing the 3R principle (reduce, recycle, and reuse) in waste management with the theme "Waste to Wealth."

Management strategies to mitigate pollution of waterbodies include: (a) avoid direct discharge of both point and non-point pollution sources into lakes; (b) ensure effective wastewater treatment in STPs; (c) promote the constructed wetland concept in lakes, i.e., allow secondary treated water (from STPs) to flow into the wetland portion (consisting of reed beds and an algal pond) of lakes to ensure better nutrient removal and avoid eutrophication in lakes; (d) desilt and rejuvenate lakes based on scientific principles; (e) implement proper waste management systems in the vicinity of lakes to reduce pollution problems; (f) conduct regular field surveys and water quality monitoring of STP treated water and lakes to reduce pollution and maintain a pristine ecosystem; (g) prevent the dumping of solid waste and construction and demolition (C&D) wastes into the lake bed; and (h) conduct environmental awareness programs to educate the public about the harmful effects of pollution on the aquatic environment.

16.4.3.2 Promotion of the Circular Economy (CE) Concept

A circular economy promotes the sustainable use of natural resources through reducing, reusing, recycling, and recovering, thus reducing environmental pollution and promoting resource recovery, which is essential for achieving the 2030 Agenda for Sustainable Development Goals (SDGs). The sustained inflow of untreated or partially treated wastewater into waterbodies has contributed to nutrient enrichment in aquatic ecosystems with algal bloom and subsequent eutrophication. The excessive use of fertilizers in agricultural fields near lakes contributes to non-point source pollution, resulting in nutrient pollution (excess nitrate and phosphate) in soil and aquatic ecosystems (Table 16.5). These will finally alter the quality of the ecosystem and lower the biodiversity. Wastewater reclamation or reuse is crucial, with the loss of water resources and increased demand leading to water scarcity in this urbanizing world. Wastewaters released from urban areas and industries are rich in nutrients, which can be recovered and reused. Wastewater treatment contributes to the circular economy by reusing water, energy production, and nutrient recycling. The treated water can be utilized for crop irrigation, cleaning, recharge of waterbodies, household applications, watering green areas, and industrial purposes, thus promoting a circular economy in the water sector (Valdes Ramos et al. 2019).

Microalgae and macrophytes (aquatic plants) utilize nutrients (nitrogen and phosphorus) in wastewater for their growth and biomass production. During photosynthesis, carbon dioxide is converted into organic matter using sunlight as an energy source. Thus, both microalgae and macrophytes aid in carbon sequestration and bioremediation, which help to tackle waste management and greenhouse gas emissions. Integrating wastewater treatment and biomass (of microalgae and macrophytes) cultivation is environmentally friendly and economically viable, as is wastewater treatment and the production of value-added products. The term "circular bioeconomy" (biobased circular economy) denotes the efficient exploitation of biomass from biological resources for the sustainable production of high-value-added products (food, medicines, animal feed, and bioenergy) that help in economic development (Leong et al. 2021). Microalgae are preferred as they have a high growth rate and yield and a high lipid content, require less land and water for cultivation than commercial crops, and are highly efficient in CO_2 mitigation. Microalgal

Table 16.5 Impacts of human activities on wetlands

Natural elements	Impacts of human activities	Consequence
Biomass	(a) Extraction of natural resources from wetlands for consumption (water, fishing, fodder, meat, fuel wood, etc.) (b) Unsustainable practices include overharvesting natural resources, overfishing, poaching, introducing exotic species, and deforestation.	(a) Decrease in biodiversity of plant and animal species. (b) Imbalance in the food chain/food web. (c) A decline in native species.
Water	(a) Rigorous use of water for domestic, agricultural, and industrial processes. (b) Pollution of groundwater and surface water through agricultural runoff and effluents from industries. (c) Pollution due to heavy metals released from industrial processes. (d) Solid waste dumping along lakes (pollution due to plastics and microplastics).	(a) Water pollution. (b) Depletion of aquifer resources. (c) Drying of wetlands. (d) Enrichment of nutrients (N and P). (e) The massive growth of macrophytes. (f) Toxic algal blooms. (g) Fish death. (h) Accumulation of organic matter and plastics in the wetlands. (i) Loss of the interconnectivity of wetlands (disturbs the environmental flow).
Soil	(a) Soil pollution in wetland areas due to chemical and organic pollutants released from domestic, agricultural, and industrial areas. (b) Dumping of solid waste. (c) Land encroachment. (d) Road construction.	(a) Erosion, acidification, contamination, and salinization of soils. (b) Destruction of habitats. (c) Decline in biodiversity. (d) Disturbance of soil microflora and fauna. (e) Reduced soil fertility.
Atmosphere	(a) Urban wetlands are susceptible to air pollution by toxic compounds (fine particles, heavy metals, toxic gases, etc.) from agriculture, industry, transport, and heating systems. (b) Pollution of the atmosphere by greenhouse gases (CO_2, methane, nitrous oxide, etc.), from industry, agriculture, and transport. (c) Burning solid waste near the wetland area.	(a) Acid rain, risks to human health through inhalation, and bioaccumulation in certain species. (b) Increasing temperature. (c) Climate change. (d) Increasing particulate matter. (e) Dust formation in the atmosphere. (f) Smoke produced at a large extent.

(continued)

Table 16.5 (continued)

Natural elements	Impacts of human activities	Consequence
Biogeochemical cycles	(a) Disruption of the nitrogen and phosphorus cycles by agricultural inputs and urban wastewater. (b) Water cycle disruption through wetland loss, drought, deforestation, groundwater depletion, stormwater drains (rajkaluvae) diversion, urban flood, and global warming. (c) A carbon cycle imbalance is caused by fossil fuel consumption, untreated wastewater, and emissions.	(a) Increasing amounts of nitrogen and phosphorus in water and soil. (b) Greenhouse gas emissions in the atmosphere. (c) Risks of invasion of ecosystems by new species, decrease in soil fertility, water pollution, etc.

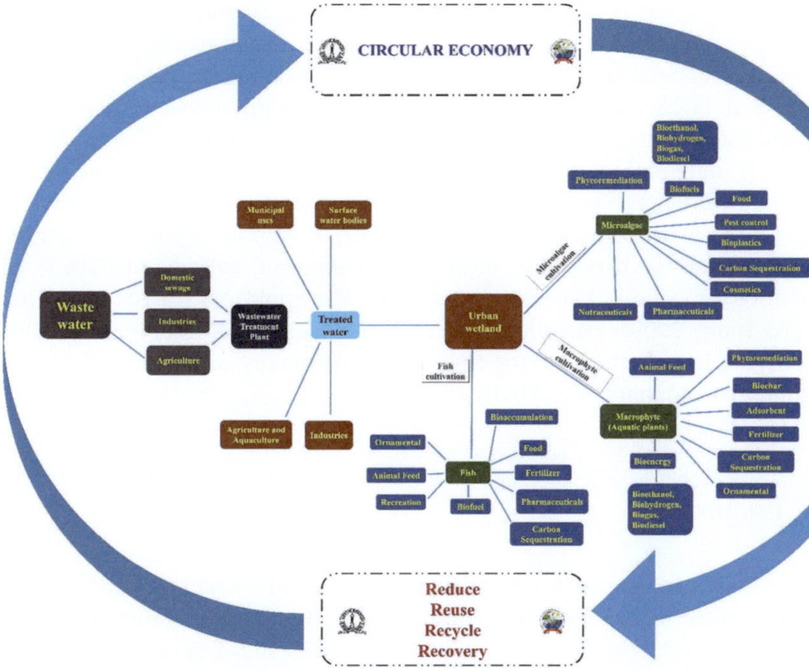

Fig. 16.5 Circular economy-based water resource management

biomass is used for producing food supplements, animal feed, and biofertilizers and is extensively used in the pharmaceutical, nutraceutical, and cosmetic industries (Díaz et al. 2022). Biofuels extracted from microalgae (third-generation biofuel sources) include biodiesel, bioethanol, biohydrogen, and biomethane (Arun et al. 2020) and are considered a solution for escalating global demands for transportation fuels (Fig. 16.5).

Macrophytes (aquatic plants) can remove both inorganic and organic pollutants from wastewater through phytoremediation, a cost-effective, sustainable method of wastewater treatment (Mustafa and Hayder 2021). The biomass of macrophytes is commercially utilized to produce animal feed, biochar, adsorbent, fertilizer, and bioenergy production, i.e., bioethanol, biohydrogen, biogas, and biodiesel (Kurniawan et al. 2021).

The natural phenomenon of sewage treatment and nutrient recovery through aquaculture and aquaponic systems is not only cost-effective but also increases food production and biofuel production, ensures nutritional requirements, purifies water, reduces pollution of the environment, and thus promotes a circular bioeconomy. Recovery of biomolecules from aquaculture, such as proteins (collagen, gelatin), polysaccharides (chitosan), lipids (omega 3), or pigments (astaxanthin or beta-carotene), can be useful for human and animal consumption, the food industry, cosmetics, or pharmaceuticals (Fraga-Corral et al. 2022). Fish waste can be used as a renewable energy source to produce biodiesel, manure compost, biogas, etc., benefiting society (Yahyaee et al. 2013). Wastewater-fed aquaculture of carp, catfish, tilapia, and aquatic plants is practiced widely in Asian countries. Aquaponics is a technology that integrates aquaculture and hydroponics and enables wastewater reuse (Das et al. 2022). Aquaponic systems utilize waste (dissolved nutrients from unconsumed fish feed and feces) from aquaculture ponds as a source of nutrients for plants. The naturally present microbes in such ponds can break down organic matter and convert the nitrogen (N) and phosphorous (P) into bioavailable forms that plants assimilate in the hydroponic system. Through the hydroponic system, vegetable crops purify water for fish. Thus, through an aquaponic system, there will be increased food production (increased crops and fish yield) and waste reduction (Graber and Junge 2009). The aquatic components like microalgae, macrophytes, and fish provide ample benefits to the environment, economy, and society, thus promoting a circular economy (Fig. 16.5).

16.5 Conclusion

The current study presents the water quality status and wastewater treatment facilities in select lakes in the Hebbal-Nagavara valley, Bengaluru, Karnataka State, India. The results highlight increased pollution of waterbodies that is likely to affect ecosystems and human well-being. As per the WQI results, only 3 lakes, Jakkur, Devasandra 1, and Thirumenahalli 1, have good water quality among the 32 monitored lakes. The major threats commonly observed were the sustained inflow of untreated or partially treated sewage, the dumping of solid waste and building debris along the shoreline, and the consequent massive growth of microalgae (or phytoplankton), invasive macrophytes, etc. This study provides baseline data, enabling decision-makers to develop appropriate strategies and policies emphasizing pollution prevention and encouraging wastewater management to protect these fragile ecosystems from further degradation. The release of untreated domestic and industrial wastewater into ecosystems has led to the contamination of surface water, soil, and groundwater, resulting in the depletion of water resources. Wastewater treatment reduces pollutant loads to safety levels that avoid eutrophication in receiving lakes. Adopting cost-effective wastewater treatment facilities and reusing treated wastewater would solve the water scarcity in the city. Thus, there is an urgent need to mitigate water pollution, encourage bioremediation (through constructed wetlands), and improve the wastewater treatment facilities in Bengaluru, which is possible only through integrated planning strategies and policies. Adopting nature-based solutions through constructed wetlands and phyco-remediation for wastewater treatment at decentralized levels would aid in wastewater treatment and the recovery of value-added products (based on reuse and recycle principles), which promote a circular economy with the reduction of environmental pollution and scope for resource recovery.

Acknowledgments We are grateful to the EIACP (ENVIS) Division; the Ministry of Environment, Forests and Climate Change; the Government of India; and the Indian Institute of Science for their sustained support for ecological research.

Compliance with Ethical Standards

Conflict of Interest The authors have no conflict of interest, either financial or non-financial.

Human and Animal Rights The research does not involve humans, animals, or tissues.

References

Abbasi SA, Tabassum-Abbasi Ponni G, Tauseef SM (2019) Potential of joyweed *Alternanthera sessilis* for rapid treatment of domestic sewage in SHEFROL® bioreactor. Int J Phytoremediation 21(2):160–169. https://doi.org/10.1080/15226514.2018.1488814

Ahmed J, Thakur A, Goyal A (2021) Industrial wastewater and its toxic effects. In: Shah MP (ed) Biological treatment of industrial wastewater. RSC, pp 1–14. https://doi.org/10.1039/9781839165399-00001

Alam AR, Hoque S (2017) Phytoremediation of industrial wastewater by culturing aquatic macrophytes, *Trapa natans* L. and *Salvinia cucullata* Roxb. Jahangirnagar Univ. J Biol Sci 6(2):19–27. https://doi.org/10.3329/jujbs.v6i2.36587

Altaf HG, Saltanat P (2014) Effect of physico-chemical conditions on the structure and composition of the phytoplankton community in Wular Lake at Lankrishipora, Kashmir. Int J Biodivers Conserv 6(1):71–84. https://doi.org/10.5897/IJBC2013.0597

Amare E, Kebede F, Mulat W (2018) Wastewater treatment by *Lemna minor* and *Azolla filiculoides* in tropical semi-arid regions of Ethiopia. Ecol Eng 120:464–473. https://doi.org/10.1016/j.ecoleng.2018.07.005

APHA (2005) Standard methods for the examination of water and wastewater, 21st edn. American Public Health Association, Washington, DC

Arun J, Gopinath KP, SundarRajan P, Felix V, Joselyn Monica M, Malolan R (2020) A conceptual review on microalgae biorefinery through thermochemical and biological pathways: bio-circular approach on carbon capture and wastewater treatment. Bioresour Technol Rep 11:100477. https://doi.org/10.1016/j.biteb.2020.100477

Atal Mission for Rejuvenation and Urban Transformation (AMRUT) (2015). https://pib.gov.in/PressReleaseIframePage.aspx?PRID=1783472. Accessed 5 July 2022

Attri SD, Singh S, Dhar A, Powar S (2022) Multi-attribute sustainability assessment of wastewater treatment technologies using combined fuzzy multi-criteria decision-making techniques. J Clean Prod 131849:1–13. https://doi.org/10.1016/j.jclepro.2022.131849

Ayache L, Boudehane ASMA (2019) Wastewater treatment by floating macrophytes (*Salvinia natans*) under Algerian semi-arid climate. Eur J Eng Sci Tech 3(1):103–110
Basavaraddi SB, Kousar H, Puttaiah ET (2014) Bacteriological contamination and remedies of ground water of Tiptur town and surrounding areas, Tumkur district, Karnataka. Int J Res Stud Biosci 2(11):107–115
Basiglini E, Pintore M, Forni C (2018) Effects of treated industrial wastewaters and temperatures on growth and enzymatic activities of duckweed (*Lemna minor* L.). Ecotoxicol Environ Saf 153:54–59. https://doi.org/10.1016/j.ecoenv.2018.01.053
Chandanshive VV, Rane NR, Tamboli AS, Gholave AR, Khandare RV, Govindwar SP (2017) Co-plantation of aquatic macrophytes *Typha angustifolia* and *Paspalum scrobiculatum* for effective treatment of textile industry effluent. J Hazard Mater 338:47–56. https://doi.org/10.1016/j.jhazmat.2017.05.021
CPCB (2021). https://cpcb.nic.in/openpdffile.php?id=UmVwb3J0RmlsZXMvMTIyOF8xNjE1MTk2MzIyX21lZGlhcGhvdG85NTY0LnBkZg. Accessed 11 Mar 2023
Das SK, Mondal B, Sarkar UK, Das BK, Borah S (2022) Understanding and approaches towards circular bio-economy of wastewater reuse in fisheries and aquaculture in India: an overview. Rev Aquac 15(3):1100–1114. https://doi.org/10.1111/raq.12758
Datta A, Singh HO, Raja SK, Dixit S (2021) Constructed wetland for improved wastewater management and increased water use efficiency in resource scarce SAT villages: a case study from Kothapally village, in India. Int J Phytoremediation 23(10):1067–1076. https://doi.org/10.1080/15226514.2021.1876627
de Groot R, Brander L, Van Der Ploeg S, Costanza R, Bernard F, Braat L, Christie M, Crossman N, Ghermandi A, Hein L, Hussain S (2012) Global estimates of the value of ecosystems and their services in monetary units. Ecosyst Serv 1(1):50–61. https://doi.org/10.1016/j.ecoser.2012.07.005
Devi WS, Singh R, Meitei NS (2015) Assessment of water quality index of Nambol River, Manipur, India. Univ J Environ Res Technol 5(3):165–172
Díaz V, Leyva-Díaz JC, Almécija MC, Poyatos JM, del Mar MM, Martín-Pascual J (2022) Microalgae bioreactor for nutrient removal and resource recovery from wastewater in the paradigm of circular economy. Bioresour Technol 363:127968. https://doi.org/10.1016/j.biortech.2022.127968
Ebrahimi A, Taheri E, Ehrampoush MH, Nasiri S, Jalali F, Soltani R, Fatehizadeh A (2013) Efficiency of constructed wetland vegetated with *Cyperus alternifolius* applied for municipal wastewater treatment. J Environ Public Health 2013:1–6. https://doi.org/10.1155/2013/815962
Fraga-Corral M, Ronza P, Garcia-Oliveira P, Pereira AG, Losada AP, Prieto MA, Quiroga MI, Simal-Gandara J (2022) Aquaculture as a circular bio-economy model with Galicia as a study case: how to transform waste into revalorized by-products. Trends Food Sci Technol 119:23–35. https://doi.org/10.1016/j.tifs.2021.11.026
Ghaitidak DM, Yadav KD (2013) Characteristics and treatment of greywater—a review. Environ Sci Pollut Res 20(5):2795–2809. https://doi.org/10.1007/s11356-013-1533-0
Graber A, Junge R (2009) Aquaponic systems: nutrient recycling from fish wastewater by vegetable production. Desalination 246(1–3):147–156. https://doi.org/10.1016/j.desal.2008.03.048
Herojeet R, Rishi MS, Lata R, Dolma K (2017) Quality characterization and pollution source identification of surface water using multivariate statistical techniques, Nalagarh Valley, Himachal Pradesh, India. Appl Water Sci 7(5):2137–2156. https://doi.org/10.1007/s13201-017-0600-y
Huong M, Costa DT, Van Hoi B (2020) Enhanced removal of nutrients and heavy metals from domestic-industrial wastewater in an academic campus of Hanoi using modified hybrid constructed wetlands. Water Sci Technol 82(10):1995–2006. https://doi.org/10.2166/wst.2020.468
Jacintha T, Rawat KS, Mishra A, Singh SK (2017) Hydrogeochemical characterization of groundwater of peninsular Indian region using multivariate statistical techniques. Appl Water Sci 7(6):3001–3013. https://doi.org/10.1007/s13201-016-0400-9

Janssen ABG, de Jager VCL, Janse JH, Kong X, Liu S, Ye Q et al (2017) Spatial identification of critical nutrient loads of large shallow lakes: implications for Lake Taihu (China). Water Res 119:276–287. https://doi.org/10.1016/j.watres.2017.04.045

Josimov-Dunđerski J, Nikolić L, Belić A, Stojanović S, Bezdan A (2011) Nutrient levels in a constructed wetland system Gložan (Vojvodina Province). Bulg J Agric Sci 17(1):31–39

Kjellstrom T, Lodh M, McMichael T, Ranmuthugala G, Shrestha R, Kingsland S (2006) Air and water pollution: burden and strategies for control. In: Jamison DT, Breman JG, Measham AR et al (eds) Disease control priorities in developing countries, 2nd edn. Oxford University Press, New York, pp 817–832

Koul B, Sharma K, Shah MP (2022) Phycoremediation: a sustainable alternative in wastewater treatment (WWT) regime. Environ Technol Innov 25:102040. https://doi.org/10.1016/j.eti.2021.102040

Kumar M (2009) Reclamation and reuse of treated municipal wastewater: an option to mitigate water stress. Curr Sci 96(7):886–889

Kurniawan SB, Ahmad A, Said NSM, Imron MF, Abdullah SRS, Othman AR, Purwanti IF, Hasan HA (2021) Macrophytes as wastewater treatment agents: nutrient uptake and potential of produced biomass utilization toward circular economy initiatives. Sci Total Environ 790:148219. https://doi.org/10.1016/j.scitotenv.2021.148219

Kutty SRM, Isa MH, Malakahmad D, Amirhossein D (2009) Nutrients removal from municipal wastewater treatment plant effluent using *Eichhornia crassipes*. World Acad Sci Eng Technol 60:826–831

Leong HY, Chang CK, Khoo KS, Chew KW, Chia SR, Lim JW, Chang JS, Show PL (2021) Waste biorefinery towards a sustainable circular bioeconomy: a solution to global issues. Biotechnol Biofuels 14(1):1–15. https://doi.org/10.1186/s13068-021-01939-5

Liao X, Luo S, Wu Y, Wang Z (2005) Comparison of nutrient removal ability between *Cyperus alternifolius* and *Vetiveria zizanioides* in constructed wetlands. Ying Yong Sheng Tai Xue Bao J Appl Ecol 16(1):156–160

Liu W, Zhang Q, Liu G (2010) Lake eutrophication associated with geographic location, lake morphology and climate in China. Hydrobiologia 644:289–299. https://doi.org/10.1007/s10750-010-0151-9

Mateo-Sagasta J, Zadeh SM, Turral H, Burke J (2017) Water pollution from agriculture: a global review. Executive summary 35. Accessed 15 March 2023

Ministry of Environment, Forest and Climate Change (MoEFCC) (2016). https://pib.gov.in/newsite/printrelease.aspx?relid=137373. Accessed 5 July 2022

Mustafa HM, Hayder G (2021) Recent studies on applications of aquatic weed plants in phytoremediation of wastewater: a review article. Ain Shams Eng J 12(1):355–365. https://doi.org/10.1016/j.asej.2020.05.009

National water policy (2012). http://jalshakti-dowr.gov.in/sites/default/files/NWP2012Eng6495132651_1.pdf. Accessed 5 July 2022

Oteng-Peprah M, Acheampong MA, DeVries NK (2018) Greywater characteristics, treatment systems, reuse strategies and user perception—a review. Water Air Soil Pollut 229(8):1–16. https://doi.org/10.1007/s11270-018-3909-8

Patel DK, Kanungo VK (2013) Treatment of domestic wastewater by potential application of a submerged aquatic plant *Hydrilla verticillata* Casp. Recent Res Sci Technol 4(10):56–61

Plastic Waste Management Rules (2016). https://kspcb.karnataka.gov.in/sites/default/files/inline-files/Plastic-Waste-Management-Rules-2016.pdf. Accessed 21 June 2022

Praveen PK, Ganguly S, Wakchaure R, Para PA, Mahajan T, Qadri K, Kamble S, Sharma R, Shekhar S, Dalai N (2016) Waterborne diseases and its effect on domestic animals and human health: a review. Int J Emerg Technol Adv Eng 6(1):242–245

Ramachandra TV, Asulabha KS, Sincy V (2021) Phosphate loading and foam formation in urban lakes. G P Glob Res J Chem 5(1):33–52

Ramachandra TV, Shivamurthy V, Subramanian AK, Varghese S, Setturu B, Aithal BH (2022b) Appraisal of environmental health and ecohydrology of free-flowing Aghanashini River, Karnataka, India. Water 14(6):1–19. https://doi.org/10.3390/w14060977

Ramachandra TV, Sincy V, Asulabha KS (2020b) Efficacy of rejuvenation of lakes in Bengaluru, India. Green Chem Technol Lett 6(1):14–26. https://doi.org/10.18510/gctl.2020.613

Ramachandra TV, Sincy V, Asulabha KS (2022a) Accounting of ecosystem services of wetlands in Karnataka state, India. JREaD 18(1&2):1–26. https://doi.org/10.3233/RED-181201

Ramachandra TV, Sincy V, Asulabha KS, Mahapatra DM, Bhat SP, Aithal BH (2018b) Optimal treatment of domestic wastewater through constructed wetlands. J Biodivers 9(1–2):81–102

Ramachandra TV, Sudarshan P, Vinay S, Asulabha KS, Varghese S (2020a) Nutrient and heavy metal composition in select biotic and abiotic components of Varthur wetlands, Bangalore, India. SN Appl Sci 2(8):1–14. https://doi.org/10.1007/s42452-020-03228-6

Ramachandra TV, Vinay S, Asulabha KS, Sincy V, Bhat S, Mahapatra DM, Aithal BH (2018a) Rejuvenation blueprint for lakes in Hebbal valley, ENVIS technical report 141. Environmental Information System, CES, Indian Institute of Science, Bangalore

Rameshbabu K, Selvanayagam M (2015) Water quality index of Kolavai lake, Chengalpet, Tamil Nadu, India. IJCC 1(1):15–18

Rezania S, Park J, Rupani PF, Darajeh N, Xu X, Shahrokhishahraki R (2019) Phytoremediation potential and control of *Phragmites australis* as a green phytomass: an overview. Environ Sci Pollut Res 26(8):7428–7441. https://doi.org/10.1007/s11356-019-04300-4

Sa'at SKM, Zaman NQ (2017) Suitability of *Ipomoea aquatica* for the treatment of effluent from palm oil mill. J Built Environ Technol Eng 2(1):39–44

Saeed T, Muntaha S, Rashid M, Sun G, Hasnat A (2018) Industrial wastewater treatment in constructed wetlands packed with construction materials and agricultural byproducts. J Clean Prod 189:442–453. https://doi.org/10.1016/j.jclepro.2018.04.115

Seroja R, Effendi H, Hariyadi S (2018) Tofu wastewater treatment using vetiver grass (*Vetiveria zizanioides*) and zeliac. Appl Water Sci 8(1):1–6. https://doi.org/10.1007/s13201-018-0640-y

Shaikh IN, Ahammed MM (2020) Quantity and quality characteristics of greywater: a review. J Environ Manag 261:110266. https://doi.org/10.1016/j.jenvman.2020.110266

Sharma S, Bhattacharya A (2017) Drinking water contamination and treatment techniques. Appl Water Sci 7(3):1043–1067. https://doi.org/10.1007/s13201-016-0455-7

Sheng-Bing HE, Li Y, Kong H-N, Zhi-Ming LIU, De-Yi WU, Zhan-Bo HU (2007) Treatment efficiencies of constructed wetlands for eutrophic landscape river water. Pedosphere 17(4):522–528. https://doi.org/10.1016/S1002-0160(07)60062-9

Singh AK, Sharma N, Farooqi H, Abdin MZ, Mock T, Kumar S (2017) Phycoremediation of municipal wastewater by microalgae to produce biofuel. Int J Phytoremediation 19(9):805–812. https://doi.org/10.1080/15226514.2017.1284758

Swachh Bharat Mission-Urban 2.0 (2021) https://sbmurban.org/storage/app/media/pdf/swachh-bharat-2.pdf. Accessed 22 June 2022

Swachh Survekshan (2023). https://www.pib.gov.in/PressReleasePage.aspx?PRID=1827969. Accessed 22 June 2022

Sydney ED, Da Silva TE, Tokarski A, Novak AD, De Carvalho JC, Woiciecohwski AL, Larroche C, Soccol CR (2011) Screening of microalgae with potential for biodiesel production and nutrient removal from treated domestic sewage. Appl Energy 88(10):3291–3294. https://doi.org/10.1016/j.apenergy.2010.11.024

The Hazardous Waste (Management and Handling) Rules (1989). https://utrenvis.nic.in/data/hazardous%20waste%20rules%201989.pdf. Accessed 13 June 2022

The Municipal Solid Wastes (Management and Handling) Rules (2000). http://toxicslink.org/docs/rulesansregulation/The-Municipal-Solid-Wastes-Management-and-Handling-Rules-2000.pdf. Accessed 23 June 2022

The National Environment Appellate Authority Act (1997). https://static.vikaspedia.in/media/files_en/energy/policy-support/environment-1/national-environment-appellate-authority-act.pdf. Accessed 21 June 2022

The National Urban Sanitation Policy (2008). http://iph.hp.gov.in/sites/default/files/Sanitation_Policy.pdf. Accessed 22 June 2022

The Water (Prevention and Control of Pollution) Act (1974). http://cpheeo.gov.in/upload/uploadfiles/files/systems_chapter2.pdf. Accessed 13 June 2022

Valdes Ramos A, Aguilera Gonzalez EN, Tobón Echeverri G, Samaniego Moreno L, Díaz Jiménez L, Carlos Hernández S (2019) Potential uses of treated municipal wastewater in a semiarid region of Mexico. Sustainability 11(8):2217. https://doi.org/10.3390/su11082217

Vymazal J (2002) The use of sub-surface constructed wetlands for wastewater treatment in The Czech Republic: 10 years experience. Ecol Eng 18(5):633–646. https://doi.org/10.1016/S0925-8574(02)00025-3

Vymazal J (2013) The use of hybrid constructed wetlands for wastewater treatment with special attention to nitrogen removal: a review of a recent development. Water Res 47(14):4795–4811. https://doi.org/10.1016/j.watres.2013.05.029

Wang M, Zhang DQ, Dong JW, Tan SK (2017) Constructed wetlands for wastewater treatment in cold climate—a review. J Environ Sci 57:293–311. https://doi.org/10.1016/j.jes.2016.12.019

WWAP (United Nations World Water Assessment Programme) (2017) The United Nations world water development report 2017. Wastewater: The untapped resource. Accessed 20 March 2023

Xu ZH, Yin XA, Yang ZF (2014) An optimisation approach for shallow lake restoration through macrophyte management. Hydrol Earth Syst Sci 18:2167–2176. https://doi.org/10.5194/hess-18-2167-2014

Yahyaee R, Ghobadian B, Najafi G (2013) Waste fish oil biodiesel as a source of renewable fuel in Iran. Renew Sust Energ Rev 17:312–319. https://doi.org/10.1016/j.rser.2012.09.025

Zarei M (2020) Wastewater resources management for energy recovery from circular economy perspective. Water-Energy Nexus 3:170–185. https://doi.org/10.1016/j.wen.2020.11.001

Zhao Y, Yang Z, Xia X, Wang F (2012) A shallow lake remediation regime with *Phragmites australis*: incorporating nutrient removal and water evapotranspiration. Water Res 46:5635–5644. https://doi.org/10.1016/j.watres.2012.07.053

Chapter 17
Performance Evaluation of Raw Rubber Treatment of Wastewater in Water Lettuce (*Pistia stratiotes*) Macrophyte Pond

H. I. Owamah, H. O. Orugba, S. Oyebisi, M. I. Alfa, and M. A. Enaboifo

Abstract This chapter presents a thorough 3-year exploration of the efficacy of water lettuce (Pistia stratiotes) macrophyte ponds for the treatment of wastewater from a typical rubber processing industry, alongside the utilization of treated effluent as biofertilizer. Initial assessment of wastewater quality involved monthly data collection and analysis over a year to establish baseline parameters, identifying several constituents surpassing permissible limits, notably conductivity, nitrates, pH, biological oxygen demand (BOD), turbidity, phosphate, total dissolved solids (TDS), and total suspended solids (TSS). Subsequently, water lettuce was introduced, covering 50% of the pond surface, and monthly monitoring of effluents ensued, assessing various chemical, physical, and biological parameters. Treatments were implemented in the ponds with retention periods ranging from 2 to 12 weeks across three inoculation phases. The findings revealed a notable reduction in most contaminants, bringing them within acceptable limits, albeit exceptions were observed for TSS and turbidity. Moreover, the treated effluent demonstrated positive effects on maize plant growth, evidenced by enhancements in plant height, stem girth, leaf area, and biomass yield. Notably, the maximum plant height of

H. I. Owamah (✉)
Department of Civil Engineering, Delta State University, Abraka, Nigeria

H. O. Orugba
Department of Chemical Engineering, Delta State University, Abraka, Nigeria

S. Oyebisi
Department of Civil Engineering, Covenant University, Ota, Ogun State, Nigeria

Department of Civil Engineering, The University of the South Pacific, Suva, Fiji

M. I. Alfa
Department of Civil Engineering, University of Jos, Jos, Plateau State, Nigeria

M. A. Enaboifo
Department of Crop Science, University of Benin, Benin, Edo State, Nigeria

S. S. Kumar, V. Kumar (eds.), *Nature-Based Solutions for Decentralised Liquid Waste Management*, Environmental Challenges and Solutions,
https://doi.org/10.1007/978-3-032-08362-3_17

117.5 ± 7.6 cm was recorded at 63 days post-planting under the second treatment regimen. The average cob weight from the second treatment group stood at 46.2 ± 6.1 g, contrasting with 21.3 ± 6.7 g from the control group.

Keywords Phytoremediation · Water lettuce · Macrophyte pond · Wastewater · Biofertilizer · Maize plant

17.1 Introduction

Water scarcity, driven by inadequate local water resources and declining quality as a result of pollution, is a mounting issue (Kivaisi 2001). Discharge of untreated wastewater into aquatic ecosystems is a major cause of water pollution in developing nations. Research has shown that more than 40% of the global population, spanning around 80 countries, is currently facing water crises, with approximately 30 of the 80 nations experiencing water scarcity for much of the year (Jiang 2009). Alongside natural scarcity, the freshwater quality in lots of developing nations is deteriorating as a result of pollution (Jiang 2009). An increase in rubber factories due to abundant virgin forests with mature rubber trees has led to an increase in water pollution from rubber factories. These factories, integral to the agro-industry, produce substantial volumes of wastewater. The indiscriminate discharge of this wastewater into soil and waterbodies poses significant environmental and health risks (Owamah et al. 2013; Owamah 2014). In Nigeria, most rubber factories release their waste into rivers and streams due to the prohibitive costs of conventional treatment systems, highlighting the need for alternative, cost-effective wastewater treatment methods. Wastewater reuse is a critical strategy for conserving water resources, particularly in water-scarce areas (Kivaisi 2001).

Baseline analyses of rubber processing wastewater reveal that several key parameters exceed permissible limits, necessitating treatment. This wastewater typically contains high levels of pollution indicators such as biochemical oxygen demand (BOD), total dissolved solids (TDS), total solids (TS), total suspended solids (TSS), ammonia, phosphate, etc. (Munandar et al. 2023; Shanker Pillai and Girish 2014; Asia and Akporhonor 2007; Girish 2014). Specifically, COD values can reach up to 3142 mg/L, BOD levels can reach 1340 mg/L, and TSS concentrations can be as high as 317 mg/L. These levels surpass permissible industrial wastewater discharge standards, underscoring the urgent need for effective treatment methods.

Efficient treatment of wastewater in the rubber processing industry is essential due to the significant environmental impact of effluents, which contain high levels of organic matter and contaminants (Ho et al. 2023; Munandar et al. 2023). Various treatment methods, such as biological processes, membrane filtration, coagulation-flocculation, and advanced oxidation processes, have been explored to address this issue, aiming for sustainable water management and resource recycling. Integrating different treatment techniques, such as electrocoagulation and membrane filtration with ozonation, can effectively remove pollutants like COD, TSS, and ammonia, enhancing treatment efficiency and reducing environmental harm (Kusworo et al. 2023a). Proper wastewater treatment ensures compliance with environmental regulations and supports the long-term sustainability of rubber processing operations by minimizing pollution and conserving water resources.

Constructed wetlands have emerged as cost-effective technologies capable of treating various forms of wastewater without the need for complex maintenance systems like activated sludge systems (Li et al. 2013). Wetland plants can uptake nitrogen and phosphorus directly, unaffected by the sewage strength, with total accumulation controlled by plant biomass (Li et al. 2013). Macrophytes are known for their rapid growth in wastewater as a result of the availability of essential nutrients. These plants make use of the nutrients for growth and metabolic functions. Papadopoulos et al. (2011) have earlier reported that the removal of fecal bacteria from septage in a full-scale duckweed-covered pond system is efficient. Duckweed is also effective in removing chromium from wastewater (Uysal 2013). Studies have emphasized the importance of selecting appropriate aquatic plants for wetlands and stabilization ponds (Owamah et al. 2013). Water lettuce macrophyte ponds have shown significant effectiveness in purifying wastewater from rubber processing industries. Research by Mahajan et al. (2023) demonstrated high removal rates of pollutants like BOD, COD, and total solids, with removal efficiencies up to 89%, 77%, and 38.5%, respectively. Gaballah et al. (2019) reported similar successes using water lettuce (*Pistia stratiotes*) in wastewater treatment, achieving removal percentages of BOD_5, NH_3, TN, and TP up to 83.5%, 97.53%, 90.3%, and 87%, respectively. These findings suggest that water lettuce macrophyte ponds can provide a valuable, cost-effective, and eco-friendly solution for treating wastewater from rubber processing industries.

Reuse of effluent from water lettuce-based ponds as biofertilizer has shown significant benefits for the growth of maize plants. Studies have demonstrated that applying rubber effluent positively influences the soil's chemical properties, nutrient uptake, and growth parameters of maize plants, enhancing plant height, leaf area, and the number of leaves. Additionally, using natural coagulants like *Moringa oleifera* seed coagulant in fishpond sediment management has resulted in enhanced plant growth, weight, and yield in maize, highlighting the potential of biofertilizers derived from organic sources to stimulate plant growth and restore soil fertility (Silva et al. 2018; Adekalu and Okunade 2001). These findings emphasize the promising role of biofertilizers in sustainable agriculture practices by effectively utilizing wastewater resources to enhance crop productivity. This chapter explores the potential of water lettuce-based ponds for treating wastewater from typical rubber processing industries and examines the benefits of reusing the effluent as a biofertilizer. Through detailed analysis and case studies, the role of macrophytes in sustainable wastewater management and agricultural productivity is highlighted.

17.1.1 Composition and Environmental Impact of Rubber Processing

Rubber, derived from *Hevea brasiliensis*, is highly valued for its natural latex, essential for producing durable plastic goods like tires and bearings (Joseph and Tharian 1994). Latex constitutes a colloidal emulsion containing rubber particles dispersed in water, surrounded by a protective protein layer that can be dislodged with soap or other agents. Additionally, fresh latex contains distinctive bright yellow spheres and lutoid particles, ranging from 2 to 10 microns in diameter (Peries and Fernando 1983). Its composition comprises 20–60% rubber, 0.3–0.7% ash, 1–2% protein, 2% lipids, 1–2% quebrachitol, 0.3% lecithin, and 33–75% water (Peries and Fernando 1983; Tan et al. 1995).

The processing of latex yields substantial wastewater, known as rubber effluents, which pose significant environmental and public health risks. A factory handling 20–30 metric tons of rubber discharges approximately 450,000 L of effluents daily (Detho et al. 2022). These effluents contain organic compounds, suspended solids, color, nitrogen, and other pollutants, contributing to eutrophication and harming aquatic ecosystems (Munandar et al. 2023). Consequently, effective wastewater treatment is imperative to mitigate these adverse impacts.

Rubber effluents exhibit diverse physical and microbiological properties, encompassing temperature, color, odor, turbidity, suspended solids, pH, organic matter, saline content, and microbial contamination. Various treatment methods have been explored for their efficacy in removing contaminants such as chemical oxygen demand (COD), total suspended solids (TSS), total Kjeldahl nitrogen (TKN), and biological oxygen demand (BOD) from rubber effluents. These methods include electrocoagulation, zeolite adsorption, upflow anaerobic sludge blanket (UASB) reactors, etc. (Zeng et al. 2021; Detho et al. 2022).

17.1.2 Wastewater Treatment Techniques in the Rubber Processing Industry

Efficient wastewater treatment is paramount in the rubber processing industry due to the substantial levels of organic matter and contaminants present in the effluents (Ho et al. 2023; Munandar et al. 2023). A range of methods, including biological processes, membrane filtration, coagulation-flocculation, and advanced oxidation processes, have been explored to address environmental concerns and promote sustainable water management practices. Studies indicate that integrating various treatment techniques, such as electrocoagulation and membrane filtration combined with ozonation, can effectively remove pollutants like chemical oxygen demand (COD), total suspended solids (TSS), and ammonia from rubber industry wastewater. This synergistic approach improves treatment efficiency while minimizing environmental impact (Kusworo et al. 2023b). Proper wastewater treatment not only ensures compliance with environmental standards but also fosters the long-term sustainability of rubber processing operations by reducing pollution and preserving water resources.

17.1.2.1 Coagulation-Flocculation in Rubber Industry Wastewater Treatment

Coagulation-flocculation plays a pivotal role in treating wastewater from the rubber industry, which is laden with elevated levels of organic matter and pollutants, including metals and inorganic substances. This process involves the addition of coagulants and flocculants to facilitate the formation of flocs, subsequently removed through sedimentation and filtration (Ho et al. 2023; Munandar et al. 2023). The efficacy of coagulation-flocculation is influenced by various factors, such as the type and dosage of chemicals, pH, temperature, and mixing rate (Solodovnik and Yakymenko 2022). Studies suggest that integrating coagulation-flocculation with other techniques, such as electrocoagulation, enhances the elimination of contaminants like chemical oxygen demand (COD) and total suspended solids (TSS) from rubber industry wastewater (Rusdianasari et al. 2021). For instance, the combined use of electrocoagulation and coagulation-flocculation achieved a remarkable 89% COD and 85% TSS removal within 80 min (Munandar et al. 2023). Various coagulants and flocculants, including natural alternatives like chitosan and sodium alginate, have shown promising results (Zaki et al. 2023). However, despite its effectiveness, coagulation-flocculation encounters challenges such as the high demand for reagents and the need for multiple treatment stages. These factors impact the cost-effectiveness and energy efficiency of the process, highlighting the necessity for ongoing research to refine these methods for industrial applications (Solodovnik and Yakymenko 2022; Rusdianasari et al. 2021).

17.1.2.2 Advanced Oxidation Processes in Rubber Industry Wastewater Treatment

Advanced oxidation processes (AOPs) stand out as highly efficient methods for treating wastewater from the rubber industry, which often contains organic contaminants resistant to conventional treatment methods (Ho et al. 2023). AOPs leverage potent oxidants, such as hydroxyl radicals, to break down persistent pollutants. These processes frequently integrate diverse techniques, including Fenton, photo-Fenton, and heterogeneous photooxidation, employing catalysts like titanium dioxide to enhance oxidation (Agrawal et al. 2023). Notably, AOPs show promise in eliminating residues from pharmaceuticals and personal care products in wastewater, owing to their robust oxidation potential (Priyanka and Varghese 2021). Recent advancements underscore the efficacy of Fenton-like and photocatalytic processes as versatile AOPs for removing organic impurities from industrial wastewater, showcasing their adaptability across various wastewater types. Ho et al. (2023)

discuss recent strides in treatment methodologies for rubber industry wastewater, encompassing biological processes, membrane processes, coagulation-flocculation, and advanced oxidation processes. In their review, Agrawal et al. (2023) examine diverse AOPs utilized for eliminating phenolic compounds and textile dyes from wastewater while outlining future research avenues aimed at enhancing catalyst properties in AOPs. Priyanka and Varghese (2021) stress the potential of AOPs reliant on free radicals with high oxidation potential for tackling stubborn pollutants in wastewater, albeit noting the challenge of generating these radicals due to their fleeting nature and instability. The primary obstacle for AOPs lies in the generation of free radicals within the treatment system, given their ephemeral nature and limited lifespan. Consequently, ongoing research endeavors target the enhancement of catalyst properties in AOPs to bolster their efficacy and applicability in industrial wastewater treatment (Agrawal et al. 2023).

17.1.2.3 Biological Processes in Rubber Industry Wastewater Treatment

Biological processes play a pivotal role in the treatment of wastewater from the rubber industry, aiming to effectively eliminate contaminants. These processes typically involve various stages such as hydrolytic acidification, aerobic activated sludge, and sedimentation to yield high-quality effluent water (Nurmalasari et al. 2022). Leveraging anaerobic reactors for methane production and organic matter removal, alongside denitrification processes, not only boosts efficiency but also reduces energy consumption. Moving bed biofilm reactors (MBBRs) have emerged as promising tools for COD and ammonia removal, underscoring the importance of microbial growth and biomass (Ninh et al. 2018). Recent insights from Ho et al. (2023) underscore the significance of biological processes in rubber wastewater treatment. Nurmalasari et al. (2022) delve into the challenges posed by elevated organic content in wastewater and explore the potential of membrane bioreactors (MBRs) in addressing these challenges. Ninh et al. (2018) elaborate on a modified model utilizing an MBBR, showcasing accelerated microbial growth and increased biomass, thereby enhancing COD and ammonia removal. The optimization of biological methods is often imperative to manage the heightened organic content in wastewater effectively. Ongoing research endeavors are directed toward integrating biological processes with other treatments to bolster overall efficiency and sustainability in rubber wastewater treatment.

17.1.2.4 Membrane Separation in Rubber Industry Wastewater Treatment

Membrane separation technology, especially nanohybrid membranes, holds significant potential for treating rubber wastewater that contains high levels of organic content and ammonia (Kusworo et al. 2023b). The incorporation of nanoparticles, such as ZnO, into polymeric membranes enhances their antifouling properties, leading to improved removal of total dissolved solids (TDS), chemical oxygen demand (COD), and ammonia (Nurmalasari et al. 2022; Kusworo et al. 2023a). These membranes increase flux rates and exhibit high selectivity for pollutants like ammonium ions, achieving rejection rates up to 87%. Combining membrane processes with ozonation further boosts selectivity and extends membrane lifespan by reducing fouling, making membrane technology a promising solution for large-scale rubber wastewater treatment.

Kusworo et al. (2021) highlight that polysulfone-silver-ZnO (PSf-sZnO) membranes, when combined with ozonation, enhance treatment efficiency by improving the removal of TDS, COD, and ammonia (N-NH_3). Kusworo et al. (2023b) report that novel PSf-sZnO membranes achieve up to 87% rejection of NH4+ ions. Nurmalasari et al. (2022) demonstrate that ozonation pretreatment enhances the performance of modified polyethersulfone (PES)-nano-ZnO membranes. Jiang et al. (2018) find that membrane processes outperform biological treatments for rubber wastewater, achieving over 99% removal of BOD and COD.

However, membrane separation requires regular maintenance and replacement to maintain performance and prevent biofouling (Susanti et al. 2019).

17.1.3 Water Lettuce Macrophyte Pond in Treatment of Wastewater from Rubber Processing Industries

Water lettuce, a floating aquatic macrophyte, shows significant potential in the phytoremediation of rubber wastewater. Phytoremediation is a sustainable and cost-effective approach that leverages the natural capabilities of plants to remove, degrade, or contain contaminants from environments such as soil, water, and air. This method involves several processes, including phytoextraction, phytodegradation, rhizofiltration, phytostabilization, phytovolatilization, rhizodegradation, and phytomining. These mechanisms enable plants to absorb, break down, or immobilize pollutants, making phytoremediation an effective and environmentally friendly method for site remediation (Ghori et al. 2016; Kafle et al. 2022).

Water lettuce, in particular, has been recognized for its effectiveness in the phytoremediation of wastewater contaminated with heavy metals and other pollutants. This plant can bioaccumulate significant amounts of contaminants, including phosphate, nitrate, nitrite, biological oxygen demand (BOD), chemical oxygen demand (COD), iron, oil, fat, zinc, and ammonia from wastewater. Through biosorption and nutrient uptake mechanisms, water lettuce can significantly reduce the concentration of these pollutants, thereby improving water quality (Zainuddin et al. 2022; Saavedra et al. 2023).

Research demonstrates the high efficacy of water lettuce in purifying wastewater from rubber processing industries. For instance, studies have shown that water lettuce can achieve removal rates of up to 89% for BOD, 77% for COD, and 38.5% for total solids (Mahajan et al. 2023). Additionally, it has been effective in removing up to 83.5% of BOD_5, 97.53% of ammonia, 90.3% of total nitrogen, and 87% of total phosphorus from wastewater, underscoring its potential as a valuable tool in environmental cleanup efforts (Gaballah et al. 2019). The effectiveness of water lettuce in treating wastewater is further enhanced by factors such as contact time, aeration, and nutrient availability, which optimize the plant's pollutant uptake and degradation capabilities (Hartono et al. 2023).

Utilizing water lettuce in the phytoremediation of rubber wastewater offers a cost-effective, eco-friendly, and efficient solution for water purification. By harnessing the natural processes of this plant, significant improvements in water quality can be achieved, contributing to sustainable wastewater management practices and environmental conservation (Yasar et al. 2018; Tang et al. 2017).

17.1.4 Reuse of Resulting Effluent as Biofertilizer for Maize Plants

Research has shown that the application of rubber effluent positively impacts soil chemical properties, nutrient uptake, and growth parameters of maize plants, resulting in improvements in plant height, leaf area, and leaf count. Additionally, the utilization of natural coagulants such as *Moringa oleifera* seed coagulant in fishpond sediment management has led to enhanced plant growth, weight, and yield in maize, indicating the potential of biofertilizers derived from organic sources to stimulate plant growth and restore soil fertility (Silva et al. 2018; Adekalu and Okunade 2001). These findings underscore the promising role of biofertilizers in sustainable agricultural practices by effectively utilizing wastewater resources to enhance crop productivity.

Studies conducted by Silva et al. (2018) evaluated the impact of fish farming effluent on maize and bean plants, revealing significant benefits for maize growth. Orhue and Osaigbovo (2011) conducted a preliminary pot trial in a greenhouse to assess the effects of rubber effluent on soil chemical properties, maize growth, and nutrient uptake. Their study concluded that rubber effluent as a biofertilizer significantly improved maize growth and soil properties, highlighting the benefits of effluent reuse from water lettuce-based ponds for maize growth. Khan et al. (2008) explored the feasibility of waste stabilization pond (WSP) technology for obtaining safe irrigation water and examined the influence of treated wastewater on maize growth parameters. The treated effluent from water lettuce-based ponds as a biofertilizer was found to enhance maize growth parameters such as plant height, leaf weight, and yield. Toua et al. (2014) investigated a treatment system based on a *Pistia stratiotes* pond to purify wastewater from the Cotton Development Company of Cameroon (CDCC) oil mill in Maroua, Cameroon, demonstrating that the final effluent used as a biofertilizer positively affected maize plant growth. Adekalu and Okunade (2001) analyzed the yield and macronutrient uptake of maize irrigated with different proportions of pond water and sewage effluent, revealing that plants benefited from the final effluent reuse as a biofertilizer. The enhanced growth was attributed to improved nutrient uptake and soil enrichment. These studies underscore the effectiveness of utilizing the final effluent from water lettuce-based ponds as a biofertilizer for maize. By harnessing these wastewater resources, it is possible to enhance crop productivity, thereby contributing to more sustainable agricultural practices.

17.2 Materials and Method

17.2.1 Sample Collection and Analysis

Two rubber processing factories, identified as Factory A and Factory B, situated near the Ikpoba River, Benin, South-South, Nigeria, were selected for the investigation. The entire research endeavor, comprising the initial monitoring of baseline wastewater quality, treatment within a macrophyte pond, and subsequent utilization of the resulting effluent for cultivating maize plants, commenced in November 2003 and concluded in March 2006. The efficacy of the water lettuce-based macrophyte pond in treating wastewater from the rubber processing industries was scrutinized over a span of 3 months for each inoculation, with adjustments made to the retention time as necessary.

17.2.2 The Making of the Water Lettuce-Based Pond Systems

A batch system of ponds was prepared to facilitate a comparative analysis with an actual field macrophyte pond. These experimental ponds were set up in plastic pots located in the screen house at the Department of Crop Science, University of Benin, Nigeria. The objective was to monitor the effectiveness of wastewater quality in these pots/ponds.

17.2.3 Collection of Data

17.2.3.1 Sampling of Effluent

From the water lettuce-based ponds, effluent samples were collected from November 25, 2004, to September 25, 2005. For the collection of these samples, new 1.5-L high-density polyethylene terephthalate (PET) screw-capped containers were utilized. Prior to sample collection, the containers and their stoppers were thoroughly washed three times with distilled water. Once the samples were collected, the containers were immediately sealed, labeled, ice-cooled, and transported to Edo State Environmental Laboratory in Benin, Nigeria.

Following Owamah et al. (2013), two types of samples were collected at each site. While a bottle was filled with effluent without the addition of acid, another was filled with effluent from the same location but acidified with a few drops of 5% HNO3 to inhibit microbial activity. Microbial analysis samples were collected with autoclave-sterilized bottles from the same sites. The non-acidified samples and microbial analysis samples were sent to the Edo State Environmental Laboratory for determination of physical parameters, anions, and total coliform bacteria. Acidified samples were delivered to the laboratory of the Department of Chemistry, Federal Polytechnic, Ado-Ekiti, for metals analysis. All effluent samples were refrigerated at 4 °C to preserve their integrity until the analyses were completed (Owamah 2014).

17.2.3.2 Analytical Procedures

The analysis of the effluent samples involved several in situ measurements and laboratory tests to assess various physicochemical and biological parameters. The specific procedures and equipment used are as follows: pH was measured in situ using a handheld digital meter (HI 9024-C, Hanna Instruments, Smithfield, RI, USA). Temperature was recorded in situ with a digital thermometer (HI 98517, Hanna Instruments). Salinity was determined in situ using a salinity meter (HI 19311, Hanna Instruments). Electrical conductivity was measured in situ with a conductivity meter (HI 2315, Hanna Instruments). Total dissolved solids (TDS) were analyzed in situ using a TDS meter (VSI 22, VSI Electronics Private Limited, Punjab, India).

Dissolved oxygen in the water samples was determined using the azide modification of Winkler's method, as described by Owamah (2014). Chloride concentration was measured by titration, following the guidelines outlined in APHA (1992, 2012). For the analysis of major anions, the ultraviolet spectrophotometer screening method was employed using a UV spectrophotometer (DR 2800, HACH, Washington, USA), adhering to APHA (1992) guidelines. To ensure the accuracy and reproducibility of the analyses, blank, standard, and pre-analyzed samples were tested after every ten samples (Owamah et al. 2013).

The total coliform bacteria count in water samples was determined using standard methods to calculate the maximum probable number (MPN), as detailed by Owamah (2014). Metal concentrations were analyzed using an atomic absorption spectrophotometer (AAS) (Sens AA 3000, GBC, Australia), according to procedures specified in APHA (2012). These rigorous methods ensured comprehensive monitoring and assessment of the water quality, contributing to the overall evaluation of the water lettuce-based macrophyte pond's effectiveness in treating rubber processing wastewater.

17.2.3.3 Treatment

Prior to introducing aquatic weeds into the ponds, the physicochemical and biological constituents of the wastewater were analyzed every month for 1 year following standard protocol (APHA 2012). The constituents analyzed are temperature, pH, electrical conductivity (EC), turbidity, TSS, hardness, TDS, COD, DO, BOD, phosphates, nitrates, iron, total coliform, *E. coli* counts, and others. This comprehensive analysis aimed to identify parameters consistently exceeding the World Health Organization (WHO) permissible limits for disposal into waterbodies. The parameters that exceeded limits were subsequently used as benchmarks for the next phase, which is wastewater quality monitoring (composite samples) and treatment in the macrophyte ponds.

Effluents from the ponds were taken biweekly and later monthly from November 25, 2004, to September 25, 2005, for determination of biological and physicochemical parameters. This was done to assess the extent of treatment provided by the ponds. The study utilized three sample crops, labeled first, second, and third inoculations. The treatments labeled first, second, and third inoculations took place from November 25, 2004, to February 25, 2005, March 25 to June 25, and July 25 to September 25, 2005, respectively.

17.2.3.4 Biofertilizer Quality of Effluent

To evaluate the quality of the biofertilizer, its application was done at 2-week intervals, starting 2 weeks before planting and continuing through 8 weeks after planting, on a small maize farm cultivated specifically for the investigation. A 10 m × 10 m plot, primarily consisting of loamy soil, was divided into eight smaller plots, each measuring 1 m × 1 m for the various treatments. Two maize seeds, Nigerian late maturity maize variety (TZESR-20), were planted in a single hole in each plot, following the method outlined by Iken and Amusa (2002). All maize seeds were checked for viability prior to planting to ensure only healthy seeds were used. The biofertilizer consisted of a mixture of effluents obtained from the 3-month retention periods of the first, second, and third inoculations.

The eight treatment (Trt) regimens were as follows:

(a) Trt 1: Effluent was applied 2 weeks prior to planting.
(b) Trt 2: Effluent was applied during planting.
(c) Trt 3: Effluent was applied 2 weeks after planting.
(d) Trt 4: Effluent was applied 4 weeks after planting.
(e) Trt 5: Effluent was applied 6 weeks after planting.
(f) Trt 6: Effluent was applied 8 weeks after planting.
(g) Trt 7: Effluent was applied 10 weeks after planting.
(h) Trt 8: Control, in which case, no effluent was applied.

Except for the control plot, where no effluent was applied, all other plots received 1 L of the effluent biofertilizer. The growing of the maize plants was under the predefined treatment conditions for 4 months, from November 1, 2005, to March 1, 2006. Leaves and roots were collected to evaluate the effects of effluent application and the timing on maize growth parameters. During this period, data were collected on biomass yield, plant height, stem girth, leaf number, and leaf area.

Nutrient chemical analyses and water composition of the water lettuce plants after treatment were done using the procedures listed in Fonkou et al. (2002). This evaluation aimed to determine the feasibility of using the resultant water lettuce biomass as feeding animal forage.

17.2.3.5 Statistical Analyses

The data were statistically evaluated using Microsoft Office Excel 2007, employing analysis of variance (ANOVA) with a significance threshold of $P < 0.05$. Mean values for the analyzed parameters were determined, and the standard deviation along with the range was determined to illustrate the level of variation in parameter values throughout the sampling months.

17.3 Results and Discussion

17.3.1 Microbial Profile and Physicochemical Analysis

Table 17.1 shows the mean microbial profile of effluent at different inoculation periods.

In the eighth week of the experiment, the highest counts of coliforms and TAPC were, respectively, recorded at 22 ± 2.6 CFU 100 mL^{-1} and 4.0 ± 0.1 CFU mL^{-1}. Conversely, the lowest counts of coliforms and TAPC, measured at 5 ± 0.1 CFU 100 mL^{-1} and 1.2 ± 0.1 CFU mL^{-1}, respectively, occurred in the 12th week. Various microorganisms, including *Escherichia coli*, diplococci, tetracocci, *Thermophilus influenza*, and *Bacillus* species, were identified in the effluent samples (refer to Table 17.1).

Table 17.1 Average microbial profile of effluent

Time taken for inoculation	Coliform count (CFU 100 mL$^{-1)}$	Total aerobic plate count (TAPC), CFU mL^{-1}	Pathogens
2 Wks	14.0 ± 2.1	2.4 ± 0.1	*Diplococci* spp.
4 Wks	10.0 ± 1.8	2.0 ± 0.3	*Influenza*
8 Wks	20.0 ± 1.9	4.3 ± 0.3	*Tetracoccus* spp.
12 Wks	4.8 ± 0.2	1.4 ± 0.2	*Escherichia coli*, *bacillus* spp.

Wks, weeks; results are shown as mean ± standard deviation of triplicate values of samples

The average physicochemical and biological parameters of the untreated composite wastewater sample collected from Factory A and Factory B (refer to Table 17.2) were compared to the WHO standard.

TDS, DO, *E. coli* count, nitrates, pH, temperature, total hardness, and total coliform were found to be within allowable limits, while PO_4, BOD, and TSS exceeded maximum permissible limits. This necessitated the monitoring and treatment of the wastewater using the water lettuce-based macrophyte pond.

17.3.2 Efficiency of Macrophyte Pond Treatment

Table 17.3 shows the efficiency of the weekly treatment in the macrophyte pond.

The maximum DO improvement (150.7%) and BOD removal (92.5%) occurred in the 12th week. TSS removal peaked in the fourth week with efficiency of 52.8%. Highest total coliform removal happened in the 12th week with an efficiency of 96.7%. NO_3 reduction was best in the fourth week at 95%, while that of PO_4 peaked in the eighth week with an efficiency of 99.3%. The highest COD and Na reductions occurred in the 12th week, with efficiencies of 85.5% and 99.4%, respectively.

The biological treatment generated effluent with residual COD and BOD ranging from 65.9 ± 8.4 mg/L to 11.6 ± 0.4 mg/L and 157.0 ± 9.9 mg/L to 20.4 ± 3.1 mg/L, respectively, within the retention period of 2 to 12 weeks. A 97.9% total coliform removal efficiency occurred at 12 weeks. The highest BOD, COD, and total coliform reduction efficiencies were, respectively, 90.6%, 85.5%, and 97.9% and found to be above most efficiencies reported in the literature for wastewater treatment in aquatic ponds (Morari and Giardini 2009).

Figures 17.1, 17.2, 17.3, 17.4, 17.5 and 17.6 illustrate contaminant removal at 2, 4, 8, and 12 weeks' retention periods for the various inoculations.

Higher efficiencies were observed mainly during the 12-week retention period. Extending the retention period beyond 12 weeks was avoided due to increased water lettuce leaf fall, potentially causing more contamination (Fonkou et al. 2002). Thus, a 12-week retention period is recommended for large-scale water lettuce-based macrophyte pond designs.

17.3.3 Impact on Maize Growth

Table 17.4 gives the influence of the rubber wastewater effluent on maize height and leaf number at different DAP.

At 42 DAP, the highest plant height (21.5 ± 2.2 cm) occurred in treatment 2, while the least (17.9 ± 3.2 cm) happened in treatment 7. For 49 DAP, the highest height of 29.7 ± 2.4 cm occurred in Trt 2, while the least (24.7 ± 1.7 cm) occurred in treatment 6. At 56 DAP, the highest height (57.1 ± 4.8 cm) occurred in Trts 2 and 5 with (44.8 ± 3.4 cm) in Trt 1. At 63 and 70 DAP, highest heights were 117.5 ± 7.6 cm

Table 17.2 Average physicochemical characteristics of the wastewater from Factories A and B

	Factory A			Factory B			
Parameters (mg/L)	First 4 months	Second 4 months	Third 4 months	First 4 months	Second 4 months	Third 4 months	WHO
DO	1.5 ± 0.1	1.5 ± 0.1	1.6 ± 0.2	1.6 ± 0.1	1.50 ± 0.0	1.7 ± 0.2	
BOD_5	203.0 ± 20.3	208.0 ± 20.2	205.8 ± 19.6	200.0 ± 20.5	207.4 ± 18.7	205.6 ± 10.2	
TSS	860.1 ± 50.5	851.0 ± 40.6	856.0 ± 42.2	854.0 ± 35.7	849.0 ± 20.6	851.0 ± 40.7	
*Total coliform and *E. Coli*	239.0 ± 15.6	242.7 ± 14.3	241.0 ± 14.7	241.0 ± 10.8	242.6 ± 25.6	241.5 ± 22.6	
NH_3	1.2 ± 0.2	1.4 ± 0.2	1.2 ± 0.2	1.4 ± 0.2	1.2 ± 0.2	1.4 ± 0.2	
Nitrates	1.3 ± 0.2	1.2 ± 0.2	1.3 ± 0.2	1.3 ± 0.1	1.3 ± 0.2	1.2 ± 0.2	50
PO_4	60.1 ± 6.1	62.2 ± 3.9	59.1 ± 6.1	62.2 ± 3.9	60.1 ± 6.1	61.2 ± 3.9	
Color	72.2 ± 11.5	67.9 ± 8.5	70.3 ± 11.9	69.8 ± 8.4	72.2 ± 11.5	67.9 ± 8.5	
**pH	9.1 ± 0.2	9.2 ± 0.2	9.1 ± 0.2	7.3 ± 0.2	9.1 ± 0.2	8.2 ± 0.2	6.5–9
EC	860.7 ± 61.2	859.6 ± 63.2	866.2 ± 43.7	859.6 ± 63.2	860.7 ± 61.2	859.6 ± 63.2	
***temperature	33.0 ± 0.2	32.9 ± 0.2	30.2 ± 0.2	32.9 ± 0.2	33.0 ± 0.2	32.9 ± 0.2	85
Turbidity	176.8 ± 11.2	177.7 ± 8.2	176.3 ± 8.6	176.8 ± 11.1	179.8 ± 8.7	175.5 ± 8.4	
COD	73.6.5	75.2 ± 4.6	76.1 ± 4.7	74.6.4	75.2 ± 4.6	74.3 ± 4.6	
Sodium	1.6 ± 0.2	1.7 ± 0.2	1.5 ± 0.2	1.7 ± 0.3	1.8 ± 0.2	1.7 ± 0.6	
Potassium	3.2 ± 0.3	2.3 ± 0.3	2.2 ± 0.3	3.3 ± 0.2	2.3 ± 0.3	2.4 ± 0.2	
Total hardness	288.2 ± 19.6	303.1 ± 25.2	304.6 ± 25.6	286.3 ± 19.4	303.1 ± 25.2	300.3 ± 23.5	500
TDS	606.5 ± 20.2	606.1 ± 23.5	607.0 ± 19.5	605.7 ± 15.4	606.0 ± 19.3	607.0 ± 24.3	
TSS	147.2 ± 11.3	144.1 ± 11.1	148.4 ± 9.8	146.0 ± 6.9	145.2 ± 7.1	146.0 ± 8.9	

Units of * and *** are CFU 100 mL^{-1} and °C, respectively; ** has no unit. Results are expressed as mean ± SD of four monthly values

Table 17.3 Average treatment efficiency

Parameters		2 wks' treatment		4 wks' treatment		8 wks' treatment		12 wks' treatment	
(mg/L)	CW (influent)	AT (effluent)	%Eff.	AT (effluent)	%Eff.	AT (effluent)	%Eff.	AT (effluent)	%Eff.
DO	1.6 ± 0.1	3.5 ± 0.1	122.6	3.9 ± 0.1	149	3.9 ± 0.2	150.3	3.9 ± 0.1	150.7
BOD_5	206 ± 5.8	156 ± 10.9	24.3	105 ± 10.3	49	43.6 ± 4.6	78.9	19.5 ± 3.2	92.5
TSS	852 ± 20.3	455 ± 6.7	46.6	10.5 ± 1.8	98.8	600 ± 50.5	29.6	402 ± 11.1	52.8
*Total coliforms	242 ± 7.4	158 ± 6.9	34.7	110 ± 8.9	58.7	8.3 ± 2.3	96.6 ± 7.4	5.0 ± 0.2	96.7
NO_3	1.0 ± 0.1	1.5 ± 0.2	45	0.1 ± 0.0	95	10.5 ± 4.4	95.0	9.8 ± 2.5	88.2
PO_4	60 ± 4.6	40 ± 10.2	33.3	0.1 ± 0.0	99.9	0.4 ± 0.1	99.3	1.0 ± 0.1	98.3
COD	74.5 ± 4.8	66 ± 8.5	11.4	42 ± 3.5	43.6	15.5 ± 2.6	79.2	10.8 ± 0.5	85.5
Sodium	1.6 ± 0.2	1.6 ± 0.1	–	1.05 ± 0.1	32.3	0.01 ± 0.0	99.4	0.01 ± 0.0	99.4
**pH	8.5 ± 0.2	8.4 ± 0.2	–	7.8 ± 0.2	–	5.6 ± 0.1	–	5.8 ± 0.1	

Results are shown as mean ± SD of values from first, second, and third inoculations; %Eff. means percent efficiency, while CW means composite wastewater; AT is after treatment; unit of * is CFU 100 mL^{-1}; ** has no unit

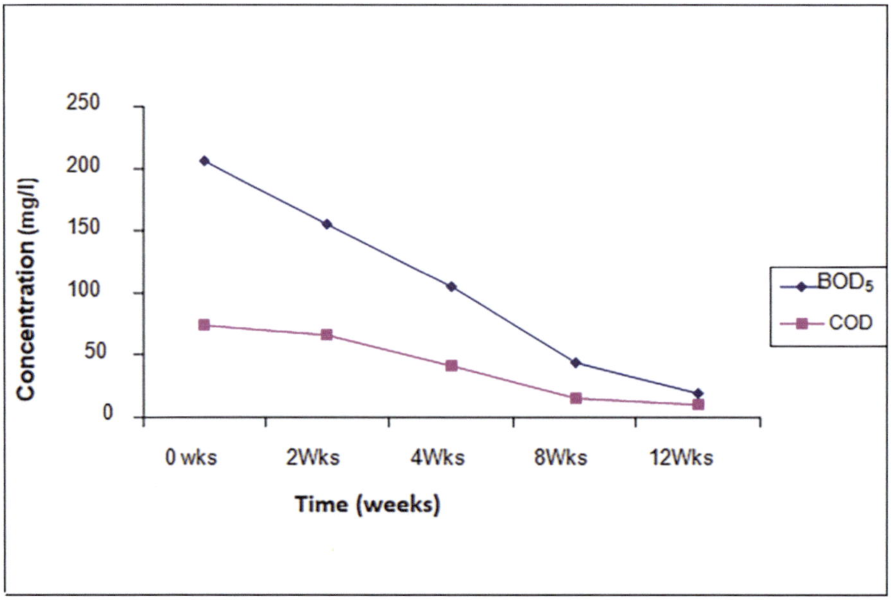

Fig. 17.1 Graph of BOD_5 and COD concentration with time at first inoculation

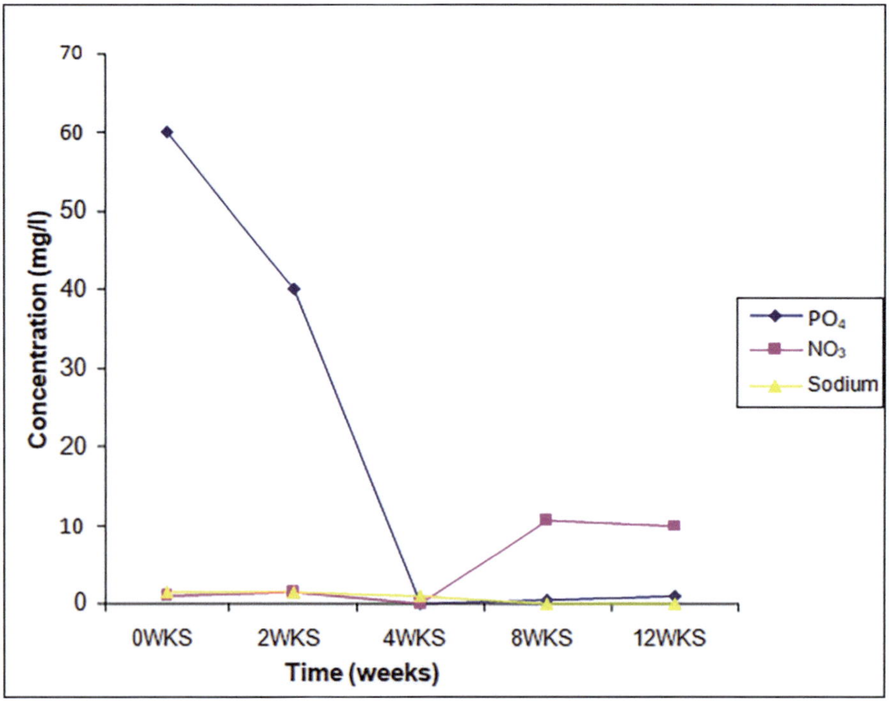

Fig. 17.2 Graph of PO_4, NO_3, and sodium concentration with time at first inoculation

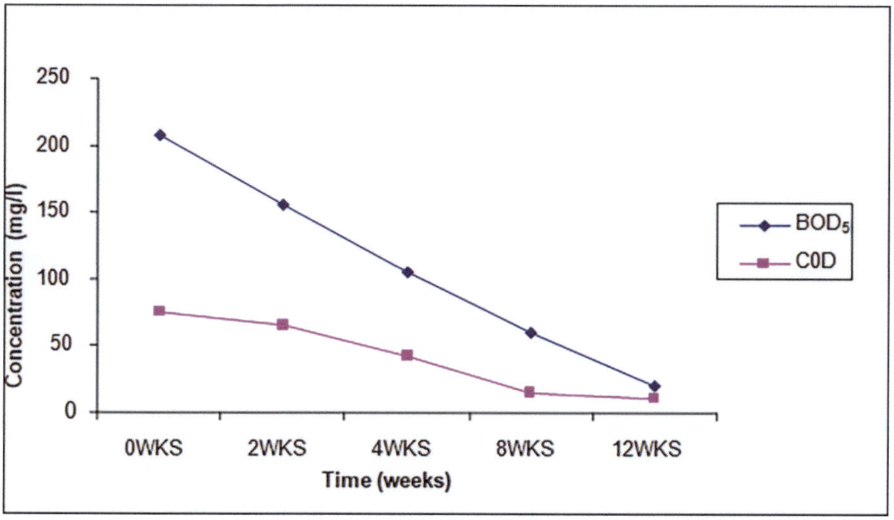

Fig. 17.3 Graph of BOD_5 and COD concentration with time at second inoculation

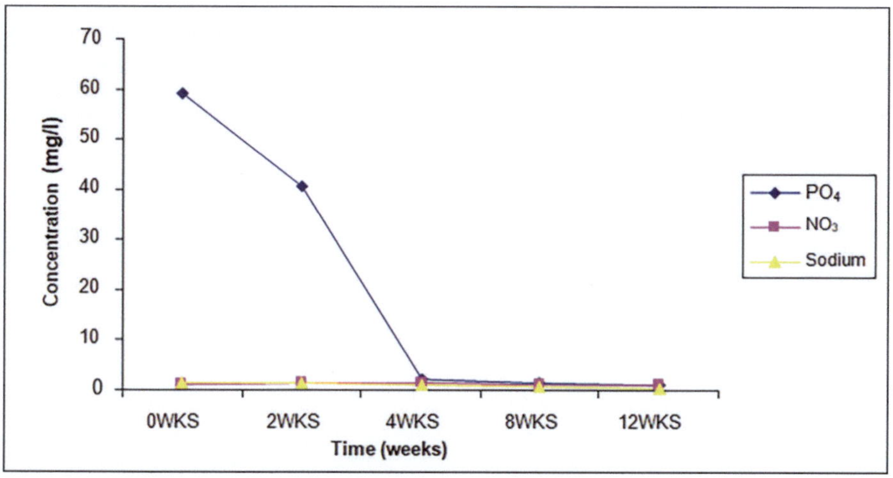

Fig. 17.4 Graph of PO_4, NO_3, and sodium concentration with time after second inoculation

and 134.0 ± 9.2 cm, respectively, with Trt 2, and the lowest heights were 72.1 ± 5.2 cm and 110.8 ± 8.6 cm, respectively, with Trt 1.

For leaf number, at 42 DAP, the maximum number of leaves (6) occurred in treatment 2 with the lowest (5) occurring in Trt 1. For 49 DAP, the maximum number (7) was obtained in Trt 6 and the least (6) in Trt 1. In 56 and 63 DAP, the maximum number of leaves (9) occurred in Trt 2 and Trt 4, respectively, while the lowest (8 and 9, respectively) were with Trts 1 and 7. At 70 DAP, the maximum number of leaves (9) occurred in Trts 4 and 6 with the least (9) in the control plot (Trt 8).

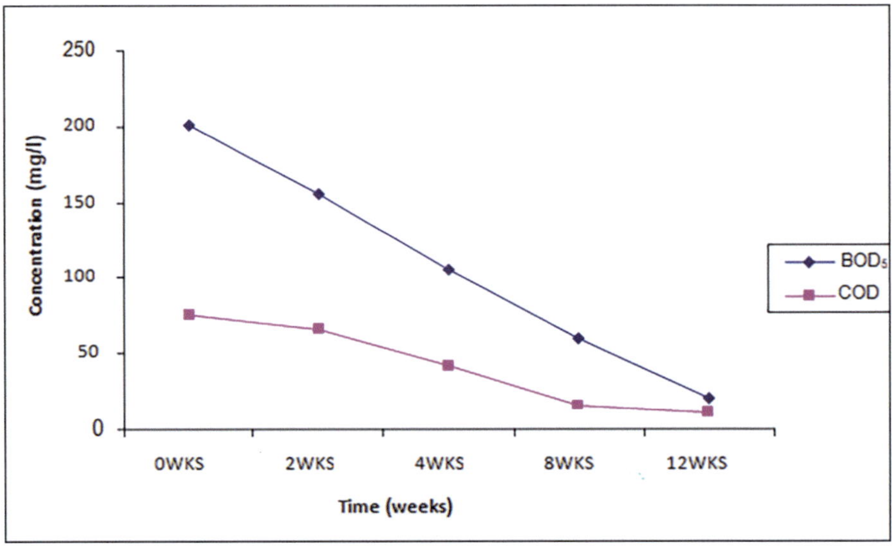

Fig. 17.5 Graph of BOD_5 and COD concentration with time at third inoculation

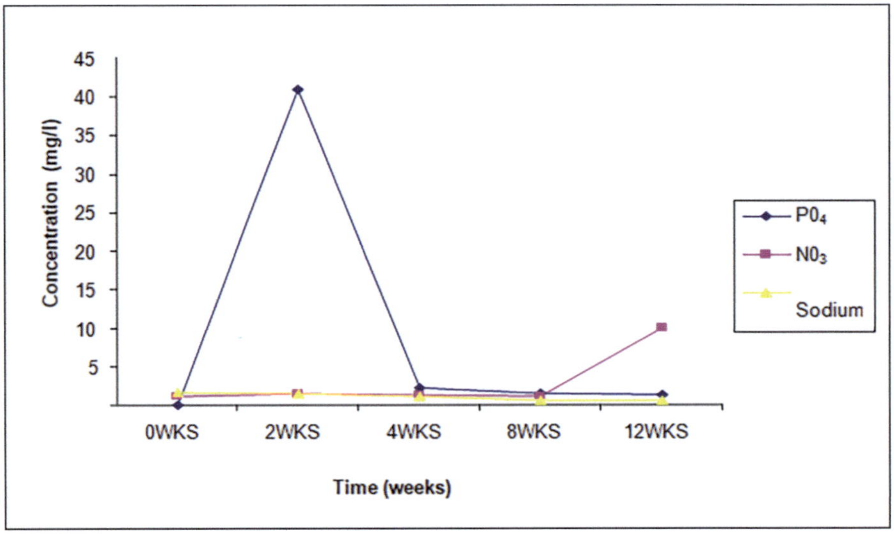

Fig. 17.6 Graph of PO_4, NO_3, and sodium concentration with time at third inoculation

The influence of treated rubber effluent on the stem girth and leaf area of maize at different DAP are shown in Table 17.5.

The highest values at 42, 49, 56, 63, and 70 DAP were 4.2 ± 0.3 cm, 4.7 ± 0.4 cm, 4.6 ± 0.2 cm, 4.4 ± 0.7 cm, and 4.5 ± 0.2 cm, respectively, and occurred in Trt 2. The least values (3.5 ± 0.1 cm, 4.5 ± 0.6 cm, 4.1 ± 0.7 cm, 3.7 ± 0.2 cm, and 3.8 ± 0.1 cm,

Table 17.4 Influence of rubber wastewater effluent on the height of maize (cm) and leaf number

[a]	(a) Maize height (cm).				
Treatment (Trt)	Time of application				
	42 DAP	49 DAP	56 DAP	63 DAP	70 DAP
Trt 1	19.9 ± 2.3	28.2 ± 2.5	44.8 ± 3.4	72.1^{b} ± 5.2	110.8 ± 8.6
Trt 2	21.5 ± 2.2	29.7 ± 2.4	57.1 ± 4.8	117.5^{a} ± 7.6	134.0 ± 9.2
Trt 3	19.4 ± 1.7	25.3 ± 2.3	48.3 ± 6.6	94.9^{ab} ± 7.9	123.3 ± 10.2
Trt 4	20.7 ± 2.8	27.1 ± 2.7	55.6 ± 5.1	115.9^{a} ± 10.3	127.1 ± 7.5
Trt 5	19.0 ± 1.9	25.7 ± 3.1	57.1 ± 4.9	106.4^{a} ± 8.6	131.3 ± 10.4
Trt 6	18.7 ± 2.3	24.7 ± 2.2	52.3 ± 4.6	111.6^{a} ± 15.1	122.0 ± 9.4
Trt 7	17.9 ± 3.2	24.7 ± 1.7	47.3 ± 5.5	103.5^{a} ± 8.3	124.1 ± 12.5
Trt 8	18.5 ± 1.7	25.6 ± 2.6	53.6 ± 7.3	94.5^{ab} ± 13.3	122.3 ± 15.8
	(b) Number of leaves.				
Treatment (Trt)	Time of application				
	42 DAP	49 DAP	56 DAP	63 DAP	70 DAP
Trt 1	6	6	8	9	9
Trt 2	7	8	9	9	10
Trt 3	6	7	9	10	10
Trt 4	7	7	9	10	10
Trt 5	6	8	9	10	10
Trt 6	6	8	9	10	10
Trt 7	6	7	8	10	10
Trt 8	6	7	9	10	9

Note: Average values of the same letter in the column are not significantly different from each other at 5% significant level. DAP, days after planting

[a]Results are shown as average ± SD of duplicate values of two different samples

respectively) occurred in Trt 8. Table 17.5 displays the influence on leaf area, with the maximum average leaf area (382.8 ± 11.6 cm^2) occurring in Trt 2 and the least (284.5 ± 6.6 cm^2) in Trt 1. Table 17.6 shows the biomass yield of maize.

17.4 Design for Size of Rubber Effluent Stabilization Pond for a Discharge of 450 m^3 per Day for 8 Weeks' and 12 Weeks' Detention Time

17.4.1 Design Criteria/Information

In formulating the design for the wastewater treatment system, several critical criteria and factors were taken into account:

(i) Soil composition: The sandy loam soil prevalent at the factory site was considered, as it can impact the construction and stability of the treatment system.

Table 17.5 Influence of the rubber wastewater effluent on the stem girth and leaf area of maize

(a) Stem girth (cm)

Treatment (Trt)	Time of application				
	42 DAP	49 DAP	56 DAP	63 DAP	70 DAP
1	$4.0^{ab} \pm 0.1$	$4.7^{ab} \pm 0.4$	$4.5^{ab} \pm 0.4$	4.5 ± 0.1	$4.4^{ab} \pm 0.3$
2	$4.1^{a} \pm 0.3$	$4.9^{a} \pm 0.3$	$4.7^{a} \pm 0.1$	4.5 ± 0.6	$4.6^{a} \pm 0.1$
3	$3.6^{ab} \pm 0.1$	$4.4^{b} \pm 0.2$	$4.4^{ab} \pm 0.1$	4.4 ± 0.4	$4.5^{ab} \pm 0.6$
4	$4.1^{a} \pm 0.4$	$4.7^{ab} \pm 1.0$	$4.5^{ab} \pm 0.8$	4.5 ± 0.7	$4.3^{ab} \pm 0.1$
5	$3.7^{ab} \pm 0.1$	$4.4^{b} \pm 0.3$	$4.5^{ab} \pm 0.4$	4.5 ± 0.2	$4.3^{ab} \pm 0.1$
6	$4.0^{ab} \pm 0.2$	$4.7^{ab} \pm 0.4$	$4.4^{ab} \pm 0.6$	4.2 ± 0.7	$4.1^{ab} \pm 0.2$
7	$3.8^{ab} \pm 0.1$	$4.4^{b} \pm 0.3$	$4.4^{ab} \pm 0.5$	4.3 ± 0.2	$4.6^{a} \pm 0.8$
8	$3.4^{b} \pm 0.2$	$4.4^{b} \pm 0.5$	$4.0^{b} \pm 0.5$	3.9 ± 0.1	$3.9^{b} \pm 0.2$

(b) Leaf area.

Treatment (Trt)	Leaf area (cm^2) at 70 DAP
1	$377.2^{ab} \pm 11.5$
2	$382.8^{a} \pm 11.8$
3	$350.0^{ab} \pm 9.2$
4	$376.1^{ab} \pm 14.2$
5	$365.7^{ab} \pm 15.1$
6	$348.6^{ab} \pm 11.7$
7	$335.2^{ab} \pm 12.2$
8	$284.8^{b} \pm 6.6$

Table 17.4 has the explanations for the treatments; results are expressed as mean ± SD of triplicate values of the same sample

Table 17.6 Cost estimates for the design period of 8 weeks

S/N	Description	Quantity	Unit	Rate $	Amount ($) × 1000
1	Clearing of site of the proposed site of WSP	16,800	m^2	3	504
2	Excavation of site for the WSP. Cost to include compaction	36,000	m^3	5	180
3	Perimeter chain link fencing for the WSP	670	m^2	10	6.7
4	5-m-wide double-leaf steel gate	2	No	2	4
Total					694.7
Contingency (10%)					69.47
Grand total					764.17

(ii) Land availability: The ample land area available for the treatment system offers flexibility in design and layout considerations.
(iii) Climatic conditions: The tropical climate of the area was factored in, as it influences evaporation rates, plant growth dynamics, and the overall efficacy of water treatment processes.

(iv) Wastewater volume: The system must accommodate the discharge of rubber effluent at a rate of 450 m^3/day, which directly influences the size and capacity requirements of the treatment units.
(v) Wastewater characteristics: A thorough understanding of the wastewater's composition is essential for devising effective treatment strategies.

Furthermore, compliance with the World Health Organization (WHO) and national standards of specific treatment targets, such as reducing pollutants to 50 mg/L for discharge into waterbodies and 500 mg/L for land application at 20 °C is required. Additionally, a security fence is to be installed 5 m outside the embankment toe to provide security and delineate the treatment area.

Taking into consideration these factors, a serial pond configuration comprising two ponds is proposed. This design facilitates sequential treatment stages aimed at meeting regulatory standards and optimizing treatment efficiency. The selection of treatment durations for 8 weeks and 12 weeks is based on specific criteria: at 8 weeks, the BOD_5 levels align with FEPA guidelines for surface water discharge, indicating successful treatment by the water lettuce system for suitable discharge. By the 12th week, the effluent's BOD_5 levels reach their lowest point, signifying the highest BOD removal efficiency based on the results obtained. Therefore, any retention time falling within the 8-week to 12-week range is considered appropriate for industrial facilities utilizing the water lettuce WSP system for effluent treatment.

17.4.2 Design Calculation

The total volume of inflow for the design period (Eq. 17.1) was determined, assuming a non-discharge condition.

$$I = i\, x\, t \tag{17.1}$$

where I is the volume (m^3) of inflow for the design period, i is the volume of inflow per day (m^3/d), and t is time (days). After calculating the inflow volume for the design period, the pond area is calculated using Eq. 17.2.

$$A = \frac{i \times t}{D} \tag{17.2}$$

where A is the pond area (m^2) and D is the pond depth (m).

17.4.2.1 Design Period of 8 Weeks Needed to Achieve Minimum FEPA Standard for BOD_5 for Stream Discharge

Based on Eq. 17.2 and assuming a pond depth of 2.5 m,

$$A = \frac{450 \times 8 \times 7}{2.5}$$

$$A = 10{,}080\,\text{m}^2$$

Using two serial ponds,

$$\text{Single pond area} = \frac{10080}{2}$$

$$= 5040\,\text{m}^2$$

Using L/B ratio of 2 (L = 2B),

$$L \times B = 5040$$

$$2B \times B = 5040$$

$$B = 50.20\,\text{m}$$

Putting the value of B as 55 m,

$$L = 2 \times B$$

$$L = 2 \times 55$$

$$L = 110\,\text{m}$$

For 0.6 m free board, use 110 × 55 × 3.1 m per pond (Fig. 17.7).

17.4.2.2 Design Period of 12 Weeks Needed to Achieve Lowest BOD_5 Remaining after Treatment

Based on Eq. 17.2 and assuming a pond depth of 2.5 m,

$$A = \frac{450 \times 12 \times 7}{2.5}$$

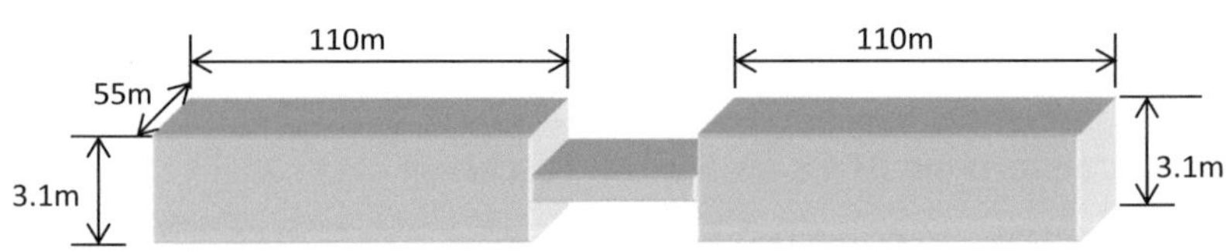

Fig. 17.7 Ponds in series for design period of 8 weeks

$$A = 15120\,\mathrm{m}^2$$

Using two serial ponds,

$$\text{Single pond area} = \frac{15120}{2}$$

$$= 7560\,\mathrm{m}^2$$

Using L/B ratio of 2 (L = 2B),

$$L \times B = 7560$$

$$2B \times B = 7560$$

$$B = 61.48\,\mathrm{m}$$

Putting the value of B as 65 m,

$$L = 2 \times B$$

$$L = 2 \times 65$$

$$L = 130\,\mathrm{m}$$

For 0.6 m free board, use 130 × 65 × 3.1 m per pond (Fig. 17.8).

17.4.3 Costing

Considerations include:

(i) Clearing of sites.
(ii) Excavation to depth not exceeding 3.5 m.
(iii) Perimeter fencing using chain link.

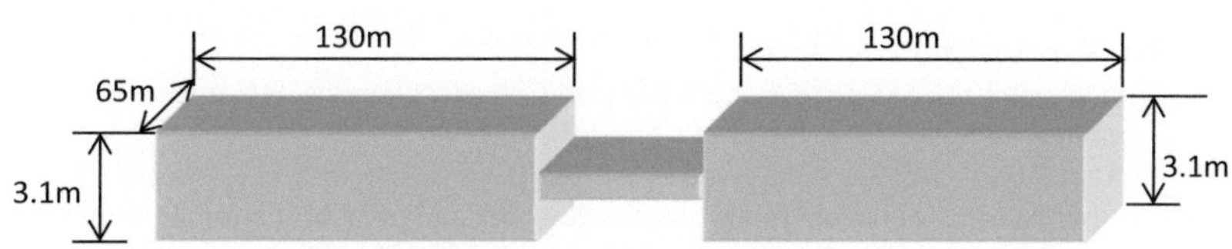

Fig. 17.8 Ponds in series for design period of 12 weeks

17.4.3.1 Cost Estimates for the Design Period of 8 Weeks Needed to Achieve Minimum FEPA Standard for BOD_5 for Stream Discharge

17.5 Conclusion

This study highlights the effectiveness and economic feasibility of employing water lettuce-based waste stabilization ponds to treat wastewater originating from rubber processing industries. By the 12th week of treatment, impressive removal rates reaching up to 90.6% for BOD, 85.5% for COD, and 97.9% for total coliform were attained. Not only does the treated effluent meet regulatory requirements but it also possesses nutritional value, qualifying it for use as a biofertilizer in maize and other crop cultivation. This approach has the potential to lower farmers' production expenses by repurposing effluent waste, although strict adherence to international and national safety standards is crucial to protect public health. Additionally, the treated effluent significantly boosts maize plant growth and yield, while the chemical composition of the resulting water lettuce biomass makes it suitable for animal feed. Given its cost-effectiveness and high purification efficiency, biological treatment utilizing water lettuce-based ponds emerges as a recommended approach for addressing industrial wastewater treatment challenges in developing nations. Furthermore, the adoption of biofertilizers presents an environmentally friendly and economically viable solution that could help tackle issues related to food insecurity.

References

Adekalu KO, Okunade DA (2001) Effects of pond water and sewage effluent treatments on yield and nutrient uptake of maize Zea mays L 120. Trop Agric 79(2):120–124

Agrawal S, Chohadia AK, Sherry P, Malhotra G, Verma K (2023) A review on wastewater treatment containing organic pollutants using advance oxidation processes. Int J Sci Res Sci Technol 10(1):50–75

APHA (1992) Standard methods for examination of water and waste-water, 18th edn. American Public Health Association, Washington, DC

APHA (2012) Standard methods for examination of water and waste-water, 22nd edn. American Public Health Association, Washington, DC

Asia IO, Akporhonor EE (2007) Characterization and physicochemical treatment of wastewater from rubber processing factory. Int J Phys Sci 2(3):61–67

Detho A, Memon AH, Aladi AF, Almohana AI, Almojil SF, Memon AA, Rosli MA, Din MFM (2022) Ammoniacal nitrogen, chemical oxygen demand, and color reduction in rubber processing industry effluent using zeolite. Desalin Water Treat 270:185–193

Fonkou T, Agendia P, Kengne I, Akoa A, Nya J (2002) Potentials of water lettuce (Pistia stratiotes) in domestic sewage treatment with macrophytic lagoon systems in Cameroon. In: Proceedings of international symposium on environmental pollution control and waste management, 7-10 January 2002, Tunis (EPCOWM'2002), pp 709–714

Gaballah MS, Ismail K, Beltagy A, Zein Eldin AM, Ismail MM (2019) Wastewater treatment potential of water lettuce (Pistia stratiotes) with modified engineering design. J Water Chem Technol 41(3):197–205. https://doi.org/10.3103/s1063455x1903010x

Ghori Z, Iftikhar H, Bhatti MF, Nasar-um-Minullah, Sharma I, Kazi AG, Ahmad P (2016) Phytoextraction. Plant metal interaction, pp 385–409. https://doi.org/10.1016/b978-0-12-803158-2.00015-1

Girish K (2014) Effect of carbon sources on the biomass build-up and degradation of rubber processing industry effluent. Int J Appl Sci Biotechnol 2(4):579–584

Hartono R, Tambun R, Nasution H (2023) The effectiveness of phytoremediation using water lettuce (Pistia stratiotes) in liquid waste treatment. In: AIP Conference Proceedings 2609(1):060001: the 6th international conference on science and technology (ICST21): challenges and opportunities for innovation research on science materials, and technology in the Covid-19 era

Ho KC, Chan MK, Chen YM, Subhramaniyum P (2023) Treatment of rubber industry wastewater review: recent advances and future prospects. J Water Process Eng 52:103559

Iken JE, Amusa NA (2002) Maize research and production in Nigeria. Afr J Biotechnol 3(6):302–307

Jiang J (2009) China's water scarcity. J Environ Manag 90:3185–3196

Jiang S-K, Zhang G-M, Yan L, Wu Y (2018) Treatment of natural rubber wastewater by membrane technologies for water reuse. Membr Water Treat 9(1):17–21

Joseph T, Tharian GK (1994) Commercial exploitation of ancillary rubber products. Econ Pol 19:413–415

Kafle A, Timilsina A, Gautam A, Adhikari K, Bhattarai A, Aryal N (2022) Phytoremediation: mechanisms, plant selection and enhancement by natural and synthetic agents. Environ Adv 8:100203. https://doi.org/10.1016/j.envadv.2022.100203

Khan MA, Shaukat SS, Khan MA (2008) Economic benefits from irrigation of maize with treated effluent of waste stabilization ponds. Pak J Bot 40(3):1091–1098

Kivaisi AK (2001) The potential for constructed wetlands for wastewater treatment and reuse in developing countries: a review. Ecol Eng 16(4):545–560. https://doi.org/10.1016/s0925-8574(00)00113-0

Kusworo TD, Dalanta F, Aryanti N, Othman NH (2021) Intensifying separation and antifouling performance of PSf membrane incorporated by GO and ZnO nanoparticles for petroleum refinery wastewater treatment. J Water Process Eng 41:102030. https://doi.org/10.1016/j.jwpe.2021.102030

Kusworo TD, Utomo DP, Candra YM, Asiyah N, Hanif MJ (2023a) Rubber wastewater treatment using integrated process ozonation-PSf-sZnO membrane filtration. AIP Conf Proc 2667:040012. https://doi.org/10.1063/5.0112320

Kusworo TD, Budiyono B, Haryani K, Utomo DP, Candra YM (2023b) Novel sulfonated zinc oxide embedded polysulfone (PSf-sZnO) membrane for rubber wastewater treatment. AIP Conf Proc 2667:040002. https://doi.org/10.1063/5.0112321

Li L, Yang Y, Tam FYN, Yang L, Mei QX, Yang F (2013) Growth characteristics of six wetland plants and their influences on domestic wastewater treatment efficiency. Ecol Eng 60:382–392

Mahajan B, Shastri S, Londhe S (2023) Waste removal efficiencies of floating macrophytes for restoration of polluted stream: an experimental analysis. Urban Sci 7(1):27. https://doi.org/10.3390/urbansci7010027

Morari F, Giardini L (2009). Municipal wastewater treatment with vertical flow constructed wetlands for irrigation reuse. Ecol Eng 35(5):643–653

Munandar A, Nabila K, Azizah RN (2023) Chemical oxygen demand (COD) and total suspended solid (TSS) removal from rubber wastewater factory using electrocoagulation technique. Indones J Environ Manag Sustain 7(1):27–31

Ninh BD, Gil NC, Baclayon DP (2018) Performance efficiency evaluation of a modified laboratory-scale process for rubber wastewater treatment using moving bed biofilm reactor. J Sci Eng Technol 6:112–126. https://doi.org/10.61569/wtcpka72

Nurmalasari E, Kusworo TD, Khairati M (2022) Rubber industry wastewater treatment using a combination of ozonation and modified PES-Nano ZnO membranes. Chemica: Jurnal Teknik Kimia 9(3). https://doi.org/10.26555/chemica.v9i3.24658

Orhue ER, Osaigbovo AU (2011) The effects of rubber effluent on some chemical properties of soil and early growth of maize (Zea Mays L.). Int J Trop Agric Food Syst 5:50–54

Owamah HI (2014) Biosorptive removal of Pb (II) and Cu (II) from wastewater using activated carbon from cassava peels. J Mater Cycles Waste Manag 16:347–358

Owamah IH, Asiagwu AK, Egboh SHO, Phil-Usiayo S (2013) Drinking water quality at Isoko north communities of The Niger Delta region, Nigeria. Toxicol Environ Chem. https://doi.org/10.1080/02772248.2013.847939

Papadopoulos F, Tsihrintzis AV, Zdragas GA (2011) Removal of faecal bacteria from septage by treating it in full-scale duckweed-covered pond system. J Environ Manag 92:3130–3135

Peries OS, Fernando DM (1983) A handbook of rubber culture and processing. Rubber Research Institute of Srilanka, pp 341–345

Priyanka MS, Varghese GK (2021) Advanced oxidation techniques for wastewater treatment. Sustainable technologies for water and wastewater treatment, 1st edn. CRC Press

Rusdianasari, Bow Y, Syakdani A, Mayasari DI (2021) The effectiveness of electrocoagulation process in rubber wastewater treatment using combination electrodes. IOP Conf Ser Earth Environ Sci 709(1):012009. https://doi.org/10.1088/1755-1315/709/1/012009

Saavedra HFF, Castillo BJS, Velasco AAA, Silva NLMB, Guaman MC (2023) Phytoremediation: an alternative to absorb metals from wastewater using plant species. NeuroQuantology 20(4):500–508

Shanker Pillai HPJ, Girish K (2014) Rubber processing industry effluent treatment using a bacterial consortium. Int J Curr Microbiol App Sci 3(10):775–782

Silva EF, Botelho HA, Venceslau AD, Magalhães DS (2018) Fish farming effluent application in the development and growth of maize and bean plants. Scientific 46(1):74–81

Solodovnik TV, Yakymenko IK (2022) Investigation of natural flocculants for use in wastewater purification processes. Water Water Purif Technol Sci Tech News 33(2):9–14. https://doi.org/10.20535/2218-93002202225986

Susanti S, Nasir S, Hermansyah H, Mataram A (2019) Treatment of wastewater from rubber industry using calcium carbide residue adsorbent and hybrid membrane UF—RO. Sriwijaya J Environ

Tan HT, Pollai KL, Pbarry DJ (1995) Possible utilization of rubber factory effluent on crop land. In: Proceedings of the Rubber Research List Malaysia. Pirs Conf. Kuala Lumpur, p 158

Tang Y, Harpenslager SF, van Kempen MML, Verbaarschot EJH, Loeffen LMJM, Roelofs JGM et al (2017) Aquatic macrophytes can be used for wastewater polishing but not for purification in constructed wetlands. Biogeosciences 14(4):755–766. https://doi.org/10.5194/bg-14-755-2017

Toua V, Clementine EMN, Jean M, Ghislain WN (2014) Treatment test of oil mill effluents by a Pistia stratiotes L., 1753 (water lettuce) pond based system, in Maroua (Far-North, Cameroon). J Ecol Nat Environ 6(11):363–372. https://doi.org/10.5897/jene2014.0477

Uysal Y (2013) Removal of chromium ions from wastewater by duckweed, Lemna minor L. by using a pilot system with continuous flow. J Hazard Mater 263:486–492

Yasar A, Zaheer A, Tabinda AB, Khan M, Mahfooz Y, Rani S, Siddiqua A (2018) Comparison of reed and water lettuce in constructed wetlands for wastewater treatment. Water Environ Res 90(2):129–135. https://doi.org/10.2175/106143017x149029682

Zainuddin NA, Din MFM, Halim KA, Salim NAA, Elias SH, Lazim ZM (2022) The phytoremediation using water hyacinth and water lettuce: correlation between sugar content, biomass growth rate, and nutrients. Jurnal Kejuruteraan 34(5):915–924

Zaki N, Hadoudi N, Charki A, Bensitel N, El Ouarghi H, Amhamdi H, Ahari M (2023) Advancements in the chemical treatment of potable water and industrial wastewater using the coagulation–flocculation process. Sep Sci Technol. https://doi.org/10.1080/01496395.2023.2219381

Zeng L, Liu G, Gao J, Du B, Lv L, Cao C, Ye W, Tong L, Wang Y (2021) A circulating ventilation system to concentrate pollutants and reduce exhaust volumes: case studies with experiments and numerical simulation for the rubber refining process. J Build Eng 35:101984

Chapter 18
Performance Assessment of a Field-Scale Constructed Wetland: A Case Study of Tepla Village, Ambala, Haryana

Shubhanjali Kori, Ravi Kumar, Pooja Ghosh, and Vivek Kumar

Abstract Wastewater (WW) management is the major concern to focus for achieving sustainable development in developing countries. Constructed wetlands are the economical and environmentally benign method, which has advantages for the economy, environment, and society. A detailed case study of rural WW treatment technology, focusing on constructed wetlands, is presented. The inlet and outlet of existing technologies in Tepla village, Ambala district, Haryana, was sampled extensively, in which *E. coli*, TDS, COD, BOD, TSS, pH, EC, and salinity were the key parameters investigated. The study found *E. coli*, TDS, and EC were beyond the permissible limit in the site, suggesting a health and environmental risk. The study found that due to behavioural issues of villagers, it was not working properly as they were polluting the inlet by throwing the solid waste directly. To tackle these difficulties, the work focuses on the incorporation of participatory rural appraisal (PRA) as a vital approach to engage and empower the community.

Keywords Constructed wetland · Wastewater treatment · Participatory rural appraisal · Rural area

18.1 Introduction

Over human history, wetlands have been acknowledged as natural resources. People like the *Marsh Arabs* who live along the Tigris and Euphrates River confluence in southern Iraq recognize the value of these water treatment resources both in their natural condition and in managed forms like rice fields, which are especially common in Southeast Asia (Al-Zahery et al. 2011). Wetlands' capacity to purify water

S. Kori (✉) · P. Ghosh · V. Kumar
Centre for Rural Development and Technology, IIT Delhi, New Delhi, India

R. Kumar
R&D Department, CLS EXP LTD, Karnal, Haryana, India

S. S. Kumar, V. Kumar (eds.), *Nature-Based Solutions for Decentralised Liquid Waste Management*, Environmental Challenges and Solutions,
https://doi.org/10.1007/978-3-032-08362-3_18

is increasingly acknowledged as a desirable wastewater treatment alternative. Wastewater from homes and industries is treated using both vertical-flow and horizontal-flow wetlands made of soil, sand, or gravel. Subterranean horizontal-flow CWs (constructed wetlands) have a high capacity for removing organic matter and suspended particles, but they have a limited ability to remove nutrients (Paing and Voisin 2005; Vymazal 2005). On the other hand, hybrid systems combine vertical flow with horizontal subsurface flow systems to provide a more effective treatment process (Vymazal 2005). They are also used as a passive method of treating diffuse pollutants, such as runoff from highways and cities and mining wastewater drainage. CWs are a kind of passive wastewater treatment technology that may efficiently bio-mineralize nutrients and organics, decrease metal acidity, and raise pH levels in wastewater (Kalin 2004). Wetland organisms contribute into environmental protection by enhancing treatment effectiveness, aiding in maintenance and management, and influencing contaminants via processes such as bioaccumulation, adsorption, and bioturbation (Li et al. 2021). Constructed wetlands benefit from the presence of macrophytes, which help in decreasing wind speed, facilitate sedimentation, provide a surface for periphyton and bacteria to grow, and absorb nutrients. In Europe and Asia, the most often seen macrophyte species is *Phragmites australis* (Vymazal 2013). Submerged macrophytes extract minerals while providing extensive surfaces towards biofilms, making them advantageous for the elimination of nitrogen in CWs and indigenous bodies of water (Zhang et al. 2016). Submerged macrophytes including the biofilms on their leaves have a significant impact on the breakdown of polycyclic aromatic hydrocarbons (PAHs) and the regulation of nitrogen cycling within CW systems. This helps to decrease the risk of pollutants for the natural environment, the aquatic organisms, and the health of humans (Qin et al. 2019). Aquatic plants flourishing in constructed wetlands enhance the efficacy and removal of cyanotoxins, pharmaceuticals, heavy metals, nanoparticles, and pesticides from aquatic systems (Kochi et al. 2020). An advantageous addition to conventional sewage treatment systems are reed beds. They often provide a less costly option to pricey wastewater treatment technologies as activated sludge processes and trickling filters (Giraldi et al. 2009; Reif et al. 2011). Moreover, there is a lack of research on comparative analyses between conventional technologies and CW systems specifically focused on water reuse (Riggio et al. 2018). The technique provides an efficient and cost-effective method of treating wastewater in both temperate and tropical climates. As a result of its ability to effectively remove pollutants from a variety of systems, CWs have gained widespread recognition as environmentally responsible solutions for the treatment of wastewater. With a high degree of efficacy in eliminating impurities and giving biological benefits, CWs provide water security, energy potential, and environmental protection. In addition, they offer advantages to biological systems. CWs that are used for the treatment of wastewater reduce the levels of total nitrogen and total phosphorus, as well as the quantities of biochemical oxygen demand and chemical oxygen demand that are present in municipal sewage. As a result, these CWs improve water quality and encourage the growth of biodiversity. This study assesses the feasibility of employing a phytoremediation procedure to recycle household wastewater including both grey and black water. Additionally, it examines people's behaviour and actions about the adoption of this technology (Fig. 18.1).

18.1.1 Study Area

The Tepla hamlet is situated in the Saha Tehsil of Ambala district in the state of Haryana, India. It is under the jurisdiction of the Saha Community Development Block. Ambala Cantt is the closest town to Tepla, located around 4 km distance. The village is accessible by public bus services. According to the statistics from 2011, there are 2803 people living in 536 households in the village of Tepla. The community consists of 1293 females and 1510 males. The female population accounts for 46.13%, while the male population accounts for 53.87% of the total population. The village general data has been shown in Table 18.1. There are a total of 1004 individuals belonging to scheduled castes, with 461 being females and 543 being men. The female population accounts for 45.92% and the male population accounts for 54.08% of the scheduled caste population. The proportion of scheduled castes in the overall population is 35.82%. The combined amount of land that has been planted or used for agriculture is 512 hectares. Approximately 109 hectares is

Fig. 18.1 CW at Tepla village, Ambala district, Haryana. (Source: https://hpwwma.org.in/Image.aspx#)

Table 18.1 Tepla village data (2011 census)

Attributes	Total	M	F
Total households	536	–	–
Total population	2803	1510	1293
Children between 0 and 6 years	327	195	132
SC	1004	543	461
ST	0	0	0
Literacy rate	81.38%	87.38%	74.59%
Total workers	844	814	30

Source: https://www.census2011.co.in/data/village/57545-tepla-haryana.html

allocated for non-agricultural use. Approximately 12 hectares is dedicated to cultivating various types of tree crops. Approximately 20 hectares of land is now unused and not being cultivated, in addition to the area that is already fallow. Approximately 2 hectares is occupied by infertile and non-arable terrain. The population density of Tepla is 427.94 people per square kilometre. The population of children aged 0–6 years in Tepla village is 327, accounting for 11.67% of the total village population. The sex ratio in Tepla village is 856, which is below the average sex ratio of 879 in the state of Haryana. The child sex ratio for Tepla, as reported by the census, is 677, which is lower than the average child sex ratio of 834 in Haryana. The literacy rate of Tepla village is greater than that of Haryana. The literacy rate of Tepla village in 2011 was 81.38%, while the literacy rate of Haryana was 75.55%. The male literacy rate of Tepla is 87.38%, while the female literacy rate is 74.59%. Tepla hamlet is mostly inhabited by individuals belonging to the scheduled caste (SC) community. The scheduled caste (SC) makes up 35.82% of the whole population of Tepla village. Tepla hamlet presently lacks any scheduled tribe (ST) population.

Out of the whole population of Tepla village, a total of 844 individuals were involved in work-related activities. About 70.50% of workers classify their employment as main employment, meaning they are employed or earn income for more than 6 months. On the other hand, 29.50% engage in marginal activity, which provides a living for less than 6 months. Out of a total of 844 individuals involved in their primary occupation, 120 were cultivators who owned or co-owned the land, while 64 were agricultural labourers (Table 18.1).

18.1.2 Participatory Rural Appraisal

The participatory rural appraisal (PRA) methodology is attuned to local conditions. It facilitates the understanding of challenges faced by rural populations by including stakeholders from the outset in the design and development of technology (Loader and Amartya 1999). Farmers/villagers are encouraged to take an active role in the decision-making process regarding the preservation, administration, and development of rural regions via the implementation of participatory rural techniques

(Menconi et al. 2017). The goal of participatory rural methods is to get an understanding of the socioeconomic and environmental repercussions of rural transformation via the active engagement of a wide range of research disciplines and stakeholders (Dougill et al. 2006). There is the potential for effective technology transfer and better water resource management to result from participatory rural methods to wastewater treatment; but, in order to ensure sustainability, the long-term partnerships and understandings are required (Gómez et al. 2008). Participatory rural approaches to wastewater treatment technologies involve considering the rural setting, natural conditions, economic levels, and environmental goals. These approaches help people become more aware of new options, make decisions about which ones to use, adapt and modify their practices, and learn and make choices. These approaches also encourage new ideas (Douthwaite et al. 2009). This analysis is used to suggest realistic solutions and evaluate their costs and impacts. The use of participatory research methods in water research gives participants more agency, democratizes the creation of information, enhances decision-making, and ushers in new environmental frontiers (Roque et al. 2022). Consequently, this study primarily aims to propose the PRA approach, since the local population lacks sufficient awareness of the technology. If an individual owns a thing, they are likely to care for it. Using the PRA technique, individuals may alter their attitudes towards technology.

18.1.3 Types of Constructed Wetlands

Based on their design and purpose, CWs are divided into four categories: free-water surface wetlands, subsurface flow wetlands, hybrid wetlands, and vertical-flow wetlands. These are discussed below.

(a) **Surface Wetlands with Free Water:** To enhance pollutant removal, free-water surface wetlands use shallow ponds or channels with emergent aquatic plants. These wetlands are good in treating pollutants that have a high surface activity, such as some medications and personal care items (Nuamah et al. 2020). The combination of plant absorption, microbial breakdown, and sedimentation processes all contributes to pollutant reduction (Saeed and Sun 2012).
(b) **Hybrid Wetlands:** Hybrid wetlands blend free-water surface and subsurface flow concepts. They are versatile in treating a wide spectrum of contaminants, making them appropriate for tackling a variety of new pollutants, such as endocrine-disrupting chemicals and microplastics (Lutterbeck et al. 2017).
(c) **Wetlands with Vertical Flow:** Vertical-flow wetlands direct water vertically through a substrate, which is often gravel or proprietary media. The regulated flow improves aeration and creates an environment favourable for microbial decomposition. Vertical-flow wetlands are good in removing pollutants like heavy metals and fertilizers (Paing and Voisin 2005).
(d) **Subsurface Flow Wetlands:** Subsurface flow wetlands utilize the movement of polluted water over a layer of sand or gravel to encourage microbial filtration

and degradation. The removal of organic contaminants and nutrients is a special use for this design. Pesticides and other industrial pollutants are particularly well-treated by subsurface flow wetlands (Jamwal et al. 2021). The available constructed wetlands of India have been discussed below.

18.1.4 Contaminant Removal Mechanisms

To remove emergent toxins, CWs use a variety of physical, chemical, and biological processes, which include sorption, microbial degradation, plant uptake, filtration, and then volatilization (Table 18.2).

18.2 Material and Methods

In India, the process of collecting detailed village information done by survey using the counting or restricted observation. However, acquiring quantitative data on agricultural activities involves prolonged periods of observation and significant exertion (Neale 1958). Survey sampling is a frequently used approach for collecting data among social scientists. Let us start by elucidating the most basic technique of survey sampling, known as simple random sampling method (Imai 1998). Prior to commencing our sample process, there is a need to ascertain the overall quantity of voters that needs to do interview. The basic idea behind the design-based approach is that the researchers' particular data collection technique—in this example, the simple random sampling method—is the sole factor that determines the statistical properties of an estimate (Imai 1998) (Thepma et al. 2016). The interviews with the villagers of around 10% of total households have been considered and analysed in Excel, which resulted that there is a need for IEC tool, which is information, education, and communication. Therefore, this technique adheres closely to the study design; also for understanding the performance of the CW, the sampling of the inlet and outlet of the technology has been taken, and the method was used as per the APHA manual (Baird et al. 2005).

18.2.1 Constructed Wetland of the Study Area

According to the Ministry of Environment and Forests (Government of India), the built-up wetland area in Haryana is 42,478 hectares, which accounts for 0.86% of the state's total land area (Panwar and Dimri 2018). This CW was made via *Haryana Pond and Wastewater Management* authority, and the results of CWs' inlet and outlet are given in Table 18.3. CWs are ecologically beneficial alternatives for the management of wastewater in countries that are still developing (underdeveloped).

Table 18.2 Study on CWs of India

SN	Study area	Wastewater type	Plants	Time period	Dimensions	Removal percentage/ efficiency	Cost	Remarks	References
1	Kothapally village of Telangana	Domestic waste from 100 households	*Typha latifolia* and *Canna indica*	1 year	20 m × 4 m × 2.5 m	COD (65%), sulphate (60%), inorganic nitrogen (67%), total coliform (80%)	$9895	Resource-scarce semi-arid tropic (SAT) villages	Datta (2021)
2	Ujjain, MP	Domestic WW	*Phragmites karka*	5 months	Surface area of 41.82 m^2	BOD (77.1%), COD (64.9%), turbidity (68.3%), suspended-solids (63%), TP (58.7%), nitrate-nitrogen (33%), fecal coliforms (96.8%), and total coliforms (95.6%)	$1200	Influent was pretreated before entering the system	Parashar et al. (2022)
3	Rural area	Grey water (household)	*Lepironia articulata*	HRT 4 min	34 × 18 × 14 cm as mini-wetlands	81.42% BOD, 84.57% COD, 39.83% AN, 54.70% SS, and 45.01% turbidity	Very low	Three containers of the same dimensions were used for treatment	Wurochekke et al. (2014)
4	Potheri village, Chennai	Dairy WW	*Typha latifolia*	12 days	50 m × 40 m × 30 m	BOD (73%), solids (75–83%), phosphorus (75.7%), COD and ammoniacal nitrogen (about 26%)		Reduce the TDS to avoid clogging the soil pores	Sudarsan et al. (2018)
5	Tribal area of Central India	Domestic sewage	*Typha latifolia* and *Commelina benghalensis*	HRT 48 h	10 × 3.5 m	BOD (79%), COD (77%), nitrate (79%), ammonia (79%), phosphate (78%), Total coliforms (64%), faecal coliforms (61%), and *E. Coli* (52%)	Very low	Average hydraulic loading rate of 250 L h^{-1}	Shukla et al. (2021)

The performance of the created wetland is shown by the dataset presented in Table 18.3. This demonstrates that the wetlands have a poor capacity for operating owing to a number of different factors, and one of those reasons has been briefly described in this paper. The techniques used to gather survey data from villages include the dissemination of questionnaires to a heterogeneous cohort of villagers, as well as the engagement of key informants and the formation of gender-inclusive or women-exclusive groups. This study examines whether conclusions made about a population are influenced by the method used to collect the data. This study juxtaposes data gathered using participatory assessment approaches with a survey of homes selected randomly from a locally conducted census that underwent meticulous revision. The community map sometimes includes households that do not really live in the specified area. In comparison, the updated official census is somewhat more prone to exclude households that really resided in the questioned region. In Fig. 18.2, the discussion with the villagers has been shown. After accounting for the survey method, it was discovered that the updated official census yields greater estimations of average household size and wealth. However, it provides lower estimations of the overall size or richness of the village compared to the community map. When comparing the survey approaches, specifically focusing on the households, it is evident that the community map consistently yields greater estimates of household size and lower estimates of wealth (Christiaensen et al. 2001) (Fig. 18.3).

Table 18.3 Water quality data of Tepla constructed wetland

TEMP	Inlet	Outlet	Average	SD
pH	7.9	7.6	7.75	0.212132
Electrical conductivity	987	866	926.5	85.55992
Total dissolved solids	2028	1900	1964	90.50967
Total alkalinity	186	108	147	55.15433
Total hardness	306	280	293	18.38478
Dissolved oxygen	10.6	8	9.3	1.838478
Biological oxygen demand (BOD_5)	16.87	10.62	13.745	4.419417
Chemical oxygen demand (COD)	130	116	123	9.899495
Carbonate ion (CO_3^{2-})	0	0	0	0
Hydrogen carbonate (HCO_3^-)	245	187	216	41.01219
Nitrate ion (NO_3^{2-})	7.5	2.2	4.85	3.747666
Phosphorus	5.8	2.5	4.15	2.333452
Sulphate	101	95	98	4.242641
Chlorine	122	86	104	25.45584
Calcium	67	55	61	8.485281
Magnesium	38	23	30.5	10.6066
Potassium	54	27	40.5	19.09188
Ammonium	47	30	38.5	12.02082
E. Coli	Present	Present	–	–
No. of samples	2			
Latitude	30°18′16.4″N			
Longitude	76°56′42.2″E			

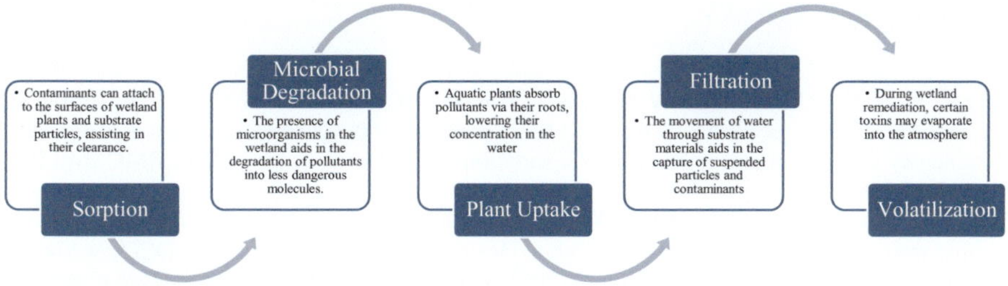

Fig. 18.2 Contaminant removal mechanisms of constructed wetland

Fig. 18.3 (**a**) Discussion with the caretaker of the CW. (**b**) Outlet of the technology. (**c**) Taking samples from the inlet. (**d**) Discussion with the villagers

18.3 Conclusion

By using the dynamic relationship that exists between plants, bacteria, and the soil matrix, clean water systems are able to successfully remove contaminants from wastewater. When compared to conventional treatment systems, CWs provide an alternative that is favourable in terms of land utilization, energy consumption, and operational requirements for the treatment of many kinds of wastewater, especially in locations that are geographically isolated and small in population. As a result, the study areas CW's ability is to successfully eliminate biochemical oxygen demand (BOD), chemical oxygen demand (COD), and total suspended solids (TSS). It has been discovered that the effectiveness of the technology has been negatively impacted as a result of the actions of the villagers; hence, suggestions have been made on the need of the villagers' involvement via PRA.

References

Al-Zahery N, Pala M, Battaglia V, Grugni V, Hamod MA, Kashani BH, Olivieri A, Torroni A, Santachiara-Benerecetti AS, Semino O (2011) In search of the genetic footprints of Sumerians: a survey of Y-chromosome and mtDNA variation in the Marsh Arabs of Iraq. BMC Evol Biol 11:288. https://doi.org/10.1186/1471-2148-11-288

Baird RB, Eaton AD, Rice EW (eds) (2005) APHA. Standard methods for the examination of water and wastewater, 23rd edn. American Public Health Association, Washington, DC/New York

Christiaensen L, Hoddinott J, Bergeron G (2001) Comparing village characteristics derived from rapid appraisals and household surveys: a tale from northern Mali. J Dev Stud 37:1–20. https://doi.org/10.1080/00220380412331321941

Dougill AJ, Fraser EDG, Holden J, Hubacek K, Prell C, Reed MS, Stagl S, Stringer LC (2006) Learning from doing participatory rural research: lessons from the Peak District National Park. J Agric Econ 57:259–275. https://doi.org/10.1111/j.1477-9552.2006.00051.x

Datta A, Singh HO, Raja SK, Dixit S (2021) Constructed wetland for improved wastewater management and increased water use efficiency in resource scarce SAT villages: a case study from Kothapally village, in India. International Journal of Phytoremediation, 23(10):1067–1076. https://doi.org/10.1080/15226514.2021.1876627

Douthwaite B, Beaulieu N, Lundy M, Peters D (2009) Understanding how participatory approaches foster innovation. Int J Agric Sustain 7:42–60. https://doi.org/10.3763/ijas.2009.0339

Giraldi D, Masciandaro G, Peruzzi E, Bianchi V, Peruzzi P, Ceccanti B, Iannelli R (2009) Hydraulic and biochemical analyses on full-scale sludge consolidation reed beds in Tuscany (Italy). Water Sci Technol 60:1209–1216. https://doi.org/10.2166/wst.2009.418

Gómez CLT, Garcia-Zamor J-C, Rojas JMC (2008) Domestic wastewater management in a rural community in Colombia. Comp Technol Transf Soc 6:212–235. https://doi.org/10.1353/ctt.0.0016

Imai K (1998) Survey sampling. Curr Sociol 46:75–87. https://doi.org/10.1177/0011392198046004005

Jamwal P, Raj AV, Raveendran L, Shirin S, Connelly S, Yeluripati J, Richards S, Rao L, Helliwell R, Tamburini M (2021) Evaluating the performance of horizontal sub-surface flow constructed wetlands: a case study from southern India. Ecol Eng 162. https://doi.org/10.1016/j.ecoleng.2021.106170

Kalin M (2004) Passive mine water treatment: the correct approach? Ecol Eng 22:299–304. https://doi.org/10.1016/j.ecoleng.2004.06.008

Kochi LY, Freitas PL, Maranho LT, Juneau P, Gomes MP (2020) Aquatic macrophytes in constructed wetlands: a fight against water pollution. Sustainability 12:9202. https://doi.org/10.3390/su12219202

Li Q, Long Z, Wang H, Zhang G (2021) Functions of constructed wetland animals in water environment protection—a critical review. Sci Total Environ 760:144038. https://doi.org/10.1016/j.scitotenv.2020.144038

Loader R, Amartya L (1999) Participatory rural appraisal: extending the research methods base. Agric Syst 62:73–85. https://doi.org/10.1016/S0308-521X(99)00056-6

Lutterbeck CA, Kist LT, Lopez DR, Zerwes FV, Machado EL (2017) Life cycle assessment of integrated wastewater treatment systems with constructed wetlands in rural areas. J Clean Prod 148:527–536. https://doi.org/10.1016/j.jclepro.2017.02.024

Menconi ME, Grohmann D, Mancinelli C (2017) European farmers and participatory rural appraisal: a systematic literature review on experiences to optimize rural development. Land Use Policy 60:1–11. https://doi.org/10.1016/j.landusepol.2016.10.007

Neale WC (1958) The limitations of Indian village survey data. J Asian Stud 17:383–402. https://doi.org/10.2307/2941423

Nuamah LA, Li Y, Pu Y, Nwankwegu AS, Haikuo Z, Norgbey E, Banahene P, Bofah-Buoh R (2020) Constructed wetlands, status, progress, and challenges. The need for critical operational reassessment for a cleaner productive ecosystem. J Clean Prod 269:122340

Paing J, Voisin J (2005) Vertical flow constructed wetlands for municipal wastewater and septage treatment in French rural area. Water Sci Technol 51(9):145–155

Panwar S, Dimri AK (2018) Trend analysis of production and productivity of major crops and its sustainability: a case study of Haryana. Indian J Agric Res. https://doi.org/10.18805/IJARe.A-5019

Parashar V, Singh S, Purohit MR, Tamhankar AJ, Singh D, Kalyanasundaram M, Lundborg CS, Diwan V (2022) Utility of constructed wetlands for treatment of hospital effluent and antibiotic resistant bacteria in resource limited settings: a case study in Ujjain. India Water Environment Research 94. https://doi.org/10.1002/wer.10783

Qin Z, Zhao Z, Liling X, Adam A, Li Y, Chen D, Mela SM, Li H (2019) The dissipation and risk alleviation mechanism of PAHs and nitrogen in constructed wetlands: the role of submerged macrophytes and their biofilms-leaves. Environ Int 131:104940. https://doi.org/10.1016/j.envint.2019.104940

Reif R, Besancon A, Le Corre K, Jefferson B, Lema JM, Omil F (2011) Comparison of PPCPs removal on a parallel-operated MBR and AS system and evaluation of effluent post-treatment on vertical flow reed beds. Water Sci Technol 63:2411–2417. https://doi.org/10.2166/wst.2011.123

Riggio VA, Ruffino B, Campo G, Comino E, Comoglio C, Zanetti M (2018) Constructed wetlands for the reuse of industrial wastewater: a case-study. J Clean Prod 171:723–732. https://doi.org/10.1016/j.jclepro.2017.10.081

Roque A, Wutich A, Quimby B, Porter S, Zheng M, Hossain MJ, Brewis A (2022) Participatory approaches in water research: a review. WIREs Water 9. https://doi.org/10.1002/wat2.1577

Saeed T, Sun G (2012) A review on nitrogen and organics removal mechanisms in subsurface flow constructed wetlands: dependency on environmental parameters, operating conditions and supporting media. J Environ Manag 112:429–448

Shukla R, Gupta D, Singh G et al (2021) Performance of horizontal flow constructed wetland for secondary treatment of domestic wastewater in a remote tribal area of Central India. Sustain Environ Res 31:13. https://doi.org/10.1186/s42834-021-00087-7

Sudarsan JS, Annadurai R, Mukhopadhyay M et al (2018) Domestic wastewater treatment using constructed wetland: an efficient and alternative way. Sustain Water Resour Manag 4:781–787. https://doi.org/10.1007/s40899-017-0164-x

Thepma A, Techakritkachorn T, Bamrungkhul S (2016) Comparison of the traditional method and the use of technology portable personal computer in the process of village survey. J Architect Plann Res Stud (JARS) 13:49–60. https://doi.org/10.56261/jars.v13i1.71589

Vymazal J (2005) Horizontal sub-surface flow and hybrid constructed wetlands systems for wastewater treatment. Ecol Eng 25:478–490. https://doi.org/10.1016/j.ecoleng.2005.07.010

Vymazal J (2013) Emergent plants used in free water surface constructed wetlands: a review. Ecol Eng 61:582–592. https://doi.org/10.1016/j.ecoleng.2013.06.023

Wurochekke AA, Harun NA, Mohamed RMSR, Kassim AHBM (2014) Constructed wetland of Lepironia Articulata for household greywater treatment. APCBEE Proc 10:103–109. https://doi.org/10.1016/J.APCBEE.2014.10.025

Zhang S, Pang S, Wang P, Wang C, Guo C, Addo FG, Li Y (2016) Responses of bacterial community structure and denitrifying bacteria in biofilm to submerged macrophytes and nitrate. Sci Rep 6:36178. https://doi.org/10.1038/srep36178

MIX
Papier aus verantwortungsvollen Quellen
Paper from responsible sources
FSC® C105338

If you have any concerns about our products,
you can contact us on
ProductSafety@springernature.com

In case Publisher is established outside the EU,
the EU authorized representative is:
Springer Nature Customer Service Center GmbH
Europaplatz 3, 69115 Heidelberg, Germany

Printed by Libri Plureos GmbH
in Hamburg, Germany